Lecture Notes in Physics

Volume 923

The Lecture Notes in Physics

The series Lecture Notes in Physics (LNP), founded in 1969, reports new developments in physics research and teaching-quickly and informally, but with a high quality and the explicit aim to summarize and communicate current knowledge in an accessible way. Books published in this series are conceived as bridging material between advanced graduate textbooks and the forefront of research and to serve three purposes:

- to be a compact and modern up-to-date source of reference on a well-defined topic
- to serve as an accessible introduction to the field to postgraduate students and nonspecialist researchers from related areas
- to be a source of advanced teaching material for specialized seminars, courses and schools

Both monographs and multi-author volumes will be considered for publication. Edited volumes should, however, consist of a very limited number of contributions only. Proceedings will not be considered for LNP.

Volumes published in LNP are disseminated both in print and in electronic formats, the electronic archive being available at springerlink.com. The series content is indexed, abstracted and referenced by many abstracting and information services, bibliographic networks, subscription agencies, library networks, and consortia.

Proposals should be sent to a member of the Editorial Board, or directly to the managing editor at Springer:

Christian Caron
Springer Heidelberg
Physics Editorial Department I
Tiergartenstrasse 17
69121 Heidelberg/Germany
christian.caron@springer.com

More information about this series at http://www.springer.com/series/5304

Andrés Santos

A Concise Course on the Theory of Classical Liquids

Basics and Selected Topics

 Springer

Andrés Santos
Departamento de Física and Instituto de
    Computación Científica Avanzada
    (ICCAEx)
Universidad de Extremadura
Badajoz, Spain

ISSN 0075-8450   ISSN 1616-6361   (electronic)
Lecture Notes in Physics
ISBN 978-3-319-29666-1   ISBN 978-3-319-29668-5   (eBook)
DOI 10.1007/978-3-319-29668-5

Library of Congress Control Number: 2016934038

Printed on acid-free paper

This Springer imprint is published by Springer Nature
The registered company is Springer International Publishing AG Switzerland

To Ángeles, María del Mar, and David

Foreword

I have been fortunate to have lived through a period during which the theory of liquids passed from being considered to be one of the unsolved problems in physics to a solved, even mature, field. When I was a student, it was felt that the gas and solid phases were understood, but the lack of a theory of liquids was believed to be a gap in our understanding of nature. I was attracted by this perceived gap and wished to play a role in changing this. In retrospect, it is amusing that an almost satisfactory theory, the van der Waals theory, was available but unappreciated. A half century ago, this theory was regarded as being of only pedagogic interest. In this theory, the contributions of the repulsive and attractive forces are treated by two separate terms. The repulsive forces in this theory are described in terms of hard-sphere interactions. Even though it was clear from the beginning that the treatment of the hard-sphere forces in the van der Waals theory, through a simple free volume, $V - Nb$, was inadequate, this was ignored and attention was directed, almost exclusively, to making empirical changes in the form of the term representing the attractive forces. Various modifications, not based on theory, were proposed on a "try this, try that" basis in order to give a more accurate value for pV/NkT at the critical point. This was misguided; it is now known that the critical point cannot be described adequately by any analytic equation. Indeed, the theory of the critical point is now regarded as a separate field. In contrast, the van der Waals treatment of the attractive forces is well founded. It is the treatment of the repulsive forces in terms of a simple free volume that is the major problem in the original form of van der Waals theory. Even though it was clear to van der Waals and his contemporaries that his expression for the free volume is a poor, even bad, approximation that grossly overestimated the hard-sphere pressure, it was retained in the various modifications that were considered. It was the treatment of the attractive forces that drew attention. The reason for retaining the free volume term of van der Waals was that only a cubic equation needs to be solved to give the volume in terms of the pressure. This, certainly, was a convenience in the days before electronic computers and avoided an iterative solution. However, from the point of view of the development of liquid state theory, it was a case of putting the cart before the horse. It is now clear that it was the nature of the gas phase at high densities, rather than the liquid phase, that was poorly understood. It was the gas phase at high densities, in particular the equation of state of the hard-sphere fluid, that was the unsolved problem. Computer simulation of hard spheres and the analytic solution of the Percus–Yevick equation for hard-spheres provided an understanding of the hard sphere gas and changed everything.

Professor Santos has presented a comprehensive treatment of liquid state theory with a clear development of integral equation theory, including the Percus–Yevick theory and the closely related hypernetted-chain and mean spherical theories. The discussion of the hard-sphere fluid is, appropriately, generous. One notable, and desirable feature of his book is the collection of photographs of some of the notables in the development of liquid state theory. Nonscientists tend to think of science as an impersonable field with scientists working in isolation. In reality, science is quite sociable with scientists interacting with each other and being stimulated through this interaction. This sociable interaction in agreeable locations is one reason that schools, such as the Warsaw schools, are so valuable and enjoyable. The large, impersonal meetings of scientific societies have their place, but, generally, it is at small meetings and schools that the real progress is made.

I look forward to placing Professor Santos' book on my bookshelf as a valuable reference and am confident that many of our peers will do this also. I am grateful to Professor Santos for writing this valuable book and for giving me the opportunity to play a small part in its production.

Provo, UT, USA   Douglas Henderson
November 2015

Preface

There exist in the market many excellent textbooks covering equilibrium statistical mechanics of liquids and dense gases. Why, then, yet another addition to the shelf? Is there any niche available for it to fill? This is perhaps a question to be answered by the reader rather than by the author. In any case, this book is not intended whatsoever to replace any of the good (some of them classical) texts on similar topics but, in the best scenario, to serve as a supplement to them. Despite the relatively small number of pages, some of the topics selected here are treated with more detail than in other books, but this is done at the expense of not addressing some other important topics. A delicate balance has been sought to have a piece of work that can be used as a textbook for a one-semester graduate-level course (perhaps by skipping some of the more advanced points), serving at the same time to the experienced researcher as a reference for some specific details.

Let me indulge myself in a little bit of personal recollection. Over more than 15 years, I had been producing, for personal use, handwritten lecture notes as a guide for (intermittent) teaching of graduate-level courses on equilibrium statistical mechanics in my university. When in the summer of 2012 Jarosław Piasecki invited me to be one of the speakers at the 5th Warsaw School of Statistical Physics (Kazimierz Dolny, June 2013), he informed me that speakers were expected to deliver six 45-min lectures to introduce a chosen subject belonging to statistical physics in a pedagogical way, inspiring further research. I decided to combine my experience as instructor of classical statistical mechanics and as researcher on simple models and approaches in liquid state theory to propose a series of lectures with the title "Playing with Marbles: Structural and Thermodynamic Properties of Hard-Sphere Systems." The lecture notes (slightly more than 90 pages long) were posted in October 2013 on the arXiv (http://arxiv.org/abs/1310.5578) and published by Warsaw University Press in the spring of 2014. This would have been the end of the story had Christian Caron, Executive Publishing Editor of Physics at Springer, not contacted me in January of 2014 to propose the extension of the Warsaw lecture notes (which he knew from the arXiv submission) to book length appropriate for the *Lecture Notes in Physics* series. After checking that there did not exist any copyright conflict with Warsaw University Press, and being aware that the lecture notes should be significantly enlarged, I accepted Christian's suggestion and presented a formal proposal. After about 2 years (much longer than anticipated!), the outcome is this monograph.

The aim of these lecture notes is to present an introduction to the equilibrium statistical mechanics of liquids and nonideal gases at a graduate-student text-book level, with emphasis on the basics and fundamentals of the field, but also with excursions into recent developments. The treatment uses classical (i.e., non-quantum) mechanics, and no special prerequisites are required, apart from standard introductory thermodynamics and statistical mechanics. Most of the content applies to any (short-range) interaction potential, any dimensionality, and (in general) any number of components. On the other hand, some specific applications deal with properties of fluids made of particles interacting via the hard-sphere potential or related potentials. Unavoidably, the selection of topics and the approach employed may be biased toward those aspects closer to the author's taste and expertise. My apologies if that bias turns out to be excessive.

While a large part of the content of this work is not that different from standard material found in well-established textbooks, some additional results published in specialized journals along the last few years are also covered. Moreover, the book includes original matter not published before, to the best of the author's knowledge. This can be found essentially as portions of Sects. 3.7–3.9, 4.5, 5.5, 6.9, 7.3, and 7.4.

An attempt has been made to preserve a pedagogical tone as much as possible. All the graphs (more than 70, many of them entirely new) have been specifically composed for the book with a uniform layout and aspect ratio. Nearly 30 tables are also included, not only for displaying diagrams or numerical values in an ordered way but also as summaries of equations and results that are obtained along the text but could be difficult to find when browsing through the pages. A list of exercises (adding to a total number higher than 200) is appended at the end of each chapter. In some cases they are just intended to fill gaps in the derivations of results presented in the text, thus stimulating the reader's self-study. In other cases, however, the exercises invite the reader to explore alternative or complementary views of the subjects under consideration.

One of the most difficult choices an author of a physics textbook must face concerns the choice of symbols and notation for mathematical and physical quantities. An imperfect balance has been attempted between avoiding repetition of symbols for different quantities as much as possible without, on the other hand, resorting to too many nonstandard, fancy, or awkward symbols. The non-exhaustive list of symbols included at the end of the front matter can alleviate the burden of this problem.

I am very much convinced that the student and the experienced researcher alike grasp more convincingly concepts, results, equations, or theories (maybe new to them) when they are able to associate faces with the names behind those concepts, results, equations, or theories. After all, science is made by beings as human (albeit with exceptional minds) as ourselves, and, therefore, the importance of the so-called human face of science cannot be overemphasized. Paying tribute to the scientists who have paved or are paving the way to the rest of us, knowing what they look like, and prompting our curiosity to know more about their scientific and personal lives (both usually being equally exciting) are issues that may perfectly belong in a "hard" monograph as much as in softer magazine articles or layman books. In agreement with that view, this book includes the photographs of more than 30 scientists, ranging from the second half of the nineteenth century to today. In some cases they are mentioned tangentially (if their main contributions overlap only partially with the content of this book), while in other cases those authors are frequently cited. Of course, not all those who have made relevant contributions to the field are represented, but the ones included here have certainly contributed to significant advances. I apologize for the absence of images of many other main contributors and practitioners.

The content of this book is organized into seven chapters. Chapters 1 and 2 present brief summaries of equilibrium thermodynamic and statistical-mechanical relations. They are mainly included to make the lecture notes as self-contained as possible and to unify the notation, but otherwise most of their content can be skipped by the knowledgeable reader.

Next, Chap. 3 describes the formal steps needed to derive the virial coefficients in the expansion of pressure in powers of density in terms of the pair interaction potential. Extensive use of diagrams is made, but several needed theorems and lemmas are justified by simple examples without formal proofs. Chapter 3 concludes with the discussion of approximate equations of state for (both one-component and multicomponent) hard-sphere fluids that are constructed by making use of the first few exact virial coefficients.

One of the core chapters of the book is Chap. 4, which starts with the definition of the reduced distribution functions and, in particular, of the radial distribution function $g(r)$ and the direct correlation function $c(r)$ and continues with the derivation of the main thermodynamic quantities in terms of $g(r)$. This includes the chemical-potential route, usually forgotten in textbooks.

Chapter 5 is perhaps a "side dish." Whereas one-dimensional systems can be seen as rather artificial, it is undoubtedly important, at least from pedagogical and illustrative perspectives, to derive their exact structural and thermophysical quantities and apply them to explicit model potentials.

The counterpart of Chap. 3 at the level of the radial distribution function makes most of Chap. 6, where the expansion of $g(r)$ in powers of density is worked out, again by diagrammatic manipulations justified with simple examples. The rest of Chap. 6 is devoted to the proposal of the hypernetted-chain and Percus–Yevick approximations, plus other approximate integral equations, and the issue of internal consistency among different thermodynamic routes in approximate theories.

Finally, Chap. 7 covers the analytical solutions of the Percus–Yevick approximation for hard spheres, sticky hard spheres, and their mixtures, derived as the simplest implementations of rational-function approximations for an auxiliary function defined in Laplace space. The latter approach is then applied to improve the Percus–Yevick solution for hard spheres and to circumvent the absence of an analytical solution of the Percus–Yevick approximation for square-well and square-shoulder potentials. Although such an approach is by now well established in the specialized literature, to the author's knowledge hardly other textbook on the subject includes the latter material.

Let me finish this already too long preface just by saying that I hope these lecture notes might be useful to students who want to be introduced to the exciting field of disordered condensed matter, to instructors who might find something profitable for their own courses, and to researchers who might need to have at hand a reference to quickly find a certain needed result.

Badajoz, Spain   Andrés Santos
December 2015

Acknowledgments

First, I want to express my gratitude to those scientists with whom I have had the privilege to collaborate in problems related to equilibrium statistical mechanics of classical fluids over the years. They are, in alphabetical order, Luis Acedo, Morad Alawneh, Mariana Bárcenas, Elena Beltrán-Heredia, Arieh Ben-Naim, J. Javier Brey, Francisco Castaño, Riccardo Fantoni, Giacomo Fiumara, Achille Giacometti, Douglas Henderson, Julio Largo, Mariano López de Haro, Stefan Luding, Miguel Ángel G. Maestre, Alexandr Malijevský, Anatol Malijevský, Gema Manzano, Gerardo Odriozola, Vitaliy Ogarko, Pedro Orea, Jarosław Piasecki, Miguel Robles, René D. Rohrmann, Francisco Romero, Luis F. Rull, Franz Saija, J. Ramón Solana, Carlos F. Tejero, and Santos B. Yuste.

Special mentions are due to some of them. I thank Mariano López de Haro, René D. Rohrmann, and Santos B. Yuste for having shared with me so many working discussion hours entangled with unforgettable personal experiences. Moreover, the two former (Mariano and René) found time to critically read the manuscript at several stages and make smart and constructive suggestions. I am especially thankful to Douglas Henderson, who immediately accepted my request and generously wrote a beautiful foreword to this little piece of work. Quite likely, these lecture notes would have never been born had Jarosław Piasecki not invited me to be one of the speakers at the 5th Warsaw School of Statistical Physics. Thank you, Jarek.

I am very grateful as well to Christian Caron, Executive Publishing Editor of Physics at Springer, who first suggested the possibility of expanding my original Warsaw lecture notes into this book and who very kindly accepted my frequent requests for deadline extensions. Suresh Kumar, from TeX Support Team of SPi Content Solutions–SPi Global, was extremely quick and efficient in helping me with some latex technical questions.

Of course, I want to acknowledge those scientists, photographers, and institutions that have granted permission to reproduce photographs. They are, in alphabetical order, AIP Emilio Segrè Visual Archives, R. Bachrach, N. F. Carnahan, CERN, E. G. D. Cohen, Cornell University, P. Cvitanović, M. E. Fisher, D. Frenkel, D. J. Henderson, Wm. G. Hoover, J. L. Lebowitz, J. K. Percus, B. Pratten, J. Russell,

University of Cambridge, M. S. Wertheim, B. Widom, Wikimedia Commons, and D. Yevick.

This is a good opportunity to express my deep appreciation to all the members of the research group (Statistical Physics in Extremadura, *SPhinX*) I belong to for their motivation and for creating the ripe environment for joyful social and scientific gatherings and discussions.

Last, but certainly not least, let me thank my wife Angeles for her continuous support and encouragement and for patiently putting up with my "Penelope-like" tendency to re-elaborate previous (and supposedly finished) parts of the text.

Contents

Acronyms

The number between parentheses refers to the page where the acronym is first introduced.

AHS	Additive hard spheres (p. 39)
BGHLL	Boublík–Grundke–Henderson–Lee–Levesque (p. 224)
BMCSL	Boublík–Mansoori–Carnahan–Starling–Leland (p. 224)
BS	Barrio–Solana (p. 79)
CS	Carnahan–Starling (p. 76)
DCF	Direct correlation function (p. 105)
EoS	Equation of state (p. 39)
GMSA	Generalized mean spherical approximation (p. 240)
H	Henderson (p. 72)
HD	Hard disk (p. 72)
HNC	Hypernetted-chain (p. 176)
HR	Hard rod (p. 146)
HS	Hard sphere (p. 37)
L	Luding (p. 74)
LDH	Linearized Debye–Hückel (p. 182)
LJ	Lennard-Jones (p. 37)
MC	Monte Carlo (p. 58)
MD	Molecular dynamics (p. 72)
MSA	Mean spherical approximation (p. 182)
NAHR	Nonadditive hard rods (p. 146)
NAHS	Nonadditive hard spheres (p. 39)
OZ	Ornstein–Zernike (p. 105)
PS	Penetrable sphere (p. 37)
PSW	Penetrable square well (p. 89)
PY	Percus–Yevick (p. 176)
RDF	Radial distribution function (p. 101)
RFA	Rational-function approximation (p. 204)
SHR	Sticky hard rod (p. 141)

Symbols

The number between parentheses refers to the page where the symbol is first introduced.

a	Helmholtz free energy per particle (p. 10)
a^{ex}	Excess Helmholtz free energy per particle (p. 79)
$a_{\mathrm{oc}}^{\mathrm{ex}}$	Excess Helmholtz free energy per particle of the one-component fluid (p. 79)
A_j	Numerical coefficients in $\Re(q)$ (p. 71)
$\mathscr{A}$	Auxiliary function related to a^{ex} (p. 82)
$\mathscr{A}_0$	Auxiliary function related to $\mathscr{A}$ (p. 83)
b_k	Rescaled virial coefficient of a hard-sphere fluid (p. 54)
$\hat{b}_k$	Rescaled virial coefficient of a hard-sphere mixture (p. 79)
$b_{k_1;k-k_1}$	Rescaled composition-independent virial coefficient in a binary AHS mixture (p. 52)
$\mathfrak{b}_\ell$	Cluster integral (p. 41)
B_k	Virial coefficient (p. 40)
$\widehat{B}_{k_1;k-k_1}$	Composition-independent virial coefficient in a binary mixture (p. 52)
$\widehat{B}_{\nu_1\nu_2\cdots\nu_k}$	Composition-independent virial coefficient in a mixture (p. 50)
$\widehat{B}_{\alpha\gamma\delta\varepsilon}^{(\mathrm{I,II,III})}$	Partial contributions to the composition-independent virial coefficient $\widehat{B}_{\alpha\gamma\delta\varepsilon}$ (p. 51)
$\mathrm{B}(a,b)$	Beta function (p. 58)
$\mathrm{B}_x(a,b)$	Incomplete beta function (p. 58)
$\mathscr{B}(r)$	"Bridge" (or "elementary") diagrams (p. 171)
$c(r)$	Direct correlation function (p. 105)
$\tilde{c}(k)$	Fourier transform of $c(r)$ (p. 106)
$c_{\alpha\gamma}(r)$	Direct correlation function in a mixture (p. 117)
$\tilde{c}_{\alpha\gamma}(k)$	Fourier transform of $c_{\alpha\gamma}(r)$ (p. 118)
$\check{\mathbf{c}}(k)$	Matrix with elements $n\sqrt{x_\alpha x_\gamma}\tilde{c}_{\alpha\gamma}(k)$ (p. 118)
C_j	Numerical coefficients in $\Re(q)$ (p. 71)
C_p	Heat capacity at constant pressure (p. 6)
C_V	Heat capacity at constant volume (p. 6)

C_N	Quantum of phase-space volume (p. 15)
$\mathscr{C}(r)$	"Chains" (or nodal diagrams) (p. 170)
d	Dimensionality of the system (p. 13)
$\widehat{D}(s)$	Determinant of the matrix $\delta_{\alpha\gamma} - \widehat{Q}_{\alpha\gamma}(s)$ (p. 133)
$D_a(z)$	Parabolic cylinder function (p. 56)
$e(r)$	Pair Boltzmann factor (p. 177)
E	Internal energy (p. 1)
$\widetilde{E}$	Most probable value of the energy in a closed system (p. 19)
$\langle E \rangle^{\mathrm{ex}}$	Excess average energy (p. 28)
$\langle E \rangle^{\mathrm{id}}$	Average energy of the ideal gas (p. 26)
$f(r)$	Mayer function (p. 34)
f_{ij}	Mayer function evaluated at $r = r_{ij}$ (p. 42)
$f_{\alpha\gamma}(r)$	Mayer function for a pair of species α and γ (p. 39)
$\tilde{f}_{\alpha\gamma}(k)$	Fourier transform of $f_{\alpha\gamma}(r)$ (p. 64)
$f_s(\mathbf{x}^s)$	s-body reduced distribution function (p. 97)
F	Helmholtz free energy (p. 4)
F^{ex}	Excess Helmholtz free energy (p. 28)
F^{id}	Helmholtz free energy of the ideal gas (p. 26)
$\widehat{F}(s)$	Auxiliary function related to the Laplace transform $\widehat{G}(s)$ (p. 207)
$g(r)$	Radial distribution function (p. 101)
$g_{\alpha\gamma}(r)$	Radial distribution function in a mixture (p. 116)
$g_s(\mathbf{r}^s)$	s-body correlation function (p. 100)
G	Gibbs free energy (p. 5)
G^{id}	Gibbs free energy of the ideal gas (p. 26)
$\widehat{G}(s)$	Laplace transform representation of $g(r)$ (pp. 127, 204, and 205)
$\widehat{G}_{\alpha\gamma}(s)$	Laplace transform representation of $g_{\alpha\gamma}(r)$ (pp. 205 and 218)
h	Planck constant (p. 15)
$h(r)$	Total correlation function (p. 102)
$h_{\alpha\gamma}(r)$	Total correlation function in a mixture (p. 117)
$h_s(\mathbf{r}^s)$	s-body cluster correlation function (p. 100)
$\tilde{h}(k)$	Fourier transform of $h(r)$ (p. 103)
$\tilde{h}_{\alpha\gamma}(k)$	Fourier transform of $h_{\alpha\gamma}(r)$ (p. 117)
$\check{h}(k)$	Matrix with elements $n\sqrt{x_\alpha x_\gamma}\tilde{h}_{\alpha\gamma}(k)$ (p. 118)
H	Enthalpy (p. 11)
$\widehat{H}(s)$	Laplace transform representation of $h(r)$ (pp. 131 and 205)
$\widehat{H}_{\alpha\gamma}(s)$	Laplace transform representation of $h_{\alpha\gamma}(r)$ (p. 218)
$H_N(\mathbf{x}^N)$	Hamiltonian of a one-component system (p. 13)
$H_N^{\mathrm{id}}(\mathbf{p}^N)$	Hamiltonian of the ideal gas (p. 26)
$H_{\{N_\nu\}}(\mathbf{x}^N)$	Hamiltonian of a mixture (p. 29)
I	Identity matrix (p. 118)
$I_x(a,b)$	Regularized beta function (p. 58)
$j_k(x)$	Spherical Bessel function (p. 205)
$J_\nu(x)$	Bessel function (p. 64)
$\widehat{J}$	Auxiliary quantity in one-dimensional mixtures (p. 134)

k	Wave number (p. 64)
k_B	Boltzmann constant (p. 10)
K	Normalization constant in $p^{(1)}(r)$ (p. 132)
$K_{\alpha\gamma}$	Normalization constant in $p_{\alpha\gamma}^{(1)}(r)$ (p. 128)
$\mathscr{K}_x^{(k)}$	kth cumulant of a random variable x (p. 19)
L	Length of a one-dimensional system (p. 125)
$L^{(k)}$	Parameters in the solution of the Percus–Yevick equation for hard spheres (p. 209) and in the rational-function approximation for hard spheres (p. 238) and square-well fluids (p. 241)
$\bar{L}^{(k)}$	Parameters in the rational-function approximation for square-well fluids (p. 241)
$\mathscr{L}$	Laplace transform (p. 205)
$\mathscr{L}^{-1}$	Inverse Laplace transform (p. 137)
m	Mass of a particle (p. 26)
m_k	Reduced kth moment of the size distribution in a hard-sphere mixture (p. 79)
M_k	kth moment of the size distribution in a hard-sphere mixture (p. 63)
n	Total number density (p. 2)
$n_s(\mathbf{r}^s)$	s-body configurational distribution function (p. 98)
n_s^{id}	s-body configurational distribution function of the ideal gas (p. 100)
n^*	Reduced number density (pp. 103 and 136)
n_v	Number density of species v (p. 2)
$n_{\alpha\gamma}$	Pair configurational distribution function in a mixture (p. 115)
N	Total number of particles (p. 1)
N_v	Number of particles of species v (p. 1)
$\widetilde{N}$	Most probable value of the number of particles in an open system (p. 22)
p	Pressure (p. 2)
p^{ex}	Excess pressure (p. 28)
p^{id}	Pressure of the ideal gas (p. 26)
$p^{(\ell)}(r)$	ℓth neighbor distribution function (p. 126)
$p_{\alpha\gamma}^{(\ell)}(r)$	ℓth neighbor distribution function in a mixture (p. 131)
$\mathbf{p}_i$	Momentum vector of particle i (p. 13)
$\mathbf{p}^N$	Set of N momentum vectors (p. 13)
$\widehat{P}^{(\ell)}(s)$	Laplace transform of $p^{(\ell)}(r)$ (p. 126)
$\widehat{P}_{\alpha\gamma}^{(\ell)}(s)$	Laplace transform of $p_{\alpha\gamma}^{(\ell)}(r)$ (p. 132)
$\widehat{\mathsf{P}}^{(1)}(s)$	Matrix of elements $\widehat{P}_{\alpha\gamma}^{(1)}(s)$ (p. 132)
$\mathscr{P}(r)$	Open "parallel" diagrams (or open "bundles") (p. 171)
$\mathscr{P}^+(r)$	"Parallel" diagrams (or "bundles") (p. 172)
$\mathscr{P}(N)$	Number probability distribution function in the grand canonical ensemble (p. 22)
$\mathscr{P}^{\mathrm{id}}(N)$	Number probability distribution function of the ideal gas in the grand canonical ensemble (p. 31)

$\mathscr{P}_N(E)$	Energy probability distribution function in the canonical ensemble (p. 19)
$\mathscr{P}_N^{\mathrm{id}}(E)$	Energy probability distribution function of the ideal gas in the canonical ensemble (p. 31)
$\mathscr{P}_N(V)$	Volume probability distribution function in the isothermal–isobaric ensemble (p. 24)
$\mathscr{P}_N^{\mathrm{id}}(V)$	Volume probability distribution function of the ideal gas in the isothermal–isobaric ensemble (p. 31)
q	Size ratio (p. 70)
q_0	Special value of the size ratio (p. 69)
$\widehat{Q}(s)$	Auxiliary functions related to $\widehat{P}_{\alpha\gamma}^{(1)}(s)$ (p. 133)
$\mathscr{Q}_N$	Configuration integral (p. 27)
$\mathscr{Q}_{\{N_\nu\}}$	Configuration integral of a mixture (p. 92)
r_0	Location of the minimum of the interaction potential (p. 56)
r_0^*	Location of the minimum of the interaction potential in reduced units (p. 57)
r_{ij}	Distance between the centers of particles i and j (p. 33)
$\mathbf{r}_i$	Position vector of particle i (p. 13)
$\mathbf{r}^N$	Set of N position vectors (p. 13)
R	Auxiliary quantity in one-dimensional mixtures (p. 133)
$\Re(q)$	Auxiliary function in $b_{2;2}(q)$ (p. 71)
s	Parameter of the generalized Lennard-Jones potential (p. 56)
S	Entropy (p. 1)
S^{id}	Entropy of the ideal gas (p. 26)
$\widetilde{S}(k)$	Structure factor (p. 103)
$\widetilde{S}_{\alpha\gamma}(k)$	Structure factor in a mixture (p. 117)
$\widetilde{\mathsf{S}}(k)$	Matrix with elements $\widetilde{S}_{\alpha\gamma}(k)$ (p. 118)
$S^{(k)}$	Parameters in the solution of the Percus–Yevick equation for hard spheres (p. 209) and in the rational-function approximation for hard spheres (p. 238) and square-well fluids (p. 241)
$\mathscr{S}[\rho_N]$	Gibbs entropy functional (p. 15)
T	Temperature (p. 2)
T^*	Reduced temperature (p. 55)
T_B	Boyle temperature (p. 57)
T_B^*	Reduced Boyle temperature (p. 57)
u	Internal energy per particle (p. 10)
u^{ex}	Excess internal energy per particle (p. 112)
$u_{\mathrm{ext}}(\mathbf{r})$	External potential (p. 158)
U_ℓ	Cluster function (p. 44)
v_d	Volume of a d-dimensional sphere of unit diameter (p. 53)
V	Volume (p. 1)
$\widetilde{V}$	Most probable value of the volume in the isothermal–isobaric system (p. 25)
V_0	Arbitrary volume scale factor (p. 15)

$\mathscr{V}_{a,b}(r)$ Intersection volume of two spheres of radii a and b whose centers are separated a distance r (p. 65)

$\bar{\mathscr{V}}_{a,b}(r)$ Relevant part of $\mathscr{V}_{a,b}(r)$ (p. 65)

w Auxiliary function in the Percus–Yevick solution for sticky hard spheres (p. 227)

$w(r)$ Shifted cavity function (p. 183)

$\widetilde{w}(k)$ Fourier transform of $w(r)$ (p. 183)

W_N N-body functions in the expansion of Ξ in powers of fugacity (p. 42)

x_ν Mole fraction of species ν (p. 1)

$x(\sigma)$ Continuous size distribution in a polydisperse hard-sphere system (p. 87)

$\mathbf{x}^N$ Point in the N-body phase space (p. 13)

X Generic thermodynamic quantity (p. 41)

X_k Coefficient in the expansion of X in powers of density (p. 41)

$\overline{X}_\ell$ Coefficient in the expansion of X in powers of fugacity (p. 41)

$y(r)$ Cavity function (p. 104)

$y_{\alpha\gamma}(r)$ Cavity function in a mixture (p. 117)

z Fugacity (p. 17)

$\hat{z}$ Rescaled fugacity (p. 28)

Z Compressibility factor (p. 10)

Z_{oc} Compressibility factor of the one-component fluid (p. 78)

$\mathscr{Z}_N$ Partition function (p. 18)

$\mathscr{Z}_N^{\mathrm{id}}$ Partition function of the ideal gas (p. 26)

$\mathscr{Z}_{\{N_\nu\}}$ Partition function of a mixture (p. 29)

α Opposite of chemical potential divided by thermal energy (p. 17)

α_p Thermal expansivity (p. 7)

$\alpha_\ell(\mathbf{r}_1, \mathbf{r}_2)$ Coefficients in the expansion of $n_2(\mathbf{r}_1, \mathbf{r}_2)$ in powers of fugacity (p. 163)

β Inverse temperature parameter (p. 10)

γ Pressure divided by thermal energy (p. 23)

$\gamma_k(\mathbf{r}_1, \mathbf{r}_2)$ Coefficients in the expansion of $n_2(\mathbf{r}_1, \mathbf{r}_2)$ in powers of density (p. 164)

$\bar{\gamma}(r)$ Indirect correlation function (p. 181)

$\Gamma(x)$ Gamma function (p. 26)

$\delta(x)$ Dirac delta function (p. 16)

Δa^* Surplus Helmholtz free energy per particle (p. 84)

Δa_{oc}^* Surplus Helmholtz free energy per particle of the one-component fluid (p. 85)

$\Delta b_{3;1}$ Rescaled partial contribution to $\widehat{B}_{3;1}$ (p. 70)

Δp^* Surplus pressure (p. 84)

Δp_{oc}^* Surplus pressure of the one-component fluid (p. 85)

ΔE Energy tolerance (p. 16)

Δ_N Isothermal–isobaric partition function (p. 23)

Δ_N^{id} Isothermal–isobaric partition function of the ideal gas (p. 26)

$\Delta_{\{N_\nu\}}$	Isothermal–isobaric partition function of a mixture (p. 29)
ε	Energy scale of the interaction potential (p. 37)
ζ	One-particle partition function (p. 26)
η	Packing fraction (pp. 54 and 78)
η_{cp}	Close-packing value of η (p. 74)
$\theta(\mathbf{r})$	Boltzmann factor associated with the external potential $u(\mathbf{r})$ (p. 158)
$\theta_k(x)$	Reverse Bessel polynomials (p. 205)
$\Theta(x)$	Heaviside step function (p. 38)
ϑ_i	Species of particle i (p. 115)
κ_T	Isothermal compressibility (p. 7)
Λ	Thermal de Broglie wavelength (p. 27)
Λ_α	Thermal de Broglie wavelength in a mixture (p. 30)
$\Lambda^{(k)}$	Parameters in the solution of the Percus–Yevick equation for hard spheres (p. 209) and sticky hard spheres (p. 225)
$\Lambda_{\alpha\gamma}(s)$	Auxiliary functions in the solution of the Percus–Yevick equation for sticky-hard-sphere mixtures (p. 220)
$\Lambda_{\alpha\gamma}^{(k)}$	Parameters in the solution of the Percus–Yevick equation for sticky-hard-sphere mixtures (p. 220)
μ	Chemical potential (p. 9)
$\overline{\mu}$	Species-averaged chemical potential (p. 5)
μ^{ex}	Excess chemical potential (p. 28)
μ^{id}	Chemical potential of the ideal gas (p. 26)
μ_ν	Chemical potential of species ν (p. 2)
μ_ν^{ex}	Excess chemical potential of species ν (p. 119)
ξ	Coupling parameter (p. 110)
Ξ	Grand partition function (p. 20)
Ξ^{id}	Grand partition function of the ideal gas (p. 26)
ϖ_k	Parameter defined as a combination of moments of the size distribution (p. 82)
$\Pi_{a,b}(x)$	Boxcar function (p. 16)
$\rho_N(\mathbf{x}^N)$	Phase-space probability density (p. 14)
$\rho_{\{N_\nu\}}(\mathbf{x}^N)$	Phase-space probability density of a mixture (p. 29)
σ	Length scale of the interaction potential (p. 37)
σ_α	Diameter of a sphere of species α (p. 39)
$\sigma_{\alpha\gamma}$	Closest distance between two spheres of species α and γ (p. 39)
$\varsigma_\eta, \varsigma_p$	Parameters combining the second and third virial coefficients of a hard-sphere mixture (p. 84)
$\Sigma_{\alpha\gamma}(s)$	Auxiliary functions in the solution of the Percus–Yevick equation for sticky-hard-sphere mixtures (p. 220)
$\overline{\Sigma}_{\alpha\gamma}$	Parameters in the solution of the Percus–Yevick equation for sticky-hard-sphere mixtures (p. 221)
τ	Inverse stickiness parameter (p. 38)
$\tau_{\alpha\gamma}$	Inverse stickiness parameter in a mixture (p. 219)
$\widehat{\Upsilon}$	Laplace transform of $\phi(r)e^{-\beta\phi(r)}$ (p. 130)

$\phi(r)$	Interaction potential (p. 33)
$\phi^*(r^*)$	Reduced interaction potential (p. 57)
$\phi_{\alpha\gamma}(r)$	Interaction potential for a pair of species α and γ (p. 39)
$\Phi_N(\mathbf{r}^N)$	Total potential energy (p. 27)
$\Phi_{\{N_v\}}(\mathbf{r}^N)$	Total potential energy of a mixture (p. 115)
$\varphi_k(x)$	Auxiliary mathematical function (p. 210)
χ_T	Isothermal susceptibility (p. 10)
$\chi_{T,k}$	Coefficient in the expansion of χ_T in powers of density (p. 185)
$\psi(r)$	Potential of mean force (p. 165)
$\Psi_\ell(r)$	Partial contribution to $g(r)$ in the one-dimensional sticky-hard-rod fluid (p. 145)
$\overline{\Psi}_\ell(r)$	Partial contribution to $g(r)$ in the three-dimensional hard-sphere and square-well fluids (p. 207)
$\Psi_\ell^{(j)}(r)$	Partial contribution to $g(r)$ in the one-dimensional square-well fluid (p. 137)
$\Psi_\ell^{(j_1,j_2)}(r)$	Partial contribution to $g_{\alpha\gamma}(r)$ in the one-dimensional nonadditive hard-rod fluid mixture (p. 151)
$\omega_{\Delta E}$	Microcanonical partition function (p. 16)
$\omega_{\Delta E}^{id}$	Microcanonical partition function of the ideal gas (p. 26)
Ω	Grand potential (p. 6)
Ω^{id}	Grand potential of the ideal gas (p. 26)
$\widehat{\Omega}(s)$	Laplace transform of $e^{-\beta\phi(r)}$ (p. 129)
$\widehat{\Omega}_{\alpha\gamma}(s)$	Laplace transform of $e^{-\beta_{\alpha\gamma}\phi(r)}$ (p. 132)
$\langle\cdots\rangle$	Statistical ensemble average (p. 14)

List of Photographs

The number between parentheses refers to the page where the photograph is reproduced.

Chapter 1
Summary of Thermodynamic Potentials

This chapter provides a brief overview of some of the most important thermodynamic relations that may appear in the book. It also serves to fix notation and nomenclature. Special attention is given to the thermodynamic potential appropriate to each set of possible thermodynamic variables. For generality, the relations are referred to multicomponent fluid systems (mixtures) rather than to pure systems. Otherwise, the treatment is restricted to simple classical fluids and does not include magnetic or electric properties, external forces, surface properties, quantum effects, etc.

1.1 Entropy: Isolated Systems

In a reversible process, the first and second laws of thermodynamics in a fluid mixture can be combined as [1, 2]

$$T\mathrm{d}S = \mathrm{d}E + p\mathrm{d}V - \sum_\nu \mu_\nu \mathrm{d}N_\nu \;, \tag{1.1}$$

where S is the entropy, E is the internal energy, V is the volume of the fluid, and N_ν is the number of particles of species ν. The total number of particles (N) and the mole fraction of each species (x_ν) are defined as

$$N = \sum_\nu N_\nu \;, \quad x_\nu = \frac{N_\nu}{N} \;. \tag{1.2}$$

© Springer International Publishing Switzerland 2016
A. Santos, *A Concise Course on the Theory of Classical Liquids*,
Lecture Notes in Physics 923, DOI 10.1007/978-3-319-29668-5_1

The partial and total number densities are

$$n_\nu = \frac{N_\nu}{V}, \quad n = \frac{N}{V} = \sum_\nu n_\nu. \tag{1.3}$$

The quantities S, E, V, and N_ν are *extensive*, i.e., they scale with the size of the system. The coefficients of the differentials in (1.1) are the conjugate *intensive* quantities: the absolute temperature (T), the pressure (p), and the chemical potentials (μ_ν).

Equation (1.1) shows that the *natural* variables of the entropy are E, V, and $\{N_\nu\}$, i.e., $S(E, V, \{N_\nu\})$. This implies that S is the right thermodynamic potential in *isolated* systems: at given E, V, and $\{N_\nu\}$, S is maximal in equilibrium. The respective partial derivatives give the intensive quantities:

$$\frac{1}{T} = \left(\frac{\partial S}{\partial E}\right)_{V,\{N_\nu\}}, \tag{1.4a}$$

$$\frac{p}{T} = \left(\frac{\partial S}{\partial V}\right)_{E,\{N_\nu\}}, \tag{1.4b}$$

$$\frac{\mu_\nu}{T} = -\left(\frac{\partial S}{\partial N_\nu}\right)_{E,V,\{N_{\gamma \neq \nu}\}}. \tag{1.4c}$$

The extensive nature of S, E, V, and $\{N_\nu\}$ implies the extensivity condition

$$S(\lambda E, \lambda V, \{\lambda N_\nu\}) = \lambda S(E, V, \{N_\nu\}), \tag{1.5}$$

where λ is any positive number. Application of Euler's homogeneous function theorem yields

$$S(E, V, \{N_\nu\}) = E\left(\frac{\partial S}{\partial E}\right)_{V,\{N_\nu\}} + V\left(\frac{\partial S}{\partial V}\right)_{E,\{N_\nu\}} + \sum_\nu N_\nu \left(\frac{\partial S}{\partial N_\nu}\right)_{E,V,\{N_{\gamma \neq \nu}\}}. \tag{1.6}$$

Using (1.4), we obtain the identity

$$\boxed{TS = E + pV - \sum_\nu \mu_\nu N_\nu.} \tag{1.7}$$

This is the so-called *fundamental equation of thermodynamics*. Differentiating (1.7) and subtracting (1.1) one arrives at the Gibbs–Duhem relation (see Figs. 1.1 and 1.2), i.e.,

$$SdT - Vdp + \sum_\nu N_\nu d\mu_\nu = 0. \tag{1.8}$$

Equation (1.1) can be inverted to express dE as a linear combination of dS, dV, and $\{dN_\nu\}$. Therefore, S, V, and $\{N_\nu\}$, are the natural variables of the internal energy $E(S, V, \{N_\nu\})$, so that

$$T = \left(\frac{\partial E}{\partial S}\right)_{V,\{N_\nu\}} , \qquad (1.9a)$$

$$p = -\left(\frac{\partial E}{\partial V}\right)_{S,\{N_\nu\}} , \qquad (1.9b)$$

$$\mu_\nu = \left(\frac{\partial E}{\partial N_\nu}\right)_{S,V,\{N_{\gamma \neq \nu}\}} . \qquad (1.9c)$$

As a consequence, in an *adiabatic* system (i.e., at fixed S, V, and $\{N_\nu\}$), the internal energy is minimal in equilibrium.

1.2   Helmholtz Free Energy: Closed Systems

From a practical point of view, it is usually more convenient to choose the temperature instead of the internal energy or the entropy as a control variable. In that case, the adequate thermodynamic potential is no longer either the entropy or the internal energy, respectively, but the Helmholtz free energy F (after Hermann L. F. von Helmholtz, see Fig. 1.3). This thermodynamic potential is defined from S or E through the *Legendre transformation*

$$F(T, V, \{N_v\}) \equiv E - TS = -pV + \sum_v \mu_v N_v ,\qquad(1.10)$$

where in the last step use has been made of (1.7). From (1.1) we obtain

$$dF = -SdT - pdV + \sum_v \mu_v dN_v ,\qquad(1.11)$$

so that

$$S = -\left(\frac{\partial F}{\partial T}\right)_{V,\{N_v\}} ,\qquad(1.12a)$$

$$p = -\left(\frac{\partial F}{\partial V}\right)_{T,\{N_v\}} ,\qquad(1.12b)$$

$$\mu_v = \left(\frac{\partial F}{\partial N_v}\right)_{T,V,\{N_{\gamma \neq v}\}} .\qquad(1.12c)$$

Fig. 1.3 Hermann Ludwig Ferdinand von Helmholtz (1821–1894) (Photograph from Wikimedia Commons, http://commons. wikimedia.org/wiki/File: Hermann_von_Helmholtz. jpg)

The Helmholtz free energy is the adequate thermodynamic potential in a closed system, that is, a system that cannot exchange mass with the environment but can exchange energy. At fixed T, V, and $\{N_v\}$, F is minimal in equilibrium.

1.3   Gibbs Free Energy: Isothermal–Isobaric Systems

If, instead of the volume, the independent thermodynamic variable is pressure, we need to perform an extra Legendre transformation from F to define the free enthalpy or Gibbs free energy (after Josiah W. Gibbs, see Fig. 1.1) as

$$G(T, p, \{N_v\}) \equiv F + pV = \sum_v \mu_v N_v \ . \tag{1.13}$$

The second equality shows that the chemical potential μ_v can be interpreted as the contribution of each particle of species v to the total Gibbs free energy. Thus, the Gibbs free energy per particle can be viewed as a species-averaged chemical potential,

$$\frac{G}{N} = \sum_v x_v \mu_v \equiv \bar{\mu} \ . \tag{1.14}$$

The differential relations for G become

$$dG = -SdT + Vdp + \sum_v \mu_v dN_v \ , \tag{1.15}$$

$$S = -\left(\frac{\partial G}{\partial T}\right)_{p,\{N_v\}} \ , \tag{1.16a}$$

$$V = \left(\frac{\partial G}{\partial p}\right)_{T,\{N_v\}} \ , \tag{1.16b}$$

$$\mu_v = \left(\frac{\partial G}{\partial N_v}\right)_{T,p,\{N_{y \neq v}\}} \ . \tag{1.16c}$$

Needless to say, G is minimal in equilibrium if one fixes T, p, and $\{N_v\}$.

1.4   Grand Potential: Open Systems

In an open system, not only energy but also particles can be exchanged with the environment. In that case, we need to replace $\{N_\nu\}$ by $\{\mu_\nu\}$ as independent variables and define the grand potential Ω from F via a new Legendre transformation:

$$\Omega(T, V, \{\mu_\nu\}) \equiv F - \sum_\nu \mu_\nu N_\nu = -pV. \qquad (1.17)$$

Interestingly, the second equality shows that $-\Omega/V$ is not but the pressure, except that it must be seen as a function of temperature and the chemical potentials. Now we have

$$d\Omega = -SdT - pdV - \sum_\nu N_\nu d\mu_\nu , \qquad (1.18)$$

$$S = -\left(\frac{\partial \Omega}{\partial T}\right)_{V,\{\mu_\nu\}} , \qquad (1.19a)$$

$$p = -\left(\frac{\partial \Omega}{\partial V}\right)_{T,\{\mu_\nu\}} = -\frac{\Omega}{V} , \qquad (1.19b)$$

$$N_\nu = -\left(\frac{\partial \Omega}{\partial \mu_\nu}\right)_{T,V,\{\mu_\gamma \neq \nu\}} . \qquad (1.19c)$$

It must be stressed that all the equilibrium thermodynamic potentials are *point* functions. This means that, in contrast to the case of path functions (like work and heat), their magnitudes depend on the thermodynamic state only and not on how the system reaches that state.

1.5   Response Functions and Maxwell Relations

We have seen that the thermodynamic variables $E \leftrightarrow T$ (or $S \leftrightarrow T$), $V \leftrightarrow p$, and $N_\nu \leftrightarrow \mu_\nu$ appear as extensive $\leftrightarrow$ intensive conjugate pairs. Depending on the thermodynamic potential of interest, one of the members of the pair acts as independent variable and the other one is obtained by differentiation, as (1.4), (1.9), (1.12), (1.16), and (1.19) show. Those relations are also displayed in Table 1.1. If an additional derivative is taken, one then obtains the so-called *response* functions. For example, the heat capacities at constant volume and at constant pressure are defined

Table 1.1 Thermodynamic variables as derived from different thermodynamic potentials

Variable	Thermodynamic potentials $S(E,V,\{N_\nu\})$	$E(S,V,\{N_\nu\})$	$F(T,V,\{N_\nu\})$	$G(T,p,\{N_\nu\})$	$\Omega(T,V,\{\mu_\nu\})$
T	$\left(\dfrac{\partial S}{\partial E}\right)_{V,\{N_\nu\}}^{-1}$	$\left(\dfrac{\partial E}{\partial S}\right)_{V,\{N_\nu\}}$	✓	✓	✓
p	$T\left(\dfrac{\partial S}{\partial V}\right)_{E,\{N_\nu\}}$	$-\left(\dfrac{\partial E}{\partial V}\right)_{S,\{N_\nu\}}$	$-\left(\dfrac{\partial F}{\partial V}\right)_{T,\{N_\nu\}}$	✓	$-\dfrac{\Omega}{V}$
μ_ν	$-T\left(\dfrac{\partial S}{\partial N_\nu}\right)_{E,V,\{N_\gamma\}}$	$\left(\dfrac{\partial E}{\partial N_\nu}\right)_{S,V,\{N_\gamma\}}$	$\left(\dfrac{\partial F}{\partial N_\nu}\right)_{T,V,\{N_\gamma\}}$	$\left(\dfrac{\partial G}{\partial N_\nu}\right)_{T,p,\{N_\gamma\}}$	✓
E	✓		$F+TS$	$G+TS-pV$	$\Omega+TS$ $+\sum_\nu \mu_\nu N_\nu$
S		✓	$-\left(\dfrac{\partial F}{\partial T}\right)_{V,\{N_\nu\}}$	$-\left(\dfrac{\partial G}{\partial T}\right)_{p,\{N_\nu\}}$	$-\left(\dfrac{\partial \Omega}{\partial T}\right)_{V,\{\mu_\nu\}}$
V	✓	✓	✓	$\left(\dfrac{\partial G}{\partial p}\right)_{T,\{N_\nu\}}$	✓
N_ν	✓	✓	✓	✓	$-\left(\dfrac{\partial \Omega}{\partial \mu_\nu}\right)_{T,p,\{\mu_\gamma\}}$

The check marks denote the independent variables for each potential

as

$$C_V \equiv \left(\frac{\partial E}{\partial T}\right)_{V,\{N_\nu\}} = T\left(\frac{\partial S}{\partial T}\right)_{V,\{N_\nu\}} = T\left(\frac{\partial^2 F}{\partial T^2}\right)_{V,\{N_\nu\}}, \qquad (1.20a)$$

$$C_p \equiv T\left(\frac{\partial S}{\partial T}\right)_{p,\{N_\nu\}} = -T\left(\frac{\partial^2 G}{\partial T^2}\right)_{p,\{N_\nu\}}. \qquad (1.20b)$$

Another response function is the thermal expansivity

$$\alpha_p \equiv \frac{1}{V}\left(\frac{\partial V}{\partial T}\right)_{p,\{N_\nu\}} = \frac{1}{V}\left(\frac{\partial^2 G}{\partial T \partial p}\right)_{\{N_\nu\}} = -\frac{1}{V}\left(\frac{\partial S}{\partial p}\right)_{T,\{N_\nu\}}. \qquad (1.21)$$

The equivalence between the second and fourth terms in (1.21) is an example of a *Maxwell relation* (see Fig. 1.4), i.e., the equivalence between the two second derivatives of a thermodynamic potential with respect to two given variables. Another simple Maxwell relation is

$$\left(\frac{\partial p}{\partial \mu_\nu}\right)_{T,V,\{\mu_{\gamma}\neq\nu\}} = -\frac{1}{V}\left(\frac{\partial \Omega}{\partial \mu_\nu}\right)_{T,V,\{\mu_{\gamma}\neq\nu\}} = \frac{N_\nu}{V} = n_\nu. \qquad (1.22)$$

An important response function is the isothermal compressibility. It is defined as

Fig. 1.4 James Clerk
Maxwell (1831–1879)
(Photograph from Wikimedia
Commons, https://commons.
wikimedia.org/wiki/File:
James_Clerk_Maxwell_big.
jpg)

$$\kappa_T \equiv -\frac{1}{V}\left(\frac{\partial V}{\partial p}\right)_{T,\{N_\nu\}} = -\frac{1}{V}\left(\frac{\partial^2 G}{\partial p^2}\right)_{T,\{N_\nu\}}. \tag{1.23}$$

Equivalently, the inverse isothermal compressibility is given by

$$\kappa_T^{-1} \equiv -V\left(\frac{\partial p}{\partial V}\right)_{T,\{N_\nu\}} = V\left(\frac{\partial^2 F}{\partial V^2}\right)_{T,\{N_\nu\}}. \tag{1.24}$$

The isothermal compressibility can also be expressed in other alternative ways. Starting from the mathematical identity

$$\left(\frac{\partial x}{\partial y}\right)_z \left(\frac{\partial y}{\partial x}\right)_x \left(\frac{\partial z}{\partial x}\right)_y = -1 , \tag{1.25}$$

and particularizing to a one-component system, one has

$$\begin{aligned} -1 &= \left(\frac{\partial N}{\partial p}\right)_{T,V} \left(\frac{\partial p}{\partial V}\right)_{T,N} \left(\frac{\partial V}{\partial N}\right)_{T,p} \\ &= \left(\frac{\partial N}{\partial p}\right)_{T,V} \left(-\frac{1}{V\kappa_T}\right)\frac{V}{N} , \end{aligned} \tag{1.26}$$

where in the second step use has been made of the first equality of (1.24) as well as of the property $(\partial V/\partial N)_{T,p} = V/N$ (as a consequence of the intensive nature of T and p). Thus, (1.26) implies

$$\kappa_T = \frac{1}{N}\left(\frac{\partial N}{\partial p}\right)_{T,V}. \tag{1.27}$$

Comparison with the first equality of (1.23) shows that what actually matters in κ_T is the relative increase of density $n = N/V$ upon an increase of pressure, i.e.,

$$\kappa_T = \frac{1}{n} \left(\frac{\partial n}{\partial p} \right)_T . \tag{1.28}$$

In (1.23) the increase of density takes place by decreasing the volume at constant number of particles, while in (1.27) it takes place by increasing the number of particles at constant volume.

Yet another equivalent form for κ_T is obtained from the chain rule relation

$$\left(\frac{\partial N}{\partial p} \right)_{T,V} = \left(\frac{\partial N}{\partial \mu} \right)_{T,V} \left(\frac{\partial \mu}{\partial p} \right)_{T,V} = \left(\frac{\partial N}{\partial \mu} \right)_{T,V} \frac{1}{n} , \tag{1.29}$$

where in the second step the result (1.22) for a one-component system of chemical potential μ has been used. Therefore, (1.27) yields

$$\kappa_T = \frac{1}{nN} \left(\frac{\partial N}{\partial \mu} \right)_{T,V} . \tag{1.30}$$

Equations (1.27) and (1.30) are not difficult to extend to the case of mixtures. In the same spirit as in (1.28), we note that, being an intensive quantity, p depends on V and $\{N_\nu\}$ only through the number densities $\{n_\nu\}$. Thus,

$$\kappa_T^{-1} = -V \left(\frac{\partial p}{\partial V} \right)_{T,\{N_\nu\}} = \sum_\alpha n_\alpha \left(\frac{\partial p}{\partial n_\alpha} \right)_{T,\{n_{\nu \neq \alpha}\}} = \sum_\alpha N_\alpha \left(\frac{\partial p}{\partial N_\alpha} \right)_{T,V,\{N_{\nu \neq \alpha}\}} . \tag{1.31}$$

This generalizes (1.27). Next, in analogy with (1.29), we use

$$\left(\frac{\partial p}{\partial N_\alpha} \right)_{T,V,\{N_{\nu \neq \alpha}\}} = \sum_\gamma \left(\frac{\partial p}{\partial \mu_\gamma} \right)_{T,V,\{\mu_{\nu \neq \gamma}\}} \left(\frac{\partial \mu_\gamma}{\partial N_\alpha} \right)_{T,V,\{N_{\nu \neq \alpha}\}}$$

$$= \sum_\gamma n_\gamma \left(\frac{\partial \mu_\gamma}{\partial N_\alpha} \right)_{T,V,\{N_{\nu \neq \alpha}\}} , \tag{1.32}$$

where in the second step use has been made of (1.22) again. Insertion of (1.32) into (1.31) yields

$$\kappa_T^{-1} = V \sum_\alpha \sum_\gamma n_\alpha n_\gamma \left(\frac{\partial \mu_\gamma}{\partial N_\alpha} \right)_{T,V,\{N_{\nu \neq \alpha}\}} , \tag{1.33}$$

what represents the multicomponent generalization of (1.30).

Since the isothermal compressibility has dimensions of inverse pressure, it is convenient to define the isothermal susceptibility

$$\chi_T = nk_BT\kappa_T \tag{1.34}$$

as a closely related dimensionless quantity.

To conclude this chapter, and for further use, we introduce the inverse temperature parameter β, the compressibility factor Z (not to be confused with the isothermal compressibility κ_T), the Helmholtz free energy per particle a, and the internal energy per particle u:

$$\beta \equiv \frac{1}{k_BT}, \quad Z \equiv \frac{\beta p}{n}, \quad a \equiv \frac{F}{N}, \quad u \equiv \frac{E}{N}. \tag{1.35}$$

Here, k_B is the Boltzmann constant. In terms of the quantities defined in (1.35), (1.12) can be rewritten as

$$u = \frac{\partial(\beta a)}{\partial \beta}, \tag{1.36a}$$

$$Z = n\frac{\partial(\beta a)}{\partial n}, \tag{1.36b}$$

$$\mu_\nu = \frac{\partial(na)}{\partial n_\nu}. \tag{1.36c}$$

In (1.36a) and (1.36b) the intensive potential a is seen as a function of $(\beta, n, \{x_\nu\})$, while it is seen as a function of $(\beta, \{n_\nu\})$ in (1.36c). The fundamental equation of thermodynamics (1.7) is equivalent to

$$\sum_\nu x_\nu \beta \mu_\nu = \frac{\partial(n\beta a)}{\partial n} = \beta a + Z. \tag{1.37}$$

Moreover, from (1.36a) and (1.36b) we can easily derive the Maxwell relation

$$n\frac{\partial u}{\partial n} = \frac{\partial Z}{\partial \beta}. \tag{1.38}$$

Also, (1.28) and (1.34) can be combined to yield

$$\chi_T^{-1} = \frac{\partial(nZ)}{\partial n}. \tag{1.39}$$

Exercises

1.1 The enthalpy is defined by the Legendre transformation $H = E + pV$. What are then the "natural" variables of H? Add a new column in Table 1.1 with the relations corresponding to H.

1.2 Construct a thermodynamic potential whose natural variables are S, V, and $\{\mu_\nu\}$. Add a new column in Table 1.1 with the relations corresponding to this new potential.

1.3 Prove that it is not possible to construct a thermodynamic potential whose natural variables are T, p, and $\{\mu_\nu\}$. Hint: Use the Gibbs–Duhem relation (1.8).

1.4 Check (1.20)–(1.24).

1.5 Check the steps leading to (1.33).

1.6 Prove (1.36).

References

1. M.W. Zemansky, *Heat and Thermodynamics* (McGraw-Hill, New York, 1981)
2. H.B. Callen, *Thermodynamics and an Introduction to Thermostatistics* (Wiley, New York, 1985)

Chapter 2
Summary of Equilibrium Statistical Ensembles

In this chapter a summary of the main equilibrium ensembles is presented, essentially to fix part of the notation that will be needed later on. The phase-space probability density associated with each ensemble is derived by maximization of the Gibbs entropy under the appropriate constraints. For simplicity, most of this chapter is restricted to one-component systems, although the extension to mixtures is straightforward and is presented in the last section.

2.1 Phase Space

Let us consider a *classical* system made of N *identical* (and hence *indistinguishable*) point particles enclosed in a volume V in d dimensions. In classical mechanics, the dynamical state of the system is characterized by the N position vectors $\{\mathbf{r}_1, \mathbf{r}_2, \ldots, \mathbf{r}_N\}$ and the N momentum vectors $\{\mathbf{p}_1, \mathbf{p}_2, \ldots, \mathbf{p}_N\}$. In what follows, we will employ the following short-hand notation

- $\mathbf{r}^N = \{\mathbf{r}_1, \mathbf{r}_2, \ldots, \mathbf{r}_N\}$, $\qquad d\mathbf{r}^N = d\mathbf{r}_1 d\mathbf{r}_2 \cdots d\mathbf{r}_N$,
- $\mathbf{p}^N = \{\mathbf{p}_1, \mathbf{p}_2, \ldots, \mathbf{p}_N\}$, $\qquad d\mathbf{p}^N = d\mathbf{p}_1 d\mathbf{p}_2 \cdots d\mathbf{p}_N$,
- $\mathbf{x}^N = \{\mathbf{r}^N, \mathbf{p}^N\}$, $\qquad d\mathbf{x}^N = d\mathbf{r}^N d\mathbf{p}^N$.

Thus, the whole *microscopic* state of the system (*microstate*) is represented by a single point $\mathbf{x}^N$ in the $(2d \times N)$-dimensional *phase space* (see Fig. 2.1). The time evolution of the microstate $\mathbf{x}^N$ is governed by the Hamiltonian of the system $H_N(\mathbf{x}^N)$ through the classical Hamilton's equations [1].

Henceforth, and in order to make contact with thermodynamics, we will generally assume that the number of particles N and the volume V are so large that specific

© Springer International Publishing Switzerland 2016
A. Santos, *A Concise Course on the Theory of Classical Liquids*,
Lecture Notes in Physics 923, DOI 10.1007/978-3-319-29668-5_2

Fig. 2.1 Sketch of the phase
space of a system of N
identical particles. The
horizontal axis represents the
$d \times N$ position variables (d
components for each
particle), while the *vertical
axis* represents the $d \times N$
momentum variables. A
differential phase-space
volume $d\mathbf{x}^N$ around a point
$\mathbf{x}^N$ is represented

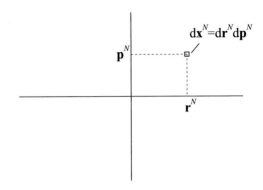

quantities (i.e., extensive quantities per particle or per unit volume) are independent
of N or V. This is equivalent to formally taking the so-called *thermodynamic limit*,
whereby

$$\left. \begin{array}{c} N \to \infty \\ V \to \infty \end{array} \right\} \text{ with a finite ratio } N/V \; . \tag{2.1}$$

Given the practical impossibility of describing the system at a microscopic level,
a statistical description is needed. Thus, we define the phase-space probability
density $\rho_N(\mathbf{x}^N)$ such that $\rho_N(\mathbf{x}^N)d\mathbf{x}^N$ is the probability that the microstate of the
system lies inside an infinitesimal (hyper)volume $d\mathbf{x}^N$ around the phase-space point
$\mathbf{x}^N$. The ensemble average of a certain dynamical variable $A_N(\mathbf{x}^N)$ is

$$\langle A \rangle = \int d\mathbf{x}^N A_N(\mathbf{x}^N)\rho_N(\mathbf{x}^N) \; . \tag{2.2}$$

Here, it is understood that the total number of particles (N) is fixed and the position
integral for each particle runs over a fixed volume (V) of the system. Otherwise, the
expression for the ensemble average may involve summation over the number of
particles and/or integration over the system volume [see (2.30) and (2.44) below].

2.2   Gibbs Entropy Functional

The concept of a phase-space probability density is valid both out of equilibrium
(where, in general, it changes with time according to the Liouville theorem [2, 3])
and in equilibrium (where it is stationary). In the latter case $\rho_N(\mathbf{x}^N)$ can be obtained
for isolated, closed, open, … systems by following logical steps and starting from
the *equal a priori probability postulate* for isolated systems [4]. Here we follow an
alternative (but equivalent) method based on information-theory arguments [3, 5, 6].

Let us define the Gibbs entropy *functional*

$$\mathscr{S}[\rho_N] = -k_B \int d\mathbf{x}^N \, \rho_N(\mathbf{x}^N) \ln\left[c_N \rho_N(\mathbf{x}^N)\right] \, ,$$

(2.3)

where

$$c_N \equiv N! h^{dN}$$

(2.4)

is the quantum of phase-space volume. In (2.4) h is the Planck constant, the coefficient h^{dN} being introduced to comply with Heisenberg's uncertainty principle and also to preserve the non-dimensional character of the argument of the logarithm. Moreover, the factorial $N!$ accounts for the fact that two apparently different microstates which only differ on the particle labels are physically the same microstate, thus avoiding the Gibbs paradox [7]. The factorial $N!$ must be removed from c_N if the particles are *distinguishable*.

Equation (2.3) applies to systems with a fixed number of particles N. On the other hand, if the system is allowed to exchange particles with the environment, microstates with different N exist, so that one needs to define a phase-space probability density $\rho_N(\mathbf{x}^N)$ for each $N \geq 0$. In that case, the entropy functional becomes

$$\mathscr{S}[\rho_N] = -k_B \sum_{N=0}^{\infty} \int d\mathbf{x}^N \, \rho_N(\mathbf{x}^N) \ln\left[c_N \rho_N(\mathbf{x}^N)\right] \, .$$

(2.5)

Analogously, if the number of particles N is fixed but the volume V occupied by the particles can vary (formally) from zero to infinity, the phase-space probability density $\rho_N(\mathbf{x}^N)$ depends on V. It is defined such that $\rho_N(\mathbf{x}^N) d\mathbf{x}^N dV$ is the probability that the particles occupy a volume between V and $V + dV$ and the microstate lies inside an infinitesimal (hyper)volume $d\mathbf{x}^N$ around the phase-space point $\mathbf{x}^N$. The corresponding entropy functional is then

$$\mathscr{S}[\rho_N] = -k_B \int_0^{\infty} dV \int d\mathbf{x}^N \, \rho_N(\mathbf{x}^N) \ln\left[c_N V_0 \rho_N(\mathbf{x}^N)\right] \, ,$$

(2.6)

where V_0 is an arbitrary volume scale factor (needed to keep the correct dimensions).

Now, the basic postulate consists in asserting that, out of all possible phase-space probability distribution functions ρ_N consistent with given *constraints* (which define the *ensemble* of accessible microstates), the *equilibrium* function ρ_N^{eq} is the one that *maximizes* the entropy functional $\mathscr{S}[\rho_N]$. Once ρ_N^{eq} is known, connection with thermodynamics is made through the identification of $S = \mathscr{S}[\rho_N^{eq}]$ as the equilibrium entropy.

2.3   Microcanonical Ensemble: Isolated Systems

The microcanical ensemble describes an isolated system and thus it is characterized by fixed values of V, N, E (the latter with a tolerance ΔE, in accordance with the uncertainty principle). Therefore, the basic constraint is the normalization condition

$$\int_{E \leq H_N(\mathbf{x}^N) \leq E + \Delta E} d\mathbf{x}^N \, \rho_N(\mathbf{x}^N) = 1 \,. \tag{2.7}$$

Maximization of the entropy functional (2.3) just says that $\rho_N(\mathbf{x}^N) = $ const for all the accessible microstates $E \leq H_N(\mathbf{x}^N) \leq E + \Delta E$. Thus,

$$\rho_N(\mathbf{x}^N) = \begin{cases} [C_N \omega_{\Delta E}(E, V, N)]^{-1} \,, & E \leq H_N(\mathbf{x}^N) \leq E + \Delta E \,, \\ 0 \,, & \text{otherwise} \,, \end{cases}$$

$$= \frac{\Pi_{E, E + \Delta E} \left(H_N(\mathbf{x}^N) \right)}{C_N \omega_{\Delta E}(E, V, N)} \,, \tag{2.8}$$

where $\Pi_{a,b}(x)$ is the boxcar function, which is equal to 1 for $a \leq x \leq b$ and 0 otherwise. The normalization function

$$\boxed{\omega_{\Delta E}(E, V, N) = \frac{1}{C_N} \int_{E \leq H_N(\mathbf{x}^N) \leq E + \Delta E} d\mathbf{x}^N} \tag{2.9}$$

is the phase-space volume comprised between the hyper-surfaces $H_N(\mathbf{x}^N) = E$ and $H_N(\mathbf{x}^N) = E + \Delta E$, in units of C_N. We will refer to the dimensionless quantity $\omega_{\Delta E}$ as the *microcanonical partition function*. It is interesting to note that, taking into account the representation

$$\delta(x - a) = \lim_{\Delta a \to 0} \frac{\Pi_{a, a + \Delta a}(x)}{\Delta a} \tag{2.10}$$

of the Dirac delta function, the microcanonical partition function can be rewritten as

$$\omega_{\Delta E}(E, V, N) \approx \frac{\Delta E}{C_N} \int d\mathbf{x}^N \, \delta(H_N(\mathbf{x}^N) - E) \,. \tag{2.11}$$

By insertion of (2.8) into (2.3) one immediately sees that $\omega_{\Delta E}(E, V, N)$ is directly related to the equilibrium entropy as

$$\boxed{S(E, V, N) = k_B \ln \omega_{\Delta E}(E, V, N) \,.} \tag{2.12}$$

In this expression, the specific value of ΔE becomes irrelevant in the thermodynamic limit (as long as $\Delta E \ll E$).

Equation (2.12) means that entropy is proportional to the logarithm of the number of microstates with energy E (within an allowance ΔE). This is usually referred to as the Boltzmann entropy. An alternative definition of entropy in the microcanonical ensemble is [8]

$$S(E, V, N) = k_B \ln \bar{\omega}(E, V, N) , \tag{2.13a}$$

$$\bar{\omega}(E, V, N) \equiv \frac{1}{C_N} \int_{0 \leq H_N(\mathbf{x}^N) \leq E} d\mathbf{x}^N , \tag{2.13b}$$

where now the so-called Gibbs entropy is proportional to the logarithm of the number of microstates with an energy smaller than or equal to E. For "normal" systems, like classical liquids, energy does not have an upper bound and the function $\omega_{\Delta E}(E)$ grows so rapidly with E that $\ln \bar{\omega}(E) \approx \ln \omega_{\Delta E}(E)$ in the thermodynamic limit, and hence both definitions (2.12) and (2.13) become fully equivalent in that limit [8]. Such an equivalence, however, does not hold for *small* systems or for systems where energy has an upper bound E_{max}. In the latter case, the function $\omega_{\Delta E}(E)$ decreases with increasing energy as E_{max} is approached from below, while the cumulative function $\bar{\omega}(E)$ monotonically increases with E. As a consequence, the thermodynamic relation (1.4a) can give rise to *negative* absolute temperatures [9, 10] if the Boltzmann entropy is used, while the Gibbs entropy always predicts positive-definite temperatures. The question of which definition of entropy (Boltzmann's versus Gibbs's) is more adequate for small systems or when the energy is bounded is still open [11–17]. On the other hand, since we will always deal here with classical normal liquids in the thermodynamic limit, (2.12) can be safely adopted for the microcanonical entropy.

Making use of (1.4) (see also Table 1.1), the thermodynamic variables conjugate to E, V, and N can be obtained from $\omega_{\Delta E}$ as

$$\beta \equiv \frac{1}{k_B T} = \frac{\partial}{\partial E} \ln \omega_{\Delta E}(E, V, N) , \tag{2.14a}$$

$$\beta p = \frac{\partial}{\partial V} \ln \omega_{\Delta E}(E, V, N) , \tag{2.14b}$$

$$\alpha \equiv -\beta \mu = \frac{\partial}{\partial N} \ln \omega_{\Delta E}(E, V, N) . \tag{2.14c}$$

The inverse temperature parameter β has dimensions of inverse energy and is usually employed in statistical-mechanical formulas more frequently than the temperature T itself [see (1.35)]. Analogously, α is a dimensionless parameter defined as the opposite of the chemical potential scaled with the thermal energy $k_B T$. The parameter α is usually preferred over μ in statistical-mechanical formal expressions. Its exponential defines the *fugacity*

$$z \equiv e^{-\alpha} \equiv e^{\beta \mu} . \tag{2.15}$$

2.4   Canonical Ensemble: Closed Systems

Now the system can have *any* value of the total energy E. However, we are free to prescribe a given value of the *average* energy $\langle E \rangle = \langle H_N \rangle$. Therefore, the constraints in the canonical ensemble are

$$\int dx^N \, \rho_N(x^N) = 1 \, , \tag{2.16a}$$

$$\int dx^N \, H_N(x^N)\rho_N(x^N) = \langle E \rangle \, . \tag{2.16b}$$

The maximization of the entropy functional (2.3) subject to the constraints (2.16) can be carried out through the Lagrange multiplier method with the result

$$\rho_N(x^N) = \frac{e^{-\beta H_N(x^N)}}{C_N \mathscr{Z}_N(\beta, V)} \, , \tag{2.17}$$

where β is the Lagrange multiplier associated with the constraint on $\langle E \rangle$ and the *canonical partition function* $\mathscr{Z}_N$ is determined from the normalization condition as

$$\boxed{\mathscr{Z}_N(\beta, V) = \frac{1}{C_N} \int dx^N \, e^{-\beta H_N(x^N)} \, .} \tag{2.18}$$

Multiplying both sides of (2.18) by $1 = \int dE \, \delta(H_N(x^N) - E)$ and using (2.11), the partition function can alternatively be written as

$$\mathscr{Z}_N(\beta, V) = \frac{1}{\Delta E} \int dE \, e^{-\beta E} \omega_{\Delta E}(E, V, N) \, . \tag{2.19}$$

Substitution of (2.17) into (2.3) and use of (2.16) yields

$$S = k_B \left(\ln \mathscr{Z}_N + \beta \langle E \rangle \right) \, . \tag{2.20}$$

Comparison with (1.10) (where now the internal energy corresponds to $\langle E \rangle$) allows one to identify $\beta = 1/k_B T$ and

$$\boxed{F(T, V, N) = -k_B T \ln \mathscr{Z}_N(\beta, V) \, .} \tag{2.21}$$

Thus, the Lagrange multiplier β acquires a physical meaning as the inverse temperature parameter already defined in (1.35) and (2.14a). Besides, in the canonical ensemble the connection with thermodynamics is conveniently established via the Helmholtz free energy rather than via the entropy.

As an average of a phase-space dynamical variable, the internal energy can be directly obtained from $\ln \mathscr{Z}_N$ as

$$\langle E \rangle = -\frac{\partial \ln \mathscr{Z}_N}{\partial \beta} \,. \tag{2.22}$$

More in general, the energy moments are

$$\langle E^k \rangle = \frac{(-1)^k}{\mathscr{Z}_N} \frac{\partial^k \mathscr{Z}_N}{\partial \beta^k} \,. \tag{2.23}$$

In particular, energy fluctuations in a closed system are measured by the variance

$$\langle E^2 \rangle - \langle E \rangle^2 = \frac{\partial^2 \ln \mathscr{Z}_N}{\partial \beta^2} = k_B T^2 C_V \,, \tag{2.24}$$

where in the last step use has been made of (1.20a). Since both the internal energy $\langle E \rangle$ and the heat capacity C_V are extensive quantities (i.e., $\langle E \rangle \propto N$, $C_V \propto N$), (2.24) implies that the *relative* standard deviation $\sqrt{\langle E^2 \rangle - \langle E \rangle^2}/\langle E \rangle$ scales with $N^{-1/2}$. Therefore, in the thermodynamic limit (2.1) the energy fluctuations become negligible and the canonical ensemble becomes equivalent to the microcanonical one.

Using (2.23), it is possible to generalize (2.24) as

$$\mathscr{K}_E^{(k)} = (-1)^k \frac{\partial^k \ln \mathscr{Z}_N}{\partial \beta^k} \,, \tag{2.25}$$

where $\mathscr{K}_x^{(k)}$ denotes the kth *cumulant* of a random variable x. The first few cumulants ($2 \leq k \leq 6$) are $\mathscr{K}_x^{(2)} = \langle (\delta x)^2 \rangle$, $\mathscr{K}_x^{(3)} = \langle (\delta x)^3 \rangle$, $\mathscr{K}_x^{(4)} = \langle (\delta x)^4 \rangle - 3\langle (\delta x)^2 \rangle^2$, $\mathscr{K}_x^{(5)} = \langle (\delta x)^5 \rangle - 10\langle (\delta x)^3 \rangle\langle (\delta x)^2 \rangle$, and $\mathscr{K}_x^{(6)} = \langle (\delta x)^6 \rangle - 15\langle (\delta x)^4 \rangle\langle (\delta x)^2 \rangle - 10\langle (\delta x)^3 \rangle^2 + 30\langle (\delta x)^2 \rangle^3$, where $\delta x \equiv x - \langle x \rangle$.

The microcanonical$\leftrightarrow$canonical ensemble equivalence can be further explored by considering the energy probability density function in the canonical ensemble,

$$\mathscr{P}_N(E; \beta, V) = \int d\mathbf{x}^N \, \delta(H_N(\mathbf{x}^N) - E)\rho_N(\mathbf{x}^N) = \frac{e^{-\beta E}\omega_{\Delta E}(E, V, N)}{\mathscr{Z}_N(\beta, V)} \,, \tag{2.26}$$

where (2.11) has been used again. While $\omega_{\Delta E}(E, V, N)$ is a rapidly increasing function of E (in classical systems with no upper bound for energy), $e^{-\beta E}$ is a rapidly decreasing function. Thus, $\mathscr{P}_N(E)$ presents an extremely sharp peak at a certain value $E = \widetilde{E}$. The extremal condition $\partial \ln \mathscr{P}_N(E)/\partial E|_{E=\widetilde{E}} = 0$ implies that $\widetilde{E}$ is implicitly given by

$$\beta = \frac{\partial \ln \omega_{\Delta E}(E, V, N)}{\partial E}\bigg|_{E=\widetilde{E}} \,. \tag{2.27}$$

Comparison with (2.14a) shows that, at given T, N, and V, the most probable energy $\widetilde{E}$ in a closed system coincides with the unique (except for the energy tolerance ΔE) energy value in an isolated system.

From (1.12b) and (1.12c) (see also Table 1.1), we note that the pressure and the chemical potential are obtained from the partition function as

$$\beta p = \frac{\partial}{\partial V} \ln \mathscr{Z}_N(\beta, V) , \tag{2.28a}$$

$$\alpha \equiv -\beta\mu = \frac{\partial}{\partial N} \ln \mathscr{Z}_N(\beta, V) . \tag{2.28b}$$

2.5   Grand Canonical Ensemble: Open Systems

In an open system neither the energy nor the number of particles are determined but we can choose to fix their average values. As a consequence, the constraints are

$$\sum_{N=0}^{\infty} \int d\mathbf{x}^N \, \rho_N(\mathbf{x}^N) = 1 , \tag{2.29a}$$

$$\sum_{N=0}^{\infty} \int d\mathbf{x}^N \, H_N(\mathbf{x}^N)\rho_N(\mathbf{x}^N) = \langle E \rangle , \tag{2.29b}$$

$$\sum_{N=0}^{\infty} N \int d\mathbf{x}^N \, \rho_N(\mathbf{x}^N) = \langle N \rangle . \tag{2.29c}$$

In general, given a dynamical variable $A_N(\mathbf{x}^N)$, its grand canonical ensemble average is

$$\langle A \rangle = \sum_{N=0}^{\infty} \int d\mathbf{x}^N \, A_N(\mathbf{x}^N)\rho_N(\mathbf{x}^N) . \tag{2.30}$$

The solution to the maximization problem of the entropy functional (2.5) with the constraints (2.29) is

$$\rho_N(\mathbf{x}^N) = \frac{e^{-\alpha N}e^{-\beta H_N(\mathbf{x}^N)}}{C_N \, \Xi(\beta, V, \alpha)} , \tag{2.31}$$

where α and β are Lagrange multipliers and the *grand partition function* is

$$\Xi(\beta, V, \alpha) = \sum_{N=0}^{\infty} \frac{e^{-\alpha N}}{C_N} \int d\mathbf{x}^N \, e^{-\beta H_N(\mathbf{x}^N)} . \tag{2.32}$$

From (2.18), the grand partition function can be rewritten as

$$\Xi(\beta, V, \alpha) = \sum_{N=0}^{\infty} e^{-\alpha N} \mathscr{Z}_N(\beta, V) . \qquad (2.33)$$

Inserting (2.31) into (2.5), it is straightforward to check that the equilibrium entropy becomes

$$S = k_B \left(\ln \Xi + \beta \langle E \rangle + \alpha \langle N \rangle \right) . \qquad (2.34)$$

From comparison with the first equality of (1.17) we can identify $\beta = 1/k_B T$, $\alpha = -\beta\mu$, and

$$\Omega(T, V, \mu) = -k_B T \ln \Xi(\beta, V, \alpha) . \qquad (2.35)$$

As happened in the canonical ensemble, the Lagrange multiplier β coincides with the inverse temperature parameter defined by (1.35) and (2.14a). Analogously, the multiplier α is not but the scaled chemical potential defined by (2.14c).

The average energy and number of particles can be obtained from the grand partition function as

$$\langle E \rangle = -\frac{\partial \ln \Xi}{\partial \beta} , \qquad (2.36a)$$

$$\langle N \rangle = -\frac{\partial \ln \Xi}{\partial \alpha} . \qquad (2.36b)$$

As for the pressure, according to (1.17) or (1.19b), we simply have

$$\beta p V = \ln \Xi(\beta, V, \alpha) . \qquad (2.37)$$

Similarly to (2.23), the moments associated with the energy and the number of particles are

$$\langle E^k \rangle = \frac{(-1)^k}{\Xi} \frac{\partial^k \Xi}{\partial \beta^k} , \qquad (2.38a)$$

$$\langle N^k \rangle = \frac{(-1)^k}{\Xi} \frac{\partial^k \Xi}{\partial \alpha^k} . \qquad (2.38b)$$

Consequently, the fluctuation relations become

$$\left\langle E^2 \right\rangle - \left\langle E \right\rangle^2 = \frac{\partial^2 \ln \varXi}{\partial \beta^2} = k_B T^2 C_V \,, \tag{2.39a}$$

$$\left\langle N^2 \right\rangle - \left\langle N \right\rangle^2 = \frac{\partial^2 \ln \varXi}{\partial \alpha^2} = -\frac{\partial \left\langle N \right\rangle}{\partial \alpha} \,. \tag{2.39b}$$

Recalling that $\alpha = -\beta \mu$ and taking into account the thermodynamic identity (1.30), we can write

$$\boxed{\left\langle N^2 \right\rangle - \left\langle N \right\rangle^2 = n k_B T \left\langle N \right\rangle \kappa_T \,.} \tag{2.40}$$

Since the isothermal compressibility is an intensive quantity, the *relative* standard deviation $\sqrt{\langle N^2 \rangle - \langle N \rangle^2} / \langle N \rangle$ scales with $\langle N \rangle^{-1/2}$ and thus decays in the thermodynamic limit. In that limit the microcanonical, canonical, and grand canonical ensembles become equivalent. On the other hand, as one approaches the vapor–liquid critical point the isothermal compressibility diverges (critical opalescence phenomenon) and so do the density fluctuations in a finite-volume cell.

As in (2.25), the cumulants of energy and number of particles in the grand canonical ensemble are

$$\mathscr{K}_E^{(k)} = (-1)^k \frac{\partial^k \ln \varXi}{\partial \beta^k} \,, \tag{2.41a}$$

$$\mathscr{K}_N^{(k)} = (-1)^k \frac{\partial^k \ln \varXi}{\partial \alpha^k} \,. \tag{2.41b}$$

This generalizes (2.39) to $k \geq 3$.

In analogy with (2.26), we can define the number probability distribution function

$$\mathscr{P}(N; \beta, V, \alpha) = \int \mathrm{d}\mathbf{x}^N \, \rho_N(\mathbf{x}^N) = \frac{\mathrm{e}^{-\alpha N} \mathscr{Z}_N(\beta, V)}{\varXi(\beta, V, \alpha)} \,. \tag{2.42}$$

This function is the product of a rapidly increasing function ($\mathscr{Z}_N$) and a rapidly decreasing function ($\mathrm{e}^{-\alpha N}$) of N, what gives rise to a sharp maximum at a value $N = \widetilde{N}$ given by the implicit condition

$$\alpha = \left. \frac{\partial \ln \mathscr{Z}_N(\beta, V)}{\partial N} \right|_{N=\widetilde{N}} \,. \tag{2.43}$$

The agreement with (2.28b) reinforces the canonical↔grand canonical ensemble equivalence for large systems (thermodynamic limit).

2.6 Isothermal–Isobaric Ensemble: Isothermal–Isobaric Systems

In this ensemble, the volume is a fluctuating quantity and only its average value is fixed. Thus, similarly to the grand canonical ensemble, the constraints are

$$\int_0^\infty dV \int d\mathbf{x}^N \rho_N(\mathbf{x}^N) = 1 , \tag{2.44a}$$

$$\int_0^\infty dV \int d\mathbf{x}^N H_N(\mathbf{x}^N) \rho_N(\mathbf{x}^N) = \langle E \rangle , \tag{2.44b}$$

$$\int_0^\infty dV\, V \int d\mathbf{x}^N \rho_N(\mathbf{x}^N) = \langle V \rangle . \tag{2.44c}$$

Not surprisingly, the solution to the maximization problem of the Gibbs entropy functional (2.6) is

$$\rho_N(\mathbf{x}^N) = \frac{e^{-\gamma V} e^{-\beta H_N(\mathbf{x}^N)}}{V_0 C_N \Delta_N(\beta, \gamma)} , \tag{2.45}$$

where γ and β are again Lagrange multipliers, and the *isothermal–isobaric partition function* is

$$\Delta_N(\beta, \gamma) = \frac{1}{V_0 C_N} \int_0^\infty dV\, e^{-\gamma V} \int d\mathbf{x}^N e^{-\beta H_N(\mathbf{x}^N)} . \tag{2.46}$$

Again, use of (2.18) allows us to write

$$\boxed{\Delta_N(\beta, \gamma) = \frac{1}{V_0} \int_0^\infty dV\, e^{-\gamma V} \mathscr{Z}_N(\beta, V) .} \tag{2.47}$$

Taking into account (2.6), the entropy becomes

$$S = k_B \left(\ln \Delta_N + \beta \langle E \rangle + \gamma \langle V \rangle \right) . \tag{2.48}$$

From comparison with (1.13) we conclude that $\beta = 1/k_B T$,

$$\gamma = \beta p , \tag{2.49}$$

and

$$\boxed{G(T, p, N) = -k_B T \ln \Delta_N(\beta, \gamma) .} \tag{2.50}$$

As before, the Lagrange multipliers are related to thermodynamic quantities: β is the inverse temperature parameter and γ is the pressure p divided by the thermal energy $k_B T$.

The average energy and volume are

$$\langle E \rangle = -\frac{\partial \ln \Delta_N}{\partial \beta} ,$$ (2.51a)

$$\langle V \rangle = -\frac{\partial \ln \Delta_N}{\partial \gamma} .$$ (2.51b)

From here one can get the Maxwell relation

$$\frac{\partial \langle E \rangle}{\partial \gamma} = \frac{\partial \langle V \rangle}{\partial \beta} .$$ (2.52)

Equations (2.51) are complemented by

$$\alpha \equiv -\beta \mu = \frac{\ln \Delta_N}{N} ,$$ (2.53)

which follows from the property $\mu = G/N$ for one-component systems.

The energy and volume fluctuations are characterized by

$$\langle E^2 \rangle - \langle E \rangle^2 = \frac{\partial^2 \ln \Delta_N}{\partial \beta^2} = k_B T^2 C_V ,$$ (2.54a)

$$\langle V^2 \rangle - \langle V \rangle^2 = \frac{\partial^2 \ln \Delta_N}{\partial \gamma^2} = -\frac{1}{\beta} \left(\frac{\partial \langle V \rangle}{\partial p} \right)_{\beta, N} = k_B T \langle V \rangle \kappa_T .$$ (2.54b)

Equations (2.40) and (2.54b) are equivalent. Both show that the density fluctuations are proportional to the isothermal compressibility and decrease as the size of the system increases. In (2.40) the volume is constant, so that the density fluctuations are due to fluctuations in the number of particles, while the opposite happens in (2.54b).

Again, the cumulants can be obtained as

$$\mathscr{K}_E^{(k)} = (-1)^k \frac{\partial^k \ln \Delta_N}{\partial \beta^k} ,$$ (2.55a)

$$\mathscr{K}_V^{(k)} = (-1)^k \frac{\partial^k \ln \Delta_N}{\partial \gamma^k} .$$ (2.55b)

The volume probability distribution function is

$$\mathscr{P}_N(V; \beta, \gamma) = \int d\mathbf{x}^N \rho_N(\mathbf{x}^N) = \frac{e^{-\gamma V} \mathscr{Z}_N(\beta, V)}{V_0 \Delta_N(\beta, \gamma)} .$$ (2.56)

Table 2.1 Summary of statistical ensembles

Quantity	Statistical ensembles			
	Microcanonical	Canonical	Grand canonical	Isothermal–isobaric
$\rho_N(\mathbf{x}^N)$	$\dfrac{\Pi_{E,E+\Delta E}\left(H_N(\mathbf{x}^N)\right)}{N!h^{dN}\omega_{\Delta E}(E,V,N)}$	$\dfrac{e^{-\beta H_N(\mathbf{x}^N)}}{N!h^{dN}\,\mathscr{Z}_N(\beta,V)}$	$\dfrac{e^{-\alpha N}e^{-\beta H_N(\mathbf{x}^N)}}{N!h^{dN}\,\Xi(\beta,V,\alpha)}$	$\dfrac{e^{-\gamma V}e^{-\beta H_N(\mathbf{x}^N)}}{V_0 N!h^{dN}\,\Delta_N(\beta,\gamma)}$
Partition fcn.				
Symbol	$\omega_{\Delta E}(E,V,N)$	$\mathscr{Z}_N(\beta,V)$	$\Xi(\beta,V,\alpha)$	$\Delta_N(\beta,\gamma)$
Expression	$\displaystyle\int_{H_N=E}^{H_N=E+\Delta E}\frac{d\mathbf{x}^N}{N!h^{dN}}$	$\displaystyle\int\frac{d\mathbf{x}^N}{N!h^{dN}}e^{-\beta H_N(\mathbf{x}^N)}$	$\displaystyle\sum_{N=0}^{\infty}e^{-\alpha N}\mathscr{Z}_N(\beta,V)$	$\displaystyle\int_0^\infty\frac{dV}{V_0}e^{-\gamma V}$ $\times\,\mathscr{Z}_N(\beta,V)$
Potential	$S=k_B\ln\omega_{\Delta E}$	$F=-k_BT\ln\mathscr{Z}_N$	$\Omega=-k_BT\ln\Xi$	$G=-k_BT\ln\Delta_N$
$\beta\equiv\dfrac{1}{k_BT}$	$\dfrac{\partial\ln\omega_{\Delta E}}{\partial E}$	✓	✓	✓
$\gamma\equiv\beta p$	$\dfrac{\partial\ln\omega_{\Delta E}}{\partial V}$	$\dfrac{\partial\ln\mathscr{Z}_N}{\partial V}$	$\dfrac{\ln\Xi}{V}$	✓
$\alpha\equiv-\beta\mu$	$\dfrac{\partial\ln\omega_{\Delta E}}{\partial N}$	$\dfrac{\partial\ln\mathscr{Z}_N}{\partial N}$	✓	$\dfrac{\ln\Delta_N}{N}$
$E,\langle E\rangle$	✓	$-\dfrac{\partial\ln\mathscr{Z}_N}{\partial\beta}$	$-\dfrac{\partial\ln\Xi}{\partial\beta}$	$-\dfrac{\partial\ln\Delta_N}{\partial\beta}$
$N,\langle N\rangle$	✓	✓	$-\dfrac{\partial\ln\Xi}{\partial\alpha}$	✓
$V,\langle V\rangle$	✓	✓	✓	$-\dfrac{\partial}{\partial\gamma}\ln\Delta_N$

The check marks denote the control variables in each ensemble

As expected, $\mathscr{P}_N(V)$ has a sharp peak at $V=\widetilde{V}$, where

$$\gamma=\left.\frac{\partial\ln\mathscr{Z}_N(\beta,V)}{\partial V}\right|_{V=\widetilde{V}}. \tag{2.57}$$

Now, comparison with (2.28a) shows the canonical↔isothermal–isobaric ensemble equivalence in the thermodynamic limit.

A summary of the main relations for the four ensembles considered in this chapter can be found in Table 2.1.

2.7 Ideal Gas

The exact evaluation of the partition functions (2.9), (2.18), (2.33), and (2.47) is in general a formidable task due to the involved dependence of the Hamiltonian on the coordinates of the particles. However, in the case of non-interacting particles (ideal gas), the Hamiltonian depends only on the momenta:

$$H_N(\mathbf{x}^N) \rightarrow H_N^{id}(\mathbf{p}^N) = \sum_{i=1}^{N} \frac{p_i^2}{2m} , \tag{2.58}$$

where m is the mass of a particle. In this case the N-body Hamiltonian is just the sum over all the particles of the one-body Hamiltonian $p_i^2/2m$ and the exact statistical-mechanical results can be easily obtained.

The expressions for the partition function, the thermodynamic potential, and the first derivatives of the latter for each one of the four ensembles considered above are listed in Table 2.2. In those expressions, $\Gamma(x)$ is the well-known gamma function,

$$\zeta(\beta, V) \equiv \frac{V}{[\Lambda(\beta)]^d} \tag{2.59}$$

Table 2.2 Physical quantities of an ideal gas

| Quantity | Statistical ensembles | | | |
	Microcanonical	Canonical	Grand canonical	Isothermal–isobaric
Partition fcn.				
Symbol	$\omega_{\Delta E}^{id}(E, V, N)$	$\mathscr{Z}_N^{id}(\beta, V)$	$\Xi^{id}(\beta, V, \alpha)$	$\Delta_N^{id}(\beta, \gamma)$
Expression	$\dfrac{\left[V(2\pi mE/h^2)^{d/2} \right]^N}{N! \Gamma(dN/2)} \dfrac{\Delta E}{E}$	$\dfrac{[\zeta(\beta, V)]^N}{N!}$	$\exp\left[e^{-\alpha} \zeta(\beta, V) \right]$	$\dfrac{\gamma^{-(N+1)}}{V_0 [\Lambda(\beta)]^{dN}}$
Potential				
Symbol	$\dfrac{S^{id}(E, V, N)}{Nk_B}$	$\dfrac{F^{id}(T, V, N)}{Nk_B T}$	$\dfrac{\Omega^{id}(T, V, \mu)}{k_B T}$	$\dfrac{G^{id}(T, p, N)}{Nk_B T}$
Expression	$\ln\left[\dfrac{V}{N}\left(\dfrac{4\pi mE}{dNh^2} \right)^{d/2} \right]$ $+ \dfrac{d+2}{2}$	$\ln \dfrac{N}{\zeta(\beta, V)} - 1$	$-e^{-\alpha} \zeta(\beta, V)$	$\ln \dfrac{p[\Lambda(\beta)]^d}{k_B T}$
T	$\dfrac{2}{d} \dfrac{E}{Nk_B}$	✓	✓	✓
p^{id}	$\dfrac{2}{d} \dfrac{E}{V}$	$\dfrac{N}{V} k_B T$	$k_B T e^{-\alpha} \dfrac{\zeta(\beta, V)}{V}$	✓
μ^{id}	$-\dfrac{2}{d} \dfrac{E}{N} \ln\left[\dfrac{V}{N}\left(\dfrac{4\pi mE}{dNh^2} \right)^{d/2} \right]$	$k_B T \ln \dfrac{N}{\zeta(\beta, V)}$	✓	$k_B T \ln \dfrac{p[\Lambda(\beta)]^d}{k_B T}$
$E, \langle E \rangle^{id}$	✓	$\dfrac{d}{2} Nk_B T$	$\dfrac{d}{2} k_B T e^{-\alpha} \zeta(\beta, V)$	$\dfrac{d}{2} Nk_B T$
$N, \langle N \rangle$	✓	✓	$e^{-\alpha} \zeta(\beta, V)$	✓
$V, \langle V \rangle$	✓	✓	✓	$\dfrac{Nk_B T}{p}$

The check marks denote the control variables in each ensemble

is the one-particle partition function and

$$\Lambda(\beta) \equiv \frac{h}{\sqrt{2\pi m/\beta}} \tag{2.60}$$

is the thermal de Broglie wavelength. When obtaining the thermodynamic potentials from the logarithm of the corresponding partition function, the thermodynamic limit $(N \to \infty)$ has been taken. This allows us to use the Stirling approximation $\ln N! \approx N(\ln N - 1)$ and the limit $N^{-1} \ln(\Delta E/E) \to 0$.

Note that the expressions for the thermodynamic potentials and the thermodynamic variables (temperature, pressure, chemical potential, internal energy, number of particles, and volume) in a given ensemble are fully equivalent to those in any other ensemble. This a manifestation of the ensemble equivalence in the thermodynamic limit, the only difference lying in the choice of independent and dependent variables.

2.8 Interacting Systems

Of course, particles do interact in real systems, so the Hamiltonian has the generic form

$$H_N(\mathbf{x}^N) = H_N^{\mathrm{id}}(\mathbf{p}^N) + \Phi_N(\mathbf{r}^N) , \tag{2.61}$$

where Φ_N denotes the *total* potential energy. Since the interactions among the particles depend on the *relative* positions of the particles only, the potential energy function is invariant under translations, i.e.,

$$\Phi_N(\mathbf{r}_1 + \mathbf{a}, \mathbf{r}_2 + \mathbf{a}, \ldots, \mathbf{r}_N + \mathbf{a}) = \Phi_N(\mathbf{r}_1, \mathbf{r}_2, \ldots, \mathbf{r}_N), \tag{2.62}$$

for any arbitrary displacement vector $\mathbf{a}$.

As a consequence of the decomposition (2.61), the canonical partition function factorizes into its ideal and non-ideal parts:

$$\mathscr{Z}_N(\beta, V) = \mathscr{Z}_N^{\mathrm{id}}(\beta, V)\mathscr{Q}_N(\beta, V) , \tag{2.63}$$

where $\mathscr{Z}_N^{\mathrm{id}}$ can be found in Table 2.2 and the non-ideal part

$$\boxed{\mathscr{Q}_N(\beta, V) = V^{-N} \int d\mathbf{r}^N \, e^{-\beta\Phi_N(\mathbf{r}^N)}} \tag{2.64}$$

is the *configuration integral*. We will refer to the exponential $\exp[-\beta\Phi_N(\mathbf{r}^N)]$ in the integrand of $\mathscr{Q}_N$ as the Boltzmann factor.

In the canonical ensemble, $\mathcal{Q}_N$ is responsible for the *excess* contributions $F^{\text{ex}} = F - F^{\text{id}}$, $\langle E \rangle^{\text{ex}} = \langle E \rangle - \langle E \rangle^{\text{id}}$, $p^{\text{ex}} = p - p^{\text{id}}$, $\mu^{\text{ex}} = \mu - \mu^{\text{id}}$:

$$F^{\text{ex}}(T, V, N) = -k_B T \ln \mathcal{Q}_N(\beta, V) , \tag{2.65a}$$

$$\langle E \rangle^{\text{ex}} = -\frac{\partial \ln \mathcal{Q}_N}{\partial \beta} , \tag{2.65b}$$

$$p^{\text{ex}} = k_B T \frac{\partial \ln \mathcal{Q}_N}{\partial V} , \tag{2.65c}$$

$$\mu^{\text{ex}} = -k_B T \frac{\partial \ln \mathcal{Q}_N}{\partial N} . \tag{2.65d}$$

In general, if $A(\mathbf{r}^N)$ is a dynamical variable that depends on the particle positions only, its canonical-ensemble average is

$$\langle A \rangle = \frac{V^{-N}}{\mathcal{Q}_N(\beta, V)} \int d\mathbf{r}^N A(\mathbf{r}^N) e^{-\beta \Phi_N(\mathbf{r}^N)} . \tag{2.66}$$

The grand partition function does not factorize but can be written as

$$\Xi(\beta, V, \alpha) = 1 + \sum_{N=1}^{\infty} \frac{V^N \mathcal{Q}_N(\beta, V)}{N!} [\hat{z}(\beta, \alpha)]^N , \tag{2.67}$$

where we have taken into account that $\mathcal{Q}_0 = 1$ and have introduced the quantity

$$\hat{z}(\beta, \alpha) \equiv \frac{z(\alpha)}{[\Lambda(\beta)]^d} , \tag{2.68}$$

z being the fugacity defined by (2.15). Thus, $\hat{z}$ is a *rescaled* fugacity with dimensions of a number density. According to (2.67), we observe that the configuration integrals $\mathcal{Q}_N$ are directly related to the coefficients in the expansion of the grand partition function in powers of the quantity $\hat{z}$.

As for the isothermal–isobaric partition function, it is easy to obtain

$$\Delta_N(\beta, \gamma) = \frac{1}{V_0 N! [\Lambda(\beta)]^{dN}} \int_0^{\infty} dV \, e^{-\gamma V} V^N \mathcal{Q}_N(\beta, V) . \tag{2.69}$$

This shows that Δ_N can be seen as proportional to the Laplace transform of $V^N \mathcal{Q}_N$ with respect to volume.

2.9 Generalization to Mixtures

While so far we have restricted ourselves to one-component systems, most of the arguments and derivations can be easily generalized to mixtures. In particular, (2.8), (2.9), (2.17), (2.18), (2.31), (2.32), (2.45), and (2.46) generalize to

$$\rho_{\{N_v\}}(\mathbf{x}^N) = \frac{\Pi_{E,E+\Delta E}\left(H_{\{N_v\}}(\mathbf{x}^N)\right)}{(\prod_v N_v!)h^{dN}\omega_{\Delta E}(E, V, \{N_v\})} , \tag{2.70a}$$

$$\omega_{\Delta E}(E, V, \{N_v\}) = \frac{1}{(\prod_v N_v!)h^{dN}} \int_{E \leq H_{\{N_v\}}(\mathbf{x}^N) \leq E+\Delta E} d\mathbf{x}^N , \tag{2.70b}$$

$$\rho_{\{N_v\}}(\mathbf{x}^N) = \frac{e^{-\beta H_{\{N_v\}}(\mathbf{x}^N)}}{(\prod_v N_v!)h^{dN} \mathscr{Z}_{\{N_v\}}(\beta, V)} , \tag{2.71a}$$

$$\mathscr{Z}_{\{N_v\}}(\beta, V) = \frac{1}{(\prod_v N_v!)h^{dN}} \int d\mathbf{x}^N e^{-\beta H_{\{N_v\}}(\mathbf{x}^N)} , \tag{2.71b}$$

$$\rho_{\{N_v\}}(\mathbf{x}^N) = \frac{\prod_v e^{-\alpha_v N_v} e^{-\beta H_{\{N_v\}}(\mathbf{x}^N)}}{(\prod_v N_v!)h^{dN} \, \Xi(\beta, V, \{\alpha_v\})} , \tag{2.72a}$$

$$\Xi(\beta, V, \{\alpha_v\}) = \sum_{N_1=0}^{\infty} \sum_{N_2=0}^{\infty} \cdots \frac{\prod_v e^{-\alpha_v N_v}}{(\prod_v N_v!)h^{dN}} \int d\mathbf{x}^N e^{-\beta H_{\{N_v\}}(\mathbf{x}^N)} , \tag{2.72b}$$

$$\rho_{\{N_v\}}(\mathbf{x}^N) = \frac{e^{-\gamma V} e^{-\beta H_{\{N_v\}}(\mathbf{x}^N)}}{V_0(\prod_v N_v!)h^{dN} \Delta_{\{N_v\}}(\beta, \gamma)} , \tag{2.73a}$$

$$\Delta_{\{N_v\}}(\beta, \gamma) = \frac{1}{V_0(\prod_v N_v!)h^{dN}} \int_0^{\infty} dV e^{-\gamma V} \int d\mathbf{x}^N e^{-\beta H_{\{N_v\}}(\mathbf{x}^N)} , \tag{2.73b}$$

respectively. For instance, from (2.72) it is easy to check that, in the grand canonical ensemble, one has

$$\langle N_v \rangle = -\frac{\partial \ln \Xi}{\partial \alpha_v} , \tag{2.74a}$$

$$\langle N_{v_1} N_{v_2} \rangle - \langle N_{v_1} \rangle \langle N_{v_2} \rangle = \frac{\partial^2 \ln \Xi}{\partial \alpha_{v_2} \partial \alpha_{v_1}} = -\frac{\partial \langle N_{v_1} \rangle}{\partial \alpha_{v_2}} . \tag{2.74b}$$

Equation (2.74b) is a generalization of (2.39b).

Table 2.2 can be generalized to ideal-gas mixtures. In particular,

$$\Delta^{id}_{\{N_v\}}(\beta,\gamma) = \frac{\gamma^{-(N+1)}N!}{V_0 \prod_v N_v! \, [\Lambda_v(\beta)]^{dN_v}} , \tag{2.75a}$$

$$G^{id}(T,p,\{N_v\}) = k_B T \sum_v N_v \ln \frac{x_v p \, [\Lambda_v(\beta)]^d}{k_B T} , \tag{2.75b}$$

$$\mu^{id}_v(T,p,x_v) = k_B T \ln \frac{x_v p \, [\Lambda_v(\beta)]^d}{k_B T} , \tag{2.75c}$$

where $\Lambda_v(\beta)$ is the thermal de Broglie wavelength of species v, which is given by (2.60) with the replacement $m \to m_v$, where m_v is the mass of a particle of species v.

Exercises

2.1 Use the Lagrange multiplier method to maximize the entropy functional (2.3) with the constraint (2.7) and prove the microcanonical distribution (2.8). Derive (2.12).

2.2 Use the Lagrange multiplier method to maximize the entropy functional (2.3) with the constraints (2.16) and prove the canonical distribution (2.17). Derive (2.20).

2.3 Derive (2.23).

2.4 Check (2.25) for $3 \le k \le 6$.

2.5 Use the Lagrange multiplier method to maximize the entropy functional (2.5) with the constraints (2.29) and prove the grand canonical distribution (2.31). Derive (2.34).

2.6 How should the derivative in (2.36a) be interpreted, at constant $\alpha = -\beta\mu$ or at constant μ? Are both interpretations equivalent?

2.7 Derive (2.38b).

2.8 Use the Lagrange multiplier method to maximize the entropy functional (2.6) with the constraints (2.44) and prove the isothermal–isobaric distribution (2.45). Derive (2.48).

2.9 Derive (2.51b).

2.10 How should the derivative in (2.51a) be interpreted, at constant $\gamma = \beta p$ or at constant p? Are both interpretations equivalent?

2.11 Prove that the area and the volume of a hypersphere of radius R in k dimensions are

$$\frac{2\pi^{k/2}}{\Gamma(k/2)}R^{k-1} \; , \qquad \frac{\pi^{k/2}}{\Gamma(k/2+1)}R^{k} , \qquad (2.76)$$

respectively. Hint: Evaluate the multiple Gaussian integral $\int d\mathbf{r}^k\, e^{-r^2}$ in both Cartesian and spherical coordinates.

2.12 Making use of (2.76), prove that the microcanonical partition function for an ideal gas, $\omega_{\Delta E}^{\mathrm{id}}$, is indeed given by the expression shown in Table 2.2.

2.13 Prove (2.59).

2.14 Check the expressions of Table 2.2.

2.15 Using Table 2.2, prove that for an ideal gas the energy, number, and volume probability distribution functions (2.26), (2.42), and (2.56) reduce to

$$\mathscr{P}_N^{\mathrm{id}}(E) = \beta\frac{e^{-\beta E}(\beta E)^{dN/2-1}}{\Gamma(dN/2)} , \qquad (2.77a)$$

$$\mathscr{P}^{\mathrm{id}}(N) = e^{-\langle N\rangle}\frac{\langle N\rangle^N}{N!} , \qquad (2.77b)$$

$$\mathscr{P}_N^{\mathrm{id}}(V) = \beta p\frac{e^{-\beta pV}(\beta pV)^N}{N!} , \qquad (2.77c)$$

respectively.

2.16 Define the scaled quantities $E^* = E/\langle E\rangle = 2\beta E/dN$, $N^* = N/\langle N\rangle$, $V^* = V/\langle V\rangle = \beta pV/N$ and obtain the corresponding distributions $\mathscr{P}_N^{\mathrm{id}}(E^*) = \langle E\rangle\,\mathscr{P}_N^{\mathrm{id}}(E)$, $\mathscr{P}^{\mathrm{id}}(N^*) = \langle N\rangle\,\mathscr{P}^{\mathrm{id}}(N)$, and $\mathscr{P}_N^{\mathrm{id}}(V^*) = \langle V\rangle\,\mathscr{P}_N^{\mathrm{id}}(V)$ from (2.77). Explore the shape of those functions as N (or $\langle N\rangle$) increases.

2.17 Justify (2.70)–(2.73).

References

1. H. Goldstein, J. Safko, C.P. Poole, *Classical Mechanics* (Pearson Education, Upper Saddle River, 2013)
2. R. Balescu, *Equilibrium and Nonequilibrium Statistical Mechanics* (Wiley, New York, 1974)
3. L.E. Reichl, *A Modern Course in Statistical Physics*, 1st edn. (University of Texas Press, Austin, 1980)
4. F. Reif, *Fundamentals of Statistical and Thermal Physics* (McGraw-Hill, Boston, 1965)
5. C.E. Shannon, W. Weaver, *The Mathematical Theory of Communication* (University of Illinois Press, Urbana, 1971)

6. A. Ben-Naim, *A Farewell to Entropy: Statistical Thermodynamics Based on Information* (World Scientific, Singapore, 2008)
7. M. Baus, C.F. Tejero, *Equilibrium Statistical Physics. Phases of Matter and Phase Transitions* (Springer, Berlin, 2008)
8. R. Kubo, *Statistical Mechanics. An Advanced Course with Problems and Solutions* (Elsevier, Amsterdam, 1965)
9. L.D. Carr, Science **339**, 42 (2013)
10. S. Braun, J.P. Ronzheimer, M. Schreiber, S.S. Hodgman, T. Rom, I. Bloch, U. Schneider, Science **339**, 52 (2013)
11. V. Romero-Rochín, Phys. Rev. E **88**, 022144 (2013)
12. J. Dunkel, S. Hilbert, Nat. Phys. **10**, 67 (2014)
13. S. Hilbert, P. Hänggi, J. Dunkel, Phys. Rev. E **90**, 062116 (2014)
14. J.M.G. Vilar, J.M. Rubi, J. Chem. Phys. **140**, 201101 (2014)
15. D. Frenkel, P.B. Warren, Am. J. Phys. **83**, 163 (2015)
16. L. Ferrari, Boltzmann vs Gibbs: a finite-size match (2015), http://arxiv.org/abs/1501.04566
17. P. Hänggi, S. Hilbert, J. Dunkel, Philos. Trans. R. Soc. A **374**, 20150039 (2016)

Chapter 3
Density Expansion of the Equation of State

This chapter is mainly devoted to the formal derivation of the virial coefficients characterizing the representation of the equation of state as a series expansion in powers of density. This requires the introduction of diagrammatic techniques, the main steps being justified by simple examples without rigorous proofs. The chapter continues with the analysis of the second virial coefficient for simple model interactions and of higher-order virial coefficients for hard spheres, both one-component and multicomponent. Finally, some simple approximate equations of state for one-component and multicomponent hard-sphere liquids are described.

3.1 Pair Interaction Potential and Mayer Function

The formal results of Chap. 2 apply regardless of the specific form of the potential energy function $\Phi_N(\mathbf{r}^N)$. From now on, however, we assume that the interactions are *pairwise additive*, i.e., Φ_N can be expressed as a sum over all pairs of a certain function (interaction potential) ϕ that depends on the distance between the two particles of the pair. In mathematical terms,

$$\Phi_N(\mathbf{r}^N) = \sum_{i=1}^{N-1}\sum_{j=i+1}^{N} \phi(r_{ij}) = \frac{1}{2}\sum_{i\neq j} \phi(r_{ij}) \,, \quad r_{ij} \equiv |\mathbf{r}_i - \mathbf{r}_j| \,. \tag{3.1}$$

The pairwise additivity condition (3.1) allows us to write the global Boltzmann factor as a product of $N(N-1)/2$ pair Boltzmann factors,

$$e^{-\beta\Phi_N(\mathbf{r}^N)} = \prod_{i=1}^{N-1}\prod_{j=i+1}^{N} e^{-\beta\phi(r_{ij})} \,. \tag{3.2}$$

© Springer International Publishing Switzerland 2016
A. Santos, *A Concise Course on the Theory of Classical Liquids*,
Lecture Notes in Physics 923, DOI 10.1007/978-3-319-29668-5_3

Since the pair Boltzmann factor $e^{-\beta\phi(r)}$ is equal to unity in the ideal-gas case, a convenient way of measuring deviations from the ideal gas is by means of the function [1]

$$\boxed{f(r) \equiv e^{-\beta\phi(r)} - 1 .}$$
(3.3)

This function is known as the *Mayer function*, after Joseph E. Mayer (see Fig. 3.1) and Maria Goeppert Mayer (see Fig. 3.2). For short-range interactions, $\phi(r)$ is small if r is sufficiently large. In that region $f(r) \approx -\beta\phi(r)$, so the Mayer function has the same range as the potential.

A few examples of effective pair interaction potentials are shown in Table 3.1. The selection of those examples is not arbitrary. They represent prototypical short-range potentials characterizing the basic and most relevant features of particle

Fig. 3.1 Joseph Edward Mayer (1904–1983) (Photograph by Bachrach, http://www.bachrachportraits. com, reproduced with permission. The photo can be seen in https://photos.aip.org/ history-programs/niels-bohr-library/photos/mayer-joseph-a2)

Fig. 3.2 Maria Goeppert-Mayer (1906–1972) (Photograph from Wikimedia Commons, http://commons. wikimedia.org/wiki/File: Mayer.jpg)

Table 3.1 A sample of model interaction potentials

Interaction potential	Graph of $\phi(r)$	Graph of $f(r)$
Hard spheres $\phi_{HS}(r) = \begin{cases} \infty, & r < \sigma, \\ 0, & r > \sigma. \end{cases}$		
Penetrable spheres $\phi_{PS}(r) = \begin{cases} \varepsilon, & r < \sigma, \\ 0, & r > \sigma. \end{cases}$		
Square shoulder $\phi_{SS} = \begin{cases} \infty, & r < \sigma, \\ \varepsilon, & \sigma < r < \sigma', \\ 0, & r > \sigma'. \end{cases}$		

(continued)

Table 3.1 (continued)

Lennard-Jones $\phi_{LJ}(r) = 4\varepsilon\left[\left(\dfrac{\sigma}{r}\right)^{12} - \left(\dfrac{\sigma}{r}\right)^{6}\right]$

$\phi(r)$

$f(r)$
$e^{\beta\varepsilon} - 1$

Square well $\phi_{SW} = \begin{cases} \infty, & r < \sigma, \\ -\varepsilon, & \sigma < r < \sigma', \\ 0, & r > \sigma'. \end{cases}$

$\phi(r)$

$f(r)$
$e^{\beta\varepsilon} - 1$

Sticky hard spheres $\phi_{SHS} = \displaystyle\lim_{\sigma' \to \sigma,\, \varepsilon \to \infty} \phi_{SW}(r)$

$\phi(r)$

$-\varepsilon \sim \ln(\sigma'/\sigma - 1)$

$f(r)$

$\sim \delta(r - \sigma)$

interactions in a great deal of real fluids (both molecular and colloidal). Moreover, their mathematical forms are simple enough as to allow for an analytical (or semi-analytical) exact or approximate treatment to the thermodynamic and structural properties of the corresponding systems, thus helping us to grasp a better physical interpretation of the statistical-mechanical behavior of real fluids. In fact, most of the specific applications considered in this book refer to potentials in Table 3.1. They are succinctly described below.

- The simplest potential is the hard-sphere (HS) one, which represents impenetrable spherical particles of diameter σ. In spite of its crudeness, this interaction model not only represents a favorite playground in statistical mechanics, both in and out of equilibrium [2], but is also important from a more practical point of view. In real fluids, especially at high temperatures and moderate and high densities, the structural and thermodynamic properties are mainly governed by the repulsive forces among molecules and in this context hard-core fluids are very useful as reference systems [3–5]. Moreover, the use of the HS model in the realm of soft condensed matter [6] has become increasingly popular [7]. For instance, the effective interaction among (sterically stabilized) colloidal particles can be tuned to match almost perfectly the HS model [8].
- When the repulsive barrier is constant but finite, the resulting potential is that of penetrable spheres (PS). Now the pair potential $\phi(r)$ takes a finite value ε if the two spheres are overlapped ($r < \sigma$) and 0 otherwise. The PS model [7, 9–18], together with the Gaussian-core model [7, 15, 19], is a prototypical example of a *bounded* potential, i.e., a potential such that $|\phi(r)| < \varepsilon_{max} = $ finite. The PS model has been proposed to understand the peculiar behavior of some colloidal systems, such as micelles in a solvent or star copolymer suspensions. The particles in these colloids are constituted by a small core surrounded by several attached polymeric arms. As a consequence of their structure, two or more of these particles allow a considerable degree of overlapping with a small energy cost [7]. An ultrasoft logarithmically divergent potential for short distances has also been proposed to describe the effective interaction between star polymers in good solvents [20].
- Going back to unbounded potentials, the square-shoulder (SS) one [21] is the simplest example of a *core-softened* potential, i.e., an interaction function with a two-length scale repulsive part exhibiting a softening region where the slope changes dramatically [22].
- The most widely used potential mimicking the effective central interaction between two molecules was proposed by John E. Lennard-Jones (Fig. 3.3) in 1924. The Lennard-Jones (LJ) potential has an attractive van der Waals tail r^{-6} and is attractive for distances larger than a certain equilibrium separation ($r_0 = 2^{1/6}\sigma$). For shorter distances, it becomes increasingly repulsive.
- A piecewise-constant caricature of realistic continuous potentials (like the LJ one) is provided by the square-well (SW) potential. It contains two length scales (σ and σ') and an energy scale (ε). In particular, the choice $\sigma'/\sigma = 1.5$ is typically adopted to represent atomistic interactions.

Fig. 3.3 John Edward
Lennard-Jones (1894–1954)
(Photograph reproduced with
permission from Computer
Laboratory, University of
Cambridge, Copyright 1999,
http://www.cl.cam.ac.uk/
relics/jpegs/jones.jpg)

Fig. 3.4 Rodney James
Baxter (b. 1940) (Photograph
reproduced with permission
from Belinda Pratten,
Copyright 2013, http://www.
belindamorganpratten.com)

- In 1968, Rodney J. Baxter (Fig. 3.4) proposed to take the synchronized limit of
  an infinitely narrow ($\sigma' \to \sigma$) and infinitely deep ($\varepsilon \to \infty$) SW potential. A non-
  trivial result, usually referred to as sticky hard spheres (SHS), is obtained when
  both limits are coupled in such a way that [23]

$$\tau^{-1} \equiv 2^{d-1} \left[\left(\frac{\sigma'}{\sigma} \right)^d - 1 \right] \left(e^{\beta\varepsilon} - 1 \right) = \text{finite} , \qquad (3.4)$$

where the temperature-dependent parameter τ^{-1} measures the "stickiness" of the
interaction. In the SHS limit the Mayer function becomes

$$f_{\text{SHS}}(r) = -\Theta(\sigma - r) + \frac{\tau^{-1}}{d2^{d-1}} \sigma \delta(r - \sigma) , \qquad (3.5)$$

where the Heaviside step function is $\Theta(x - a) = 1$ if $x \geq a$ and 0 otherwise.
Strictly speaking, a one-component SHS system is not thermodynamically stable,
but a small degree of polydispersity is sufficient to restore stability [24, 25].
While the SHS model was originally introduced as a rather academic potential, it
has proved to provide an excellent starting point for the study of colloidal systems

with short-range attraction [26–30], interactions between protein molecules in solution [31], and other interesting applications [32, 33].

Of course, the pairwise additivity hypothesis (3.1) can be extended to mixtures. In that case, instead of a single potential function $\phi(r)$, there exists in general a different function $\phi_{\alpha\gamma}(r)$ for each pair of species α and γ. As an obvious consequence, there is a Mayer function

$$f_{\alpha\gamma}(r) = e^{-\beta\phi_{\alpha\gamma}(r)} - 1 \tag{3.6}$$

for each pair of species α and γ.

In particular, in the case of a HS multicomponent system,

$$\phi_{\alpha\gamma}(r) = \begin{cases} \infty, & r < \sigma_{\alpha\gamma}, \\ 0, & r > \sigma_{\alpha\gamma}, \end{cases} \tag{3.7a}$$

$$f_{\alpha\gamma}(r) = -\Theta(\sigma_{\alpha\gamma} - r). \tag{3.7b}$$

Here, $\sigma_{\alpha\gamma}$ is the closest possible distance between the center of a sphere of species α and the center of a sphere of species γ. If we call $\sigma_\alpha = \sigma_{\alpha\alpha}$ to the closest distance between two spheres of the same species α, it is legitimate to refer to σ_α as the *diameter* of a sphere of species α. However, that does not necessarily mean that two spheres of different type repel each other with a distance equal to the sum of their radii. Depending on that, one can classify HS mixtures into *additive* or *nonadditive*:

- Additive HS (AHS) mixtures: $\sigma_{\alpha\gamma} = \frac{1}{2}(\sigma_\alpha + \sigma_\gamma)$ for all pairs $\alpha\gamma$.
- Nonadditive HS (NAHS) mixtures: $\sigma_{\alpha\gamma} \neq \frac{1}{2}(\sigma_\alpha + \sigma_\gamma)$ for at least one pair $\alpha\gamma$.

Furthermore, the nonadditivity is said to be *negative* if $\sigma_{\alpha\gamma} < \frac{1}{2}(\sigma_\alpha + \sigma_\gamma)$, while it is *positive* if $\sigma_{\alpha\gamma} > \frac{1}{2}(\sigma_\alpha + \sigma_\gamma)$.

3.2 Virial Expansion

Except for one-dimensional systems with nearest-neighbor interactions (see Chap. 5), the exact evaluation by theoretical tools of the *equation of state* (EoS) expressing the pressure $p(n, T)$ for arbitrary interaction potential $\phi(r)$, density n, and temperature T is simply not possible. At a formal level, the non-ideal EoS is given by (2.65c) in the canonical ensemble, but the evaluation of the dN-order configuration integral $\mathcal{Q}_N$ defined in (2.64) is beyond our capabilities.

However, the problem can be controlled if one gives up the "arbitrary density" requirement and is satisfied with the low-density regime. In such a case, a series expansion in powers of density (*virial expansion*) is the adequate tool:

$$Z \equiv \frac{p}{nk_BT} = 1 + B_2(T)n + B_3(T)n^2 + \cdots$$

$$= 1 + \sum_{k=2}^{\infty} B_k n^{k-1} , \qquad (3.8)$$

where Z is the compressibility factor [see (1.35)] and $B_k(T)$ are the *virial* coefficients. Our main aim in this chapter is to derive expressions for the coefficients $B_k(T)$ as functions of T for any (short-range) interaction potential $\phi(r)$.

The expansion (3.8) was originally introduced by Thiesen [34] as early as in 1885 as an approximation to the EoS of low-density fluids. It was apparently independently reintroduced in 1901 by Kamerlingh Onnes (see Fig. 3.5) as a mathematical representation of experimental data on the EoS [35].

What is the basic physical idea behind the virial expansion? This is very clearly stated by E.G.D. Cohen (see Fig. 3.6) [36]:

> The virial or density expansions reduce the intractable $N(\sim10^{23})$-particle problem of a macroscopic gas in a volume V to a sum of an increasing number of tractable isolated few (1, 2, 3, ...) particle problems, where each group of particles *moves* alone in the volume V of the system.
>
> Density expansions will then appear, since the number of single particles, pairs of particles, triplets of particles, ..., in the system are proportional to n, n^2, n^3, ..., respectively, where $n = N/V$ is the number density of the particles.

Fig. 3.5 Heike Kamerlingh Onnes (1853–1926) (Photograph from Wikimedia Commons, https://upload. wikimedia.org/wikipedia/ commons/f/fa/ Kamerlingh_Onnes.jpg)

Fig. 3.6 Ezechiel "Eddie"
Godert David Cohen
(b. 1923)
(Photograph courtesy of
E.G.D. Cohen)

In order to attain (3.8), and exploiting the equivalence among the statistical ensembles in the thermodynamic limit, it is convenient to work with the grand canonical ensemble. This is because in that ensemble we already have a natural series power expansion for free: as shown in (2.67), the grand partition function Ξ is already expressed as a series in powers of fugacity.

Let us consider a generic quantity X that can be obtained from Ξ by taking its logarithm, by differentiation, etc. Then, from the expansion in (2.67) one could in principle obtain

$$X = \sum_{\ell=0}^{\infty} \bar{X}_\ell \hat{z}^\ell , \qquad (3.9)$$

where the coefficients $\bar{X}_\ell$ are related to the configuration integrals $\mathcal{Q}_N$ and depend on the specific choice of X. In particular, in the case of the average number density $n = \langle N \rangle / V$, we can write

$$n = \sum_{\ell=1}^{\infty} \ell \mathfrak{b}_\ell \hat{z}^\ell , \qquad (3.10)$$

where, because of reasons that will become apparent later, the coefficients $\mathfrak{b}_\ell$ are termed *cluster integrals*.

Now, eliminating the (rescaled) fugacity $\hat{z}$ between (3.9) and (3.10) one can express X in powers of n:

$$X = \sum_{k=0}^{\infty} X_k n^k . \qquad (3.11)$$

The first few relations are

$$X_0 = \bar{X}_0 \,, \tag{3.12a}$$

$$X_1 = \frac{\bar{X}_1}{\mathfrak{b}_1} \,, \tag{3.12b}$$

$$X_2 = \frac{\bar{X}_2}{\mathfrak{b}_1^2} - \frac{2\mathfrak{b}_2}{\mathfrak{b}_1^3}\bar{X}_1 \,, \tag{3.12c}$$

$$X_3 = \frac{\bar{X}_3}{\mathfrak{b}_1^3} - \frac{4\mathfrak{b}_2}{\mathfrak{b}_1^4}\bar{X}_2 - \left(\frac{3\mathfrak{b}_3}{\mathfrak{b}_1^4} - \frac{8\mathfrak{b}_2^2}{\mathfrak{b}_1^5}\right)\bar{X}_1 \,. \tag{3.12d}$$

$$X_4 = \frac{\bar{X}_4}{\mathfrak{b}_1^4} - \frac{6\mathfrak{b}_2}{\mathfrak{b}_1^5}\bar{X}_3 - 2\left(\frac{3\mathfrak{b}_3}{\mathfrak{b}_1^5} - \frac{10\mathfrak{b}_2^2}{\mathfrak{b}_1^6}\right)\bar{X}_2 - 2\left(\frac{2\mathfrak{b}_4}{\mathfrak{b}_1^5} + \frac{20\mathfrak{b}_2^3}{\mathfrak{b}_1^7} - \frac{15\mathfrak{b}_2\mathfrak{b}_3}{\mathfrak{b}_1^6}\right)\bar{X}_1 \,. \tag{3.12e}$$

3.3   Diagrammatic Method

Let us consider a one-component system and rewrite (2.67) as

$$\Xi = 1 + \sum_{N=1}^{\infty} \frac{z^N}{N!} \int d\mathbf{r}^N \, W_N(1, 2, \ldots, N) \,, \tag{3.13}$$

where we have introduced the N-body function

$$W_N(1, 2, \ldots, N) \equiv W_N(\mathbf{r}^N) = e^{-\beta\Phi_N(\mathbf{r}^N)}$$

$$= \prod_{1 \le i < j \le N} (1 + f_{ij}) \,, \quad f_{ij} \equiv f(r_{ij}) \,, \tag{3.14}$$

and use has been made of (2.64) and of the pairwise additivity property (3.1). The notation $f_{ij} \equiv f(r_{ij})$ should not be confused with the notation $f_{\alpha\gamma}(r)$ introduced for mixtures in (3.6).

When expanding the product in (3.14), $2^{N(N-1)/2}$ terms appear in W_N. To manage those terms, it is very convenient to represent them with *diagrams*. Each diagram contributing to W_N is made of N open circles (representing the N particles), some of them being joined by a bond (representing a factor f_{ij}). For example, the diagrams contributing to W_1-W_4 are

$$W_1(1) = 1 = \text{O} \quad , \tag{3.15a}$$

$$W_2(1,2) = 1 + f_{12}$$

$$= \text{O} \quad \text{O} \; + \; \text{O}\!\!-\!\!\text{O} \; , \tag{3.15b}$$

$$W_3(1,2,3) = (1 + f_{12})(1 + f_{13})(1 + f_{23})$$

$$= \; + \{3\} \; + \{3\} \; + \; , \tag{3.15c}$$

$$W_4(1,2,3,4) = (1 + f_{12})(1 + f_{13})(1 + f_{14})(1 + f_{23})(1 + f_{24})(1 + f_{34})$$

$$= \; + \{6\} \; + \{12\} \; + \{3\} \; + \{4\}$$

$$+\{12\} \; + \{4\} \; + \{12\} \; + \{3\}$$

$$+\{6\} \; + \; . \tag{3.15d}$$

The numerical coefficients enclosed by curly braces (e.g., {3}) in front of some diagrams refer to the number of diagrams topologically equivalent, i.e., those that differ only in the particle labels associated with each circle. For instance,

$$\{3\} \; = \; {}_1^3 \; + \; {}_1^2{}_3 \; + \; {}_2^1{}_3 \; . \tag{3.16}$$

Some of the diagrams are *disconnected* (i.e., there exists at least one particle isolated from the remaining ones), while the other ones are *connected* diagrams or *clusters* (i.e., it is possible to go from any particle to any other particle by following a path made of bonds) [1]. Therefore, in general,

$$W_N(1,2,\ldots,N) = \sum all \text{ (connected and disconnected) diagrams of } N \text{ particles.}$$

As will be seen, in our goal of obtaining the coefficients in the expansion (3.8), we will follow a *distillation* process upon which we will get rid of the least relevant diagrams at each stage, keeping only those containing more information. The first step consists in taking the logarithm of the grand partition function:

$$\ln \Xi = \sum_{\ell=1}^{\infty} \frac{\tilde{z}^\ell}{\ell!} \int d\mathbf{r}^\ell \, U_\ell(1,2,\ldots,\ell) \, , \tag{3.17}$$

where the functions $U_\ell(1, 2, \ldots, \ell)$ are called *cluster* (or Ursell) functions. They are obviously related to the functions $W_N(1, 2, \ldots, N)$. In fact, by comparing (3.13) and (3.17), one realizes that the relationship between $\{W_N\}$ and $\{U_\ell\}$ is exactly the same as that between *moments* and *cumulants* of a certain probability distribution [37]. In that analogy, Ξ plays the role of the *characteristic function* (or Fourier transform of the probability distribution) and $-i\hat{z}$ plays the role of the Fourier variable. The first few relations are

$$W_1(1) = U_1(1) , \tag{3.18a}$$

$$W_2(1, 2) = U_1(1)U_1(2) + U_2(1, 2) , \tag{3.18b}$$

$$W_3(1, 2, 3) = U_1(1)U_1(2)U_1(3) + \{3\}U_1(1)U_2(2, 3) + U_3(1, 2, 3) , \tag{3.18c}$$

$$W_4(1, 2, 3, 4) = U_1(1)U_1(2)U_1(3)U_1(4) + \{6\}U_1(1)U_1(2)U_2(3, 4)$$
$$+ \{3\}U_2(1, 2)U_2(3, 4) + \{4\}U_1(1)U_3(2, 3, 4) + U_4(1, 2, 3, 4) . \tag{3.18d}$$

Again, each numerical factor represents the number of terms equivalent (except for particle labeling) to the indicated canonical term. Using (3.15), one finds

$$U_1(1) = 1 = \circ , \tag{3.19a}$$

$$U_2(1, 2) = f_{12} = \circ\!\!-\!\!\circ , \tag{3.19b}$$

$$U_3(1, 2, 3) = \{3\}\; \triangle + \triangle , \tag{3.19c}$$

$$U_4(1, 2, 3, 4) = \{12\}\; \square + \{4\}\; \square + \{12\}\; \square + \{3\}\; \square$$

$$+ \{6\}\; \square + \square . \tag{3.19d}$$

We observe that all the disconnected diagrams have gone away. In general,

$$U_\ell(1, 2, \ldots, \ell) = \sum \text{all } \textit{connected} \text{ diagrams (i.e., ``clusters'') of } \ell \text{ particles.}$$

For later use, it is important to classify the clusters into *reducible* (or singly connected) and *irreducible* (or biconnected). The first class is made of those clusters having at least one *articulation point*, i.e., a point that, if removed together with its bonds, the resulting diagram becomes disconnected. Examples of reducible clusters are

$$\text{} \tag{3.20}$$

where the articulation points are surrounded by circles. Irreducible clusters (also called *stars*) are those clusters with no articulation point. For instance,

$$\text{} \tag{3.21}$$

The functions $W_N(\mathbf{r}^N)$ and $U_\ell(\mathbf{r}^\ell)$ depend on the position vectors of the corresponding number of particles. This is indicated by open circles in the diagrams (3.15) and (3.19). We will refer to them as *root* points. On the other hand, in (3.13) and (3.17) one has to integrate out the positions of the particles. The quantity obtained by integrating a given diagram will be represented by the same diagram, except that the integrated points will be indicated by filled circles and will be referred to as *field* points. For instance,

$$\int d\mathbf{r}_1 \int d\mathbf{r}_2 \int d\mathbf{r}_3 \ \text{} = \text{} , \tag{3.22a}$$

$$\int d\mathbf{r}_2 \int d\mathbf{r}_3 \ \text{} = \text{} , \tag{3.22b}$$

$$\int d\mathbf{r}_3 \ \text{} = \text{} . \tag{3.22c}$$

If $A(1, 2, \ldots, \ell)$ is an arbitrary translation-invariant function, then it actually depends on the relative positions of the ℓ particles only, i.e.,

$$A(\mathbf{r}_1, \mathbf{r}_2, \ldots, \mathbf{r}_\ell) = \bar{A}(\mathbf{r}_{21}, \ldots, \mathbf{r}_{\ell 1}) . \tag{3.23}$$

Thanks to this property,

$$\int d\mathbf{r}_1 \int d\mathbf{r}_2 \cdots \int d\mathbf{r}_\ell A(\mathbf{r}_1, \mathbf{r}_2, \ldots, \mathbf{r}_\ell) = V \int d\mathbf{r}_{21} \cdots \int d\mathbf{r}_{\ell 1} \bar{A}(\mathbf{r}_{21}, \ldots, \mathbf{r}_{\ell 1})$$

$$= V \int d\mathbf{r}_2 \cdots \int d\mathbf{r}_\ell A(\mathbf{r}_1, \mathbf{r}_2, \ldots, \mathbf{r}_\ell) . \tag{3.24}$$

As a consequence, every zero-root diagram factorizes into V times the corresponding one-root diagram. Thus, for instance,

$$V^{-1}\; \bullet\!\!-\!\!\bullet \; = \; \circ\!\!-\!\!\bullet \; , \tag{3.25a}$$

$$V^{-1}\; \bigwedge \; = \; \bigwedge \; = \; \bigwedge \; , \tag{3.25b}$$

$$V^{-1}\; \bigtriangleup \; = \; \bigtriangleup \; . \tag{3.25c}$$

A further factorization property applies to the reducible diagrams. The existence of at least one articulation point allows one to express the zero-root reducible diagrams as products of zero-root stars. In particular, in the case of the reducible diagrams (3.20) one has

$$V^{-1}\; \bigwedge \; = \left(\int d\mathbf{r}_{21}\; \circ\!\!-\!\!\circ \; \right)^{2} = \left(V^{-1}\; \bullet\!\!-\!\!\bullet \; \right)^{2} , \tag{3.26a}$$

$$V^{-1}\; \sqcup\!\!\sqcup \; = \left(\int d\mathbf{r}_{21}\; \circ\!\!-\!\!\circ \; \right)^{3} = \left(V^{-1}\; \bullet\!\!-\!\!\bullet \; \right)^{3} , \tag{3.26b}$$

$$V^{-1}\; \diagdown\!\!\diagup \; = \left(\int d\mathbf{r}_{21}\; \circ\!\!-\!\!\circ \; \right)^{3} = \left(V^{-1}\; \bullet\!\!-\!\!\bullet \; \right)^{3} , \tag{3.26c}$$

$$V^{-1}\; \boxtimes \; = \left(\int d\mathbf{r}_{21}\; \circ\!\!-\!\!\circ \; \right)\left(\int d\mathbf{r}_{21} \int d\mathbf{r}_{31}\; \bigtriangleup \; \right)$$

$$= \left(V^{-1}\; \bullet\!\!-\!\!\bullet \; \right)\left(V^{-1}\; \bigtriangleup \; \right) . \tag{3.26d}$$

3.4   Grand Canonical Ensemble: Expansion in Powers of Fugacity

Let us rewrite (3.17) as

$$\ln \Xi = V \sum_{\ell=1}^{\infty} \mathfrak{b}_{\ell} \hat{z}^{\ell} , \tag{3.27}$$

where the cluster integrals are defined as

$$\mathfrak{b}_{\ell} \equiv \frac{V^{-1}}{\ell!} \int d\mathbf{r}^{\ell}\, U_{\ell}(1, 2, \ldots, \ell) . \tag{3.28}$$

Taking into account (3.19), the first few cluster integrals are

$$b_1 = V^{-1} \, \bullet \qquad = 1 \,, \tag{3.29a}$$

$$b_2 = \frac{V^{-1}}{2} \, \bullet\!\!-\!\!\bullet \,, \tag{3.29b}$$

$$b_3 = \frac{V^{-1}}{3!} \left(3 \,\triangle + \,\blacktriangle \right), \tag{3.29c}$$

$$b_4 = \frac{V^{-1}}{4!} \left(12 \; \square_{\text{(path)}} + 4 \; \square + 12 \; \square + 3 \; \square \right.$$

$$\left. + 6 \; \square + \boxtimes \right). \tag{3.29d}$$

Now the numerical coefficients in front of each zero-root diagram do not need to be enclosed between braces because all the topologically equivalent "bare" diagrams become the same upon integration. We will refer to the numerical coefficient in front of a given diagram as the *degeneracy* of that diagram.

In general,

$$b_\ell(T) = \frac{V^{-1}}{\ell!} \sum \text{all } \textit{clusters} \text{ with zero roots and } \ell \text{ field points}$$

$$= \frac{1}{\ell!} \sum \text{all } \textit{clusters} \text{ with 1 root and } \ell - 1 \text{ field points} \,. \tag{3.30}$$

From the grand canonical relations (2.36b) and (2.37) we have

$$\beta p = \sum_{\ell=1}^{\infty} b_\ell \hat{z}^\ell \,, \tag{3.31a}$$

$$n = -\frac{1}{V} \frac{\partial \ln \varXi}{\partial \alpha} = \frac{\hat{z}}{V} \frac{\partial \ln \varXi}{\partial \hat{z}} = \sum_{\ell=1}^{\infty} \ell \, b_\ell \hat{z}^\ell \,, \tag{3.31b}$$

where we have taken into account that $\partial \hat{z}/\partial \alpha = -\hat{z}$, as follows from the definitions (2.15) and (2.68). Equations (3.31) express the pressure and the number density as series expansions in powers of fugacity. Note that the last term in (3.31b) was already written in (3.10).

3.5   Expansion of Pressure in Powers of Density: Virial Coefficients

Equations (3.8) and (3.31a) have the same structure as (3.11) and (3.9), respectively, with $X \to \beta p$, $X_k \to B_k$, and $\tilde{X}_\ell \to b_\ell$. Therefore, the general relations (3.12) yield

$$B_1 = 1 ,\tag{3.32a}$$

$$B_2 = -b_2 ,\tag{3.32b}$$

$$B_3 = 4b_2^2 - 2b_3 ,\tag{3.32c}$$

$$B_4 = -20b_2^3 + 18b_2b_3 - 3b_4 ,\tag{3.32d}$$

where (3.29a) has been used. The diagrams representing the virial coefficients B_k can be obtained from those representing the cluster integrals b_ℓ. Apparently, given the nonlinear relationship between both classes of coefficients, one could expect that the diagram complexity increases in the process $\{b_\ell\} \to \{B_k\}$. However, what actually happens is just the opposite. It turns out that, thanks to the factorization property of the reducible diagrams present in the cluster integrals, they exactly cancel out the nonlinear terms. As a consequence, only the irreducible diagrams (stars) survive in the virial coefficients. For instance, inserting (3.29) into (3.32), and making use of (3.26), one finds

$$B_2 = -\frac{V^{-1}}{2}\;\text{•—•}\; = -\frac{1}{2}\;\text{○—•}\; ,\tag{3.33a}$$

$$B_3 = -\frac{2}{3!}V^{-1}\triangle = -\frac{2}{3!}\triangle ,\tag{3.33b}$$

$$B_4 = -\frac{3}{4!}V^{-1}\left(3\,\square + 6\,\boxtimes + \boxtimes\right)$$

$$= -\frac{3}{4!}\left(3\,\square + 6\,\boxtimes + \boxtimes\right) .\tag{3.33c}$$

The star diagrams contributing to the virial coefficients B_2–B_5 are displayed in Table 3.2 [38]. In general,

$$
\begin{aligned}
B_k(T) &= -\frac{k-1}{k!}V^{-1}\sum \text{ all } \textit{stars} \text{ with zero roots and } k \text{ field points} \\
&= -\frac{k-1}{k!}\sum \text{ all } \textit{stars} \text{ with 1 root and } k-1 \text{ field points}.
\end{aligned}
\tag{3.34}
$$

The "distillation" process leading to (3.8) is summarized in Table 3.3 and Fig. 3.7. As shown in Fig. 3.8, the number of independent diagrams needed to evaluate the kth virial coefficient B_k grows explosively with k, what significantly hampers the computation beyond the first few coefficients, even with sophisticated numerical algorithms [39, 41, 42].

Table 3.2 Diagrams contributing to $B_2(r)$, $B_3(r)$, $B_4(r)$, and $B_5(r)$ in the expansion (3.8)

Density term	Coefficient	Diagrams
$-\dfrac{n}{2}$	$-2B_2$	
$-\dfrac{n^2}{3}$	$-3B_3$	
$-\dfrac{n^3}{8}$	$-8B_4$	
$-\dfrac{n^4}{30}$	$-30B_5$	

Table 3.3 Summary of diagrams contributing to different quantities

Quantity	Expansion in powers of	Coefficient	Diagrams	Equation
$\varXi$	Fugacity ($\hat{z}$)	$W_N/N!$	All (disconnected+clusters)	(3.13)
$\ln \varXi$	Fugacity ($\hat{z}$)	$U_\ell/\ell!$	Clusters (reducible+stars)	(3.17)
p	Fugacity ($\hat{z}$)	b_ℓ	Clusters (reducible+stars)	(3.31a)
n	Fugacity ($\hat{z}$)	ℓb_ℓ	Clusters (reducible+stars)	(3.31b)
p	Density (n)	B_k	Stars	(3.8)

Fig. 3.7 Schematic
representation of the
diagrams contributing to
different quantities

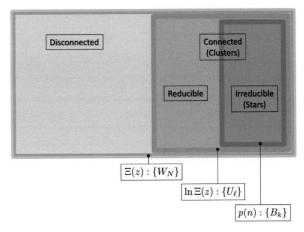

Fig. 3.8 Number of star
diagrams (in logarithmic
scale) contributing to each
virial coefficient B_k [39, 40].
The unlabeled and labeled
cases correspond to
monodisperse and
polydisperse systems,
respectively

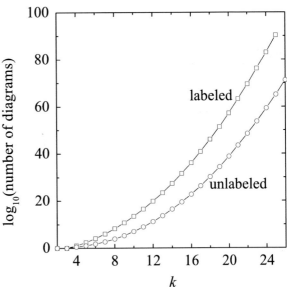

3.6 Virial Coefficients for Mixtures

Although the steps followed so far in this chapter have been restricted to one-component systems, it is not difficult to extend the final results to mixtures. In that case, the virial coefficients not only depend on temperature but also on the set of mole fractions $\{x_\nu\}$, i.e., $B_k(T) \rightarrow B_k(T, \{x_\nu\})$. The dependence of B_k on $\{x_\nu\}$ is necessarily polynomial (of degree k), so that one can define *composition-independent* virial coefficients $\widehat{B}_{\nu_1\nu_2\cdots\nu_k}(T)$ as the coefficients in the polynomial expansion of $B_k(T, \{x_\nu\})$, namely

$$B_k(T, \{x_\nu\}) = \sum_{\nu_1}\sum_{\nu_2}\cdots\sum_{\nu_k} x_{\nu_1} x_{\nu_2}\cdots x_{\nu_k}\widehat{B}_{\nu_1\nu_2\cdots\nu_k}(T) , \tag{3.35}$$

where the indices $\nu_1, \nu_2, \ldots, \nu_k$ run over all the species of the mixture. For instance, the second $(\widehat{B}_{\alpha\gamma})$, third $(\widehat{B}_{\alpha\gamma\delta})$, and fourth $(\widehat{B}_{\alpha\gamma\delta\epsilon})$ composition-independent virial coefficients are defined by

$$B_2(T) = \sum_{\alpha,\gamma} x_\alpha x_\gamma \widehat{B}_{\alpha\gamma}(T) , \tag{3.36a}$$

$$B_3(T) = \sum_{\alpha,\gamma,\delta} x_\alpha x_\gamma x_\delta \widehat{B}_{\alpha\gamma\delta}(T) , \tag{3.36b}$$

$$B_4(T) = \sum_{\alpha,\gamma,\delta,\epsilon} x_\alpha x_\gamma x_\delta x_\epsilon \widehat{B}_{\alpha\gamma\delta\epsilon}(T) . \tag{3.36c}$$

Without loss of generality, one can always construct the coefficients $\widehat{B}_{\nu_1\nu_2\cdots\nu_k}$ as invariant under any permutation of indices, i.e., $\widehat{B}_{\nu_1\nu_2\cdots\nu_k} = \widehat{B}_{\nu_2\nu_1\cdots\nu_k}$, etc.

Each coefficient $\widehat{B}_{\nu_1\nu_2\cdots\nu_k}$ is represented by the same diagrams as the one-component virial coefficient B_k, except that now the k field points represent particles of species $\nu_1, \nu_2, \ldots, \nu_k$ and, consequently, the degeneracy of each diagram is broken down (see Fig. 3.8). In particular, (3.33) now becomes

$$\widehat{B}_{\alpha\gamma} = -\frac{V^{-1}}{2}\ \alpha\ \bullet\!\!-\!\!\bullet\ \gamma, \tag{3.37a}$$

$$\widehat{B}_{\alpha\gamma\delta} = -\frac{V^{-1}}{3}\ \alpha\ \triangle\ \gamma\ , \tag{3.37b}$$

$$\widehat{B}_{\alpha\gamma\delta\epsilon} = \widehat{B}^{(I)}_{\alpha\gamma\delta\epsilon} + \widehat{B}^{(I)}_{\alpha\gamma\epsilon\delta} + \widehat{B}^{(I)}_{\alpha\delta\epsilon\gamma} + \widehat{B}^{(II)}_{\alpha\gamma\delta\epsilon} + \widehat{B}^{(II)}_{\alpha\gamma\epsilon\delta} + \widehat{B}^{(II)}_{\alpha\delta\epsilon\gamma}$$
$$+ \widehat{B}^{(II)}_{\gamma\alpha\epsilon\delta} + \widehat{B}^{(II)}_{\gamma\alpha\delta\epsilon} + \widehat{B}^{(II)}_{\delta\alpha\gamma\epsilon} + \widehat{B}^{(III)}_{\alpha\gamma\delta\epsilon} , \tag{3.37c}$$

where

$$\widehat{B}^{(I)}_{\alpha\gamma\delta\epsilon} = -\frac{V^{-1}}{8}\ \substack{\alpha\\ \delta}\ \square\ \substack{\gamma\\ \epsilon}\ , \tag{3.38a}$$

$$\widehat{B}^{(II)}_{\alpha\gamma\delta\epsilon} = -\frac{V^{-1}}{8}\ \substack{\alpha\\ \delta}\ \boxtimes\ \substack{\gamma\\ \epsilon}\ , \tag{3.38b}$$

$$\widehat{B}^{(III)}_{\alpha\gamma\delta\epsilon} = -\frac{V^{-1}}{8}\ \substack{\alpha\\ \delta}\ \boxtimes\ \substack{\gamma\\ \epsilon}\ . \tag{3.38c}$$

Note that, in contrast to $\widehat{B}_{\alpha\gamma\delta\epsilon}$ and $\widehat{B}^{(III)}_{\alpha\gamma\delta\epsilon}$, the partial contributions $\widehat{B}^{(I)}_{\alpha\gamma\delta\epsilon}$ and $\widehat{B}^{(II)}_{\alpha\gamma\delta\epsilon}$ are, in general, not invariant under a permutation of indices.

The simplest type of mixture is the *binary* one, i.e., only two components (say 1 and 2) are present. In that case,

$$\widehat{B}_{\nu_1\nu_2\cdots\nu_k} \rightarrow \widehat{B}_{\underbrace{11\cdots1}_{k_1}\underbrace{22\cdots2}_{k-k_1}} \equiv \widehat{B}_{k_1;k-k_1} \tag{3.39}$$

and thus (3.35) reduces to

$$B_k(T, x_1) = \sum_{k_1=0}^{k} \binom{k}{k_1} x_1^{k_1} x_2^{k-k_1} \widehat{B}_{k_1;k-k_1}(T) , \tag{3.40}$$

with $x_2 = 1 - x_1$. For example, the fourth and fifth virial coefficients are

$$B_4 = x_1^4 \widehat{B}_{4;0} + 4x_1^3 x_2 \widehat{B}_{3;1} + 6x_1^2 x_2^2 \widehat{B}_{2;2} + 4x_1 x_2^3 \widehat{B}_{1;3} + x_2^4 \widehat{B}_{0;4} , \tag{3.41a}$$

$$B_5 = x_1^5 \widehat{B}_{5;0} + 5x_1^4 x_2 \widehat{B}_{4;1} + 10x_1^3 x_2^2 \widehat{B}_{3;2} + 10x_1^2 x_2^3 \widehat{B}_{2;3} + 5x_1 x_2^4 \widehat{B}_{1;4} + x_2^5 \widehat{B}_{0;5} . \tag{3.41b}$$

The coefficients $\widehat{B}_{4;0}$, $\widehat{B}_{0;4}$, $\widehat{B}_{5;0}$, and $\widehat{B}_{0;5}$ are equivalent to those of the one-component system (see Table 3.2). The diagrams associated with $\widehat{B}_{3;1}$ and $\widehat{B}_{2;2}$ are shown in Table 3.4, while those associated with $\widehat{B}_{4;1}$ and $\widehat{B}_{3;2}$ are shown in Table 3.5. The remaining coefficients ($\widehat{B}_{1;3}$, $\widehat{B}_{2;3}$, and $\widehat{B}_{1;4}$) can be obtained by exchanging species 1 and 2.

In the special case of a binary AHS mixture made of particles of diameters σ_1 and σ_2, so that $\sigma_{12} = \frac{1}{2}(\sigma_1 + \sigma_2)$, it is convenient to introduce the *rescaled* composition-independent virial coefficients $b_{k_1;k-k_1}$ by the relation

$$\widehat{B}^{AHS}_{k_1;k-k_1}(\sigma_1, \sigma_2) = \left(v_d \sigma_1^d\right)^{k-1} b_{k_1;k-k_1}(\sigma_2/\sigma_1) , \tag{3.42}$$

Table 3.4 Diagrams contributing to $-8V\widehat{B}_{k_1;k_2}$ with $k_1 + k_2 = 4$

$(k_1; k_2)$	Diagrams
$(3; 1)$	
$(2; 2)$	

Filled circles and points enclosed by open circles denote species 1 and 2, respectively

Table 3.5 Diagrams contributing to $-30V\widehat{B}_{k_1;k_2}$ with $k_1 + k_2 = 5$

$(k_1;k_2)$	Diagrams
(4; 1)	
(3; 2)	

Filled circles and points enclosed by open circles denote species 1 and 2, respectively

where

$$v_d = \frac{(\pi/4)^{d/2}}{\Gamma(1 + d/2)} \tag{3.43}$$

is the volume of a d-dimensional sphere of unit diameter [see (2.76)]. For instance, $v_d = 1, \frac{\pi}{4}, \frac{\pi}{6}, \frac{\pi^2}{32},$ and $\frac{\pi^2}{60}$ for $d = 1, 2, 3, 4,$ and 5, respectively. Obviously, $\widehat{B}^{\text{AHS}}_{k_1;k-k_1}(\sigma_1,\sigma_2) = \widehat{B}^{\text{AHS}}_{k-k_1;k_1}(\sigma_2,\sigma_1)$, what implies the symmetry property

$$b_{k_1;k-k_1}(q) = q^{d(k-1)}b_{k-k_1;k_1}(1/q) . \tag{3.44}$$

If the two species have the same diameter ($\sigma_1 = \sigma_2 = \sigma$) one recovers the one-component case and therefore

$$\lim_{q\to 1} b_{k_1;k-k_1}(q) = b_k , \tag{3.45}$$

where

$$b_k \equiv \left(v_d\sigma^d\right)^{-(k-1)} B_k^{HS} \tag{3.46}$$

are the rescaled virial coefficients of a pure HS fluid. Note also that $b_{k;0} = b_k$. Other exact relations are less straightforward [43–46]:

$$\lim_{q\to 0} q^{-d(k-k_1-1)} b_{k_1;k-k_1}(q) = \frac{k-k_1}{k} b_{k-k_1} , \tag{3.47a}$$

$$\left.\frac{\partial b_{k_1;k-k_1}}{\partial q}\right|_{q=1} = d\frac{(k-k_1)(k-1)}{k} b_k , \tag{3.47b}$$

$$\left.\frac{\partial b_{k-1;1}}{\partial q}\right|_{q=0} = d\frac{k-1}{k} . \tag{3.47c}$$

3.7 Second Virial Coefficient

The second virial coefficient provides the very first correction to the ideal gas EoS. From (3.33a),

$$B_2(T) = -\frac{1}{2}\int d\mathbf{r} f(r) = -d2^{d-1} v_d \int_0^\infty dr\, r^{d-1} f(r) , \tag{3.48}$$

where in the second step we have used spherical coordinates.

In the case of the HS potential, the second virial coefficient is simply

$$B_2^{HS} = 2^{d-1} v_d\sigma^d , \tag{3.49}$$

so that $b_2 = 2^{d-1}$ and the EoS truncated after B_2 is

$$Z \equiv \frac{\beta p}{n} = 1 + 2^{d-1}\eta + \cdots , \tag{3.50}$$

where

$$\boxed{\eta \equiv n v_d\sigma^d} \tag{3.51}$$

is the packing fraction.

The HS Mayer function is independent of temperature and so are all the HS virial coefficients. In general, however, $B_2(T)$ is a function of temperature. As simple examples, the result for the purely repulsive PS and SS potentials (see Table 3.1 for definitions) are displayed in Table 3.6. The temperature dependence of B_2^{PS} and B_2^{SS} is illustrated by Fig. 3.9. In the zero-temperature limit, both B_2^{PS} and B_2^{SS} tend to HS values corresponding to diameters σ and σ', respectively. In the opposite limit $T \to \infty$, B_2^{SS} tends to HS value corresponding to diameter σ, while B_2^{PS} tends to zero.

As a simple example of the second virial coefficient for a potential with an attractive part, let us consider the SW interaction (see Table 3.1). The result is simply obtained from $B_2^{PS}(T)$ by the replacement $\varepsilon \to -\varepsilon$, as shown in Table 3.6. Table 3.6 also includes B_2^{SHS}, as obtained from B_2^{SW} by taking the SHS limit [see (3.4)].

Table 3.6 Expressions for the second virial coefficient corresponding to several interaction potentials, relative to that of a HS fluid with the same value of σ [see (3.49)]

Potential	$B_2(T)/B_2^{HS}$
PS	$1 - e^{-1/T^*}$
SS	$1 + \left[(\sigma'/\sigma)^d - 1 \right] \left(1 - e^{-1/T^*} \right)$
SW	$1 - \left[(\sigma'/\sigma)^d - 1 \right] \left(e^{1/T^*} - 1 \right)$
SHS	$1 - 2^{1-d} \tau^{-1}$
LJ (2s-s)	$\Gamma(1 - d/s) \left(\dfrac{8}{T^*} \right)^{d/2s} e^{1/2T^*} D_{d/s} \left(-\sqrt{2/T^*} \right)$

Here, $T^* \equiv k_B T / \varepsilon$ and, in the SHS case, τ^{-1} is defined by (3.4)

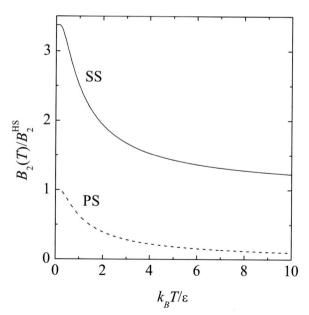

Fig. 3.9 Second virial coefficient $B_2(T)$ of PS and SS fluids, relative to that of a HS fluid with the same value of σ. In the SS case, $\sigma'/\sigma = 1.5$ and $d = 3$

The evaluation is much less straightforward in the case of continuous potentials like the LJ one. Let us consider the more general case of the LJ (2s-s) potential (with $s > d$):

$$\phi_{LJ}(r) = 4\varepsilon \left[\left(\frac{\sigma}{r} \right)^{2s} - \left(\frac{\sigma}{r} \right)^{s} \right] . \qquad (3.52)$$

The minimum is located at $r_0 = 2^{1/s}\sigma$. The conventional LJ potential (see Table 3.1) corresponds to $s = 6$.

Starting from the last equality in (3.48) and introducing the change of variable $r \to t \equiv \sqrt{8\beta\varepsilon}(\sigma/r)^s$, one obtains

$$B_2^{LJ}(T) = -2^{d-1} v_d \sigma^d \frac{d}{s} (8\beta\varepsilon)^{d/2s} \int_0^\infty dt\, t^{-d/s-1} \left(e^{-t^2/2+\sqrt{2\beta\varepsilon}t} - 1 \right) . \qquad (3.53)$$

From the properties of the parabolic cylinder function $D_a(z)$ [47, 48], it is possible to prove the integral representation

$$D_a(z) = \frac{e^{-z^2/4}}{\Gamma(-a)} \int_0^\infty dt\, t^{-a-1} \left(e^{-t^2/2-zt} - 1 \right) , \quad 0 < \mathrm{Re}(a) < 1 . \qquad (3.54)$$

Comparison between (3.53) and (3.54) yields the expression displayed in Table 3.6. That compact expression seems to have been first published in [49].

Figure 3.10 shows the temperature dependence of B_2/B_2^{HS} for (three-dimensional) SW and LJ fluids [50]. For low temperatures, the attractive part

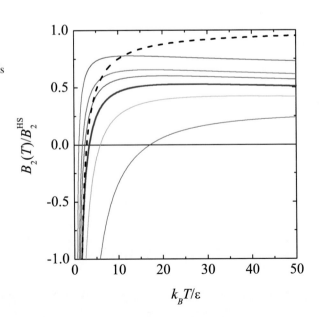

Fig. 3.10 Second virial coefficient $B_2(T)$, relative to that of a HS fluid with the same value of σ, of SW fluids with $\sigma'/\sigma = 1.5$ (*dashed line*) and of LJ (2s-s) fluids with $s = 4, 5, 6$ (*thick line*), 7, 8, and 12, *from bottom to top* (*solid lines*). The dimensionality is $d = 3$

of the potential dominates and thus $B_2 < 0$, meaning that in the low-density regime the pressure is smaller than that of an ideal gas at the same density. Reciprocally, $B_2 > 0$ for high temperatures, in which case the repulsive part of the potential prevails. The transition between both situations takes place at the so-called Boyle temperature T_B, where $B_2 = 0$. In the LJ case the Boyle temperature is $T_B^* = 2/z_0^2$, where $z = -z_0$ is the zero of $D_{d/s}(z)$, for which good approximations are known [47, 48].

Note that, while B_2^{SW} monotonically grows with temperature and asymptotically tends to the HS value, the LJ coefficient reaches a maximum (smaller than the HS value corresponding to a diameter σ) and then decreases very slowly. This reflects the fact that for very high temperatures the LJ fluid behaves practically as a HS system but with an *effective* diameter smaller than the nominal value σ.

Let us analyze the low-temperature and high-temperature limits of B_2^{LJ} with some detail. Making use of the mathematical property $D_a(z = 0) = 2^{a/2}\sqrt{\pi}/\Gamma(1/2 - a/2)$ [47, 48], we have

$$\lim_{T^* \to \infty} \frac{B_2^{LJ}(T)}{B_2^{HS}} = \frac{\sqrt{\pi}\,\Gamma(1 - d/s)}{\Gamma(1/2 - d/2s)}\left(\frac{16}{T^*}\right)^{d/2s}. \tag{3.55}$$

On the other hand, the asymptotic behavior of $D_a(-z)$ in the limit of large z is $D_a(-z) \approx \sqrt{2\pi}e^{z^2/4}z^{-a-1}/\Gamma(-a)$ [47, 48]. This implies

$$\lim_{T^* \to 0} \frac{B_2^{LJ}(T)}{B_2^{HS}} = -\frac{d}{s}2^{d/s-1/2}e^{1/T^*}\sqrt{2\pi T^*}. \tag{3.56}$$

The asymptotic property $B_2 \sim -e^{1/T^*}\sqrt{T^*}$ for $T^* \ll 1$ is independent of the parameter s. In fact, it applies beyond the LJ potential (3.52). Consider a general class of continuous potentials of the form

$$\phi(r) = \varepsilon\phi^*(r^*), \quad r^* = r/\sigma, \tag{3.57}$$

where the dimensionless function $\phi^*(r^*)$ satisfies the general properties

$$\lim_{r^* \to 0} \phi^*(r^*) = \infty, \quad \lim_{r^* \to \infty} \phi^*(r^*) = 0, \quad \phi^*(1) = 0, \quad \phi^*(r_0^*) = -1, \tag{3.58}$$

$$\left.\frac{d\phi^*(r^*)}{dr^*}\right|_{r^*=r_0^*} = 0, \quad \left.\frac{d^2\phi^*(r^*)}{dr^{*2}}\right|_{r^*=r_0^*} \equiv \phi_0^{*''} > 0, \tag{3.59}$$

$r_0^* > 1$ denoting the location of the absolute minimum of $\phi^*(r^*)$. Thus, from (3.48) we can write

$$\frac{B_2(T)}{B_2^{HS}} = -d\int_0^\infty dr^*\, r^{*d-1}e^{Y(r^*,T^*)}, \quad Y(r^*,T^*) \equiv \ln\left[e^{-\phi^*(r^*)/T^*} - 1\right]. \tag{3.60}$$

In the limit $T^* \to 0$, $e^{Y(r^*, T^*)}$ presents a sharp maximum at $r^* = r_0^*$, so that we can evaluate the integral by the Laplace method [51]. First, we expand $Y(r^*, T^*)$ around $r^* = r_0^*$ and truncate the expansion after the second-order term, i.e.,

$$Y(r^*, T^*) \approx \ln\left(e^{1/T^*} - 1\right) - \frac{\phi_0^{*''}}{2T^*} \frac{(r^* - r_0^*)^2}{1 - e^{-1/T^*}} \approx \frac{1}{T^*} - \frac{\phi_0^{*''}}{2T^*}(r^* - r_0^*)^2 , \quad (3.61)$$

where in the last step we have taken into account that $T^* \gg 1$. Next, we approximate the integral in (3.60) by a Gaussian integral, so that

$$\frac{B_2(T)}{B_2^{HS}} \approx -dr_0^{*d-1}e^{1/T^*} \int_{-\infty}^{\infty} dx\, e^{-\phi_0^{*''} x^2/2T^*} = -dr_0^{*d-1}e^{1/T^*}\sqrt{2\pi T^*/\phi_0^{*''}} . \quad (3.62)$$

In the particular case of the LJ potential (3.52), $r_0^* = 2^{1/s}$ and $\phi_0^{*''} = 2^{1-2/s}s^2$. Therefore, (3.62) becomes (3.56).

3.8 Higher-Order Virial Coefficients for Hard Spheres

The evaluation of virial coefficients beyond B_2 becomes a formidable task as the order increases and it is thus necessary to resort to numerical Monte Carlo (MC) methods to perform the multiple integrals involved. Needless to say, the computational task is much more manageable in the case of HS fluids.

3.8.1 One-Component Systems

In the one-component case, the third and fourth virial coefficients are analytically known [52, 53] and higher-order virial coefficients have been numerically evaluated [39, 41, 42, 54–58].

The third virial coefficient is [59, 60]

$$\frac{B_3}{B_2^2} = 2I_{3/4}\left(\frac{d+1}{2}, \frac{1}{2}\right) , \quad (3.63)$$

where $I_x(a, b) = B_x(a, b)/B(a, b)$ is the *regularized* beta function,

$$B_x(a, b) = \int_0^x dt\, t^{a-1}(1-t)^{b-1} , \quad B(a, b) = B_1(a, b) = \frac{\Gamma(a)\Gamma(b)}{\Gamma(a+b)} \quad (3.64)$$

being the incomplete beta function and the beta function [47, 48], respectively. The explicit expressions of B_3/B_2^2 for $d \leq 12$ can be found in Table 3.7. We note that

Table 3.7 Third virial
coefficient B_3 (in units of B_2^2)
for HS fluids with dimensions
$d = 1\text{--}12$

d	Exact	Numerical
1	1	1
2	$\dfrac{4}{3} - \dfrac{\sqrt{3}}{\pi}$	$0.782\,004\,4\cdots$
3	$\dfrac{5}{2^3}$	0.625
4	$\dfrac{4}{3} - \dfrac{3\sqrt{3}}{2\pi}$	$0.506\,340\,0\cdots$
5	$\dfrac{53}{2^7}$	$0.414\,062\,5$
6	$\dfrac{4}{3} - \dfrac{9\sqrt{3}}{5\pi}$	$0.340\,941\,3\cdots$
7	$\dfrac{289}{2^{10}}$	$0.282\,226\,6\cdots$
8	$\dfrac{4}{3} - \dfrac{279\sqrt{3}}{140\pi}$	$0.234\,613\,6\cdots$
9	$\dfrac{6413}{2^{15}}$	$0.195\,709\,2\cdots$
10	$\dfrac{4}{3} - \dfrac{297\sqrt{3}}{140\pi}$	$0.163\,728\,5\cdots$
11	$\dfrac{35\,995}{2^{18}}$	$0.137\,310\,0\cdots$
12	$\dfrac{4}{3} - \dfrac{243\sqrt{3}}{110\pi}$	$0.115\,397\,7\cdots$

B_3/B_2^2 is an irrational number (since it includes $\sqrt{3}/\pi$) if $d = $ even, while it is a
rational number if $d = $ odd. More generally [61],

$$\frac{B_3}{B_2^2} = 2 - \frac{2\Gamma(1 + d/2)}{\sqrt{\pi}} \sum_{j=0}^{j_{\max}^{(d)}} \frac{(-4)^{-j}}{(2j+1)j!\,\Gamma\left(\frac{d+1}{2} - j\right)}, \qquad (3.65)$$

where $j_{\max}^{(d)} = (d-1)/2$ if $d = $ odd and $j_{\max}^{(d)} = \infty$ if $d = $ even. In the latter case,
however, the infinite series representation (3.65) converges so rapidly that a finite
upper limit $\gtrsim d/2$ is enough for practical purposes.

The ratio B_3/B_2^2 monotonically decreases with increasing dimensionality. To
analyze this with more detail, note that $\mathrm{B}(a,b) \approx \Gamma(b)a^{-b}$ and $\mathrm{B}_x(a,b) \approx
a^{-1}x^a(1-x)^{b-1}$ if $a \gg 1$ with $b > 0$ and $0 < x < 1$ [48]. Consequently,

$$\frac{B_3}{B_2^2} \approx \frac{4}{\sqrt{\pi d/2}} \left(\frac{\sqrt{3}}{2}\right)^{d+1}, \qquad d \gg 1. \qquad (3.66)$$

The influence of the parity of d is also present in the exact evaluation of B_4, which has been carried out separately for $d =$ even [55] and $d =$ odd [53]. The results for $d \leq 12$ are shown in Table 3.8. We see that B_4/B_2^3 is always an irrational number that includes $\sqrt{3}/\pi$ and $1/\pi^2$ if $d =$ even, while it includes $\sqrt{2}/\pi$ and $\cos^{-1}(1/3)/\pi$ if $d =$ odd. Interestingly, the fourth virial coefficient becomes negative for $d \geq 8$ [52].

Equation (3.49) and Tables 3.7 and 3.8 show that $B_k = \sigma^{k-1}$ for $k = 2\text{--}4$ in the one-dimensional case. This is in fact true for any k, as will be seen in Chap. 5 [see (5.67)].

Table 3.8 Fourth virial coefficient B_4 (in units of B_2^3) for HS fluids with dimensions $d = 1\text{--}12$

d	Exact	Numerical
1	1	1
2	$2 - \dfrac{9\sqrt{3}}{2\pi} + \dfrac{10}{\pi^2}$	$0.532\,231\,8\cdots$
3	$\dfrac{2707}{4480} + \dfrac{219\sqrt{2}}{2240\pi} - \dfrac{4131\cos^{-1}(1/3)}{4480\pi}$	$0.286\,949\,5\cdots$
4	$2 - \dfrac{27\sqrt{3}}{4\pi} + \dfrac{832}{45\pi^2}$	$0.151\,846\,1\cdots$
5	$\dfrac{25\,315\,393}{32\,800\,768} + \dfrac{3\,888\,425\sqrt{2}}{16\,400\,384\pi} - \dfrac{67\,183\,425\cos^{-1}(1/3)}{32\,800\,768\pi}$	$0.075\,972\,5\cdots$
6	$2 - \dfrac{81\sqrt{3}}{10\pi} + \dfrac{38\,848}{1\,575\pi^2}$	$0.033\,363\,1\cdots$
7	$\dfrac{299\,189\,248\,759}{290\,596\,061\,184} + \dfrac{159\,966\,456\,685\sqrt{2}}{435\,894\,091\,776\pi}$ $- \dfrac{292\,926\,667\,005\cos^{-1}(1/3)}{96\,865\,353\,728\pi}$	$0.009\,864\,9\cdots$
8	$2 - \dfrac{2511\sqrt{3}}{280\pi} + \dfrac{17\,605\,024}{606\,375\pi^2}$	$-0.002\,557\,7\cdots$
9	$\dfrac{2\,886\,207\,717\,678\,787}{2\,281\,372\,811\,001\,856} + \dfrac{2\,698\,457\,589\,952\,103\sqrt{2}}{570\,343\,2027\,504\,640\pi}$ $- \dfrac{8\,656\,066\,770\,083\,523\cos^{-1}(1/3)}{2\,281\,372\,811\,001\,856\pi}$	$-0.008\,580\,8\cdots$
10	$2 - \dfrac{2673\sqrt{3}}{280\pi} + \dfrac{49\,048\,616}{1\,528\,065\pi^2}$	$-0.010\,962\,5\cdots$
11	$\dfrac{17\,357\,449\,486\,516\,274\,011}{11\,932\,824\,186\,709\,344\,256} + \dfrac{16\,554\,115\,383\,300\,832\,799\sqrt{2}}{29\,832\,060\,466\,773\,360\,640\pi}$ $- \dfrac{52\,251\,492\,946\,866\,520\,923\cos^{-1}(1/3)}{11\,932\,824\,186\,709\,344\,256\pi}$	$-0.011\,337\,2\cdots$
12	$2 - \dfrac{2187\sqrt{3}}{220\pi} + \dfrac{11\,565\,604\,768}{337\,702\,365\pi^2}$	$-0.010\,670\,3\cdots$

It is instructive at this point to insert a brief historical digression about a controversy between Boltzmann (see Fig. 3.11) and van der Waals (see Fig. 3.12) on the correct value of the fourth virial coefficient for HS systems [62]. After the kinetic theory of gases was developed, a key problem concerning the so-called "excluded volume" problem attracted the attention of nineteenth century physicists. It was realized that, since molecules have a finite size, the actual volume of the container of a gas had to be corrected in the EoS for the volume occupied by the molecules themselves. Van der Waals's arguments gave only the first order effect (second virial coefficient) of the deviation. Later on, Boltzmann and Jäger independently calculated the value of the third virial coefficient for a HS gas, correcting an error made in a former calculation by van der Waals. Without any strict formalism to guide him, since at that time the presently available expression (3.33c) for B_4 in terms of the intermolecular potential had not been derived yet, and realizing that a further step with the same method was virtually impossible, Boltzmann went on to compute in an admirable "tour de force" the fourth virial coefficient of three-dimensional HSs (see Table 3.8). Boltzmann's result turned out to be at odds with the result derived by J.J. van Laar using a method suggested by van der Waals. It took a long time and the advent of new and powerful statistical-mechanical methods

Fig. 3.11 Ludwig Eduard Boltzmann (1844–1906) (Photograph from Wikimedia Commons, https://commons. wikimedia.org/wiki/File: Boltzmann2.jpg)

Fig. 3.12 Johannes Diderik van der Waals (1837–1923) (Photograph from Wikimedia Commons, https://commons. wikimedia.org/wiki/File: Johannes_Diderik_van_der_Waals. jpg)

Table 3.9 Fifth to twelfth virial coefficient B_k, $k = 5, \ldots, 12$, (in units of B_2^{k-1}) for HS fluids with dimensions $d = 2$–8 (with missing values of B_{11} and B_{12} for $d = 4$)

d	B_5/B_2^4	B_6/B_2^5	B_7/B_2^6	B_8/B_2^7
2	0.33355604(4)[a]	0.1988446(3)[b]	0.1148763(4)[b]	0.0649896(5)[b]
3	0.11025147(6)[c]	0.03888206(10)[c]	0.01302297(12)[c]	0.00418265(17)[c]
4	0.03570438(12)[d]	0.00773280(16)[d]	0.0014308(2)[d]	0.0002905(3)[d]
5	0.01295219(16)[d]	0.00098184(19)[d]	0.0004165(2)[d]	−0.0001127(4)[d]
6	0.00752384(18)[d]	−0.0017402(2)[d]	0.0013052(3)[d]	−0.0008925(3)[d]
7	0.00707178(15)[d]	−0.00351139(16)[d]	0.00253868(17)[d]	−0.0019941(3)[d]
8	0.00743187(14)[d]	−0.00451452(7)[d]	0.00341415(7)[d]	−0.00286716(11)[d]

d	B_9/B_2^8	B_{10}/B_2^9	B_{11}/B_2^{10}	B_{12}/B_2^{11}
2	0.0362202(11)[b]	0.019952(6)[d]	0.010933(19)[d]	0.00586(5)[d]
3	0.0013096(3)[c]	0.0004032(4)[c]	0.0001206(6)[c]	0.000031(6)[c]
4	0.0000457(6)[d]	0.0000106(8)[d]		
5	0.0000789(5)[d]	−0.0000468(10)[d]	0.0000309(11)[d]	−0.000023(2)[d]
6	0.0006693(4)[d]	−0.0005294(10)[d]	0.000438(2)[d]	−0.000377(4)[d]
7	0.0016903(3)[d]	−0.0015178(5)[d]	0.0014261(8)[d]	−0.0013877(9)[d]
8	0.00260397(16)[d]	−0.0025104(4)[d]	0.0025367(6)[d]	−0.0026592(10)[d]

The numbers in parentheses indicate the statistical error in the last significant digits
[a]Kratky [54]
[b]Labík et al. [56]
[c]Schultz and Kofke [39]
[d]Zhang and Pettitt [42]

before Nijboer and van Hove [63] confirmed the correctness of Boltzmann's value, doing final justice to his outstanding degree of insight.

The MC numerical values of the virial coefficients B_5–B_{12} up to $d = 8$ [39, 42, 54, 56–58] are displayed in Table 3.9. Apart from those coefficients, B_{13} and B_{14} for $d = 2$, B_{13}–B_{16} for $d = 6$, B_{13}–B_{20} for $d = 7$, B_{13}–B_{24} for $d = 8$, B_4–B_8 for $d \geq 100$, and B_{16}, B_{32}, B_{48}, and B_{64} for $13 \leq d \leq 100$ are known [42, 52].

Several features emerge from inspection of the virial coefficients. While the odd-order coefficients B_3, B_5, B_7, and B_9 remain positive, the even-order coefficients B_4, B_6, B_8, and B_{10} become negative if $d \geq 8$, $d \geq 6$, $d \geq 5$, and $d \geq 5$, respectively. These trends are confirmed by the known virial coefficients not included in Table 3.9 [42]. Even though the known first ten and twelve virial coefficients are positive if $d = 4$ and $d = 3$ [39, 41], respectively, the behavior observed when $d \geq 5$ shows that this does not need to be necessarily the case for all the virial coefficients. It is then legitimate to speculate that, for three-dimensional HS systems, a certain high-order coefficient B_k (perhaps with $k =$ even) might become negative, alternating in sign thereafter. This scenario would be consistent with a singularity of the EoS on the (density) negative real axis that would determine the radius of convergence of the virial series [42, 57, 58, 64].

Equation (3.66), as well as the known data, show that

$$\lim_{d\to\infty} \frac{B_k}{B_2^{k-1}} = 0 , \quad k \geq 3 .$$
(3.67)

This implies that, in the density range $B_2 n = 2^{d-1}\eta \lesssim 1$ (which includes the fluid regime), the virial expansion in the high dimensionality limit can be truncated after the second virial coefficient and therefore (3.50) becomes exact in that limit [65, 66].

3.8.2   Multicomponent Systems

The evaluation of the virial coefficients becomes much more complicated for mixtures, even in the case of HS systems.

3.8.2.1   Second Virial Coefficient

In the HS case, the evaluation of $\widehat{B}_{\alpha\gamma}$ follows exactly the same steps as that of B_2 for the one-component system [see (3.49)], and thus

$$\widehat{B}_{\alpha\gamma}^{HS} = 2^{d-1} v_d \sigma_{\alpha\gamma}^d .$$
(3.68)

The net second virial coefficient B_2 is obtained by inserting (3.68) into (3.36a). In the additive case, i.e., if $\sigma_{\alpha\gamma} = \frac{1}{2} (\sigma_\alpha + \sigma_\gamma)$ for every pair (α, γ), one has

$$\widehat{B}_{\alpha\gamma}^{AHS} = \frac{v_d}{2} \sum_{k=0}^{d} \binom{d}{k} \sigma_\alpha^k \sigma_\gamma^{d-k} ,$$
(3.69a)

$$B_2^{AHS} = \frac{v_d}{2} \sum_{k=0}^{d} \binom{d}{k} M_k M_{d-k} ,$$
(3.69b)

where

$$M_k \equiv \sum_\alpha x_\alpha \sigma_\alpha^k$$
(3.70)

denotes the kth *moment* of the size distribution.

3.8.2.2 Third Virial Coefficient

For the evaluation of $\widehat{B}_{\alpha\gamma\delta}$ we proceed as follows. First, we note that [see (3.37b)]

$$\widehat{B}_{\alpha\gamma\delta} = -\frac{1}{3} \int d\mathbf{r}_1 \int d\mathbf{r}_2 f_{\alpha\gamma}(r_1) f_{\alpha\delta}(r_2) f_{\gamma\delta}(r_{12})$$

$$= -\frac{1}{3} \frac{1}{(2\pi)^d} \int d\mathbf{k} \tilde{f}_{\alpha\gamma}(k) \tilde{f}_{\alpha\delta}(k) \tilde{f}_{\gamma\delta}(k) , \qquad (3.71)$$

where

$$\tilde{f}_{\alpha\gamma}(k) \equiv \int d\mathbf{r} \, e^{-i\mathbf{k}\cdot\mathbf{r}} f_{\alpha\gamma}(r) = (2\pi)^{d/2} \int_0^\infty dr \, r^{d-1} f_{\alpha\gamma}(r) \frac{J_{d/2-1}(kr)}{(kr)^{d/2-1}} \qquad (3.72)$$

is the Fourier transform of the Mayer function. In the second equality of (3.72), $J_\nu(x)$ is the Bessel function of the first kind of order ν and use has been made of the fact that $f_{\alpha\gamma}(r)$ is an isotropic function. The inverse Fourier transformation is

$$f_{\alpha\gamma}(k) = \frac{1}{(2\pi)^d} \int d\mathbf{k} \, e^{i\mathbf{k}\cdot\mathbf{r}} \tilde{f}_{\alpha\gamma}(k) = \frac{1}{(2\pi)^{d/2}} \int_0^\infty dk \, k^{d-1} \tilde{f}_{\alpha\gamma}(k) \frac{J_{d/2-1}(kr)}{(kr)^{d/2-1}}. \qquad (3.73)$$

Equation (3.71) is valid for any isotropic potential. Now, in the case of HS systems,

$$\tilde{f}_{\alpha\gamma}^{HS}(k) = -(2\pi\sigma_{\alpha\gamma})^{d/2} k^{-d/2} J_{d/2}(k\sigma_{\alpha\gamma}) . \qquad (3.74)$$

Inserting this into (3.71), we finally get

$$\widehat{B}_{\alpha\gamma\delta}^{HS} = v_d^2 \frac{d^2}{3} 2^{5d/2-1} \Gamma(d/2) \left(\sigma_{\alpha\gamma}\sigma_{\alpha\delta}\sigma_{\gamma\delta}\right)^{d/2}$$

$$\times \int_0^\infty dk \, k^{-(1+d/2)} J_{d/2}\left(k\sigma_{\alpha\gamma}\right) J_{d/2}\left(k\sigma_{\alpha\delta}\right) J_{d/2}\left(k\sigma_{\gamma\delta}\right) . \qquad (3.75)$$

The mathematical property

$$\frac{d^2}{3} 2^{d/2} \Gamma(d/2) \int_0^\infty dx \, x^{-(1+d/2)} \left[J_{d/2}(x)\right]^3 = I_{3/4}\left(\frac{d+1}{2}, \frac{1}{2}\right) \qquad (3.76)$$

yields

$$\widehat{B}_{\alpha\alpha\alpha}^{HS} = v_d^2 2^{2d-1} \sigma_\alpha^{2d} I_{3/4}\left(\frac{d+1}{2}, \frac{1}{2}\right) , \qquad (3.77)$$

in agreement with (3.63). On the other hand, the evaluation of the integral in (3.75) for arbitrary values of $(\sigma_{\alpha\gamma}, \sigma_{\alpha\delta}, \sigma_{\gamma\delta})$ becomes much more involved, especially if d = even.

It can be proved that (3.75) is equivalent to (see also Sect. 6.5.1)

$$\widehat{B}_{\alpha\gamma\delta}^{HS} = \frac{v_d 2^{d-1}}{3} \{3\} \sigma_{\alpha\gamma}^d \, \mathscr{V}_{\sigma_{\alpha\delta}, \sigma_{\gamma\delta}}(\sigma_{\alpha\gamma}) \,, \tag{3.78}$$

where $\mathscr{V}_{a,b}(r)$ denotes the *intersection volume* of two d-dimensional spheres of radii a and b whose centers are separated by a distance r. Analogously to the convention employed in (3.15c) and (3.15d), henceforth a numerical coefficient enclosed by curly braces, such as $\{3\}$ in (3.78), refers to the number of terms equivalent, except for a permutation of indices, to the shown canonical one. For instance,

$$\{3\}\sigma_{\alpha\gamma}^d \, \mathscr{V}_{\sigma_{\alpha\delta}, \sigma_{\gamma\delta}}(\sigma_{\alpha\gamma}) \rightarrow \sigma_{\alpha\gamma}^d \, \mathscr{V}_{\sigma_{\alpha\delta}, \sigma_{\gamma\delta}}(\sigma_{\alpha\gamma}) + \sigma_{\alpha\delta}^d \, \mathscr{V}_{\sigma_{\alpha\gamma}, \sigma_{\gamma\delta}}(\sigma_{\alpha\delta})$$

$$+ \sigma_{\gamma\delta}^d \, \mathscr{V}_{\sigma_{\alpha\gamma}\sigma_{\alpha\delta}}(\sigma_{\gamma\delta}) \,. \tag{3.79}$$

The structure of $\mathscr{V}_{a,b}(r)$ is

$$\mathscr{V}_{a,b}(r) = \begin{cases} 2^d v_d \min(a^d, b^d) \,, & 0 \le r \le |a-b| \,, \\ \bar{\mathscr{V}}_{a,b}(r) \,, & |a-b| \le r \le a+b \,, \\ 0 \,, & r \ge a+b \,, \end{cases} \tag{3.80}$$

where $\bar{\mathscr{V}}_{a,b}(r)$ is the relevant, non-trivial part of $\mathscr{V}_{a,b}(r)$. In the case d = odd, $\bar{\mathscr{V}}_{a,b}(r)$ has the form of a polynomial in r of degree $2(d-1)$ divided by r^{d-2} [67]. For example, if $d = 3$ and 5,

$$\bar{\mathscr{V}}_{a,b}(r) = \frac{\pi(a+b-r)^2}{12r}\left[r^2 + 2(a+b)r - 3(b-a)^2\right] \,, \quad (d=3) \,, \tag{3.81a}$$

$$\bar{\mathscr{V}}_{a,b}(r) = \frac{\pi^2(a+b-r)^3}{480r^3}\left[3r^5 + 9(a+b)r^4 + 2(18ab - a^2 - b^2)r^3 \right.$$

$$\left. -30(b-a)^2(a+b)r^2 + 5(b-a)^4(3r+a+b)\right] \,, \quad (d=5) \,. \tag{3.81b}$$

The mathematical expression of $\bar{\mathscr{V}}_{a,b}(r)$ is much more involved if d = even. In particular, for $d = 2$ one has

$$\bar{\mathscr{V}}_{a,b}(r) = a^2 \cos^{-1}\frac{r^2 + a^2 - b^2}{2ar} + b^2 \cos^{-1}\frac{r^2 + b^2 - a^2}{2br}$$

$$-\frac{1}{2}\sqrt{2r^2(a^2 + b^2) - (b^2 - a^2)^2 - r^4} \,, \quad (d=2) \,. \tag{3.82}$$

If $a = b$, the intersection volume for any d can be expressed as [60, 68, 69]

$$\bar{\mathcal{V}}_{a,a}(r) = v_d(2a)^d I_{1-r^2/4a^2}\left(\frac{d+1}{2}, \frac{1}{2}\right)$$

$$= v_d(2a)^d\left[1 - \frac{2\Gamma(1+d/2)}{\pi^{1/2}}\sum_{j=0}^{j_{\max}^{(d)}}\frac{(-1)^j(r/2a)^{2j+1}}{(2j+1)j!\Gamma\left(\frac{d+1}{2}-j\right)}\right], \quad (3.83)$$

where $j_{\max}^{(d)}$ was defined below (3.65). If $d = $ odd, the second line of (3.83) gives $\bar{\mathcal{V}}_{a,a}(r)$ as an explicit polynomial in r of degree d. In the case $d = $ even, $\bar{\mathcal{V}}_{a,a}(r)$ is expressed as a series in powers of $r/2a \leq 1$ with a rapid convergence. Needless to say, (3.65) is recovered from (3.78) by setting $r = a = \sigma$ in (3.83).

Equations (3.75) and (3.78) apply to both AHS and NAHS mixtures. Now we restrict ourselves to additive mixtures. In that case, one always has $|\sigma_{\alpha\delta} - \sigma_{\gamma\delta}| \leq \sigma_{\alpha\gamma} \leq \sigma_{\alpha\delta} + \sigma_{\gamma\delta}$ and thus $\mathcal{V}_{\sigma_{\alpha\delta},\sigma_{\gamma\delta}}(\sigma_{\alpha\gamma}) = \bar{\mathcal{V}}_{\sigma_{\alpha\delta},\sigma_{\gamma\delta}}(\sigma_{\alpha\gamma})$. If $d = $ odd, the composition-independent coefficients $\widehat{B}_{\alpha\gamma\delta}$ turn out to have the *polynomial* structure

$$\widehat{B}_{\alpha\gamma\delta}^{\mathrm{AHS}} = v_d^2\sum_{k_\alpha=0}^{d}\sum_{\substack{k_\gamma=0 \\ (k_\alpha+k_\gamma+k_\delta=2d)}}^{d}\sum_{k_\delta=0}^{d} C_{k_\alpha k_\beta k_\gamma}\sigma_\alpha^{k_\alpha}\sigma_\gamma^{k_\gamma}\sigma_\delta^{k_\delta}, \quad (d = \mathrm{odd}), \quad (3.84)$$

where the coefficients $C_{k_\alpha k_\beta k_\gamma}$ are *rational* numbers. As a consequence, according to (3.36b),

$$B_3^{\mathrm{AHS}} = v_d^2\sum_{k_\alpha=0}^{d}\sum_{\substack{k_\gamma=0 \\ (k_\alpha+k_\gamma+k_\delta=2d)}}^{d}\sum_{k_\delta=0}^{d} C_{k_\alpha k_\beta k_\gamma}M_{k_\alpha}M_{k_\gamma}M_{k_\delta}, \quad (d = \mathrm{odd}). \quad (3.85)$$

In particular, making use of (3.78) and (3.81),

$$\widehat{B}_{\alpha\gamma\delta}^{\mathrm{AHS}} = \left(\frac{\pi}{6}\right)^2\left(\{6\}\sigma_\alpha^3\sigma_\gamma^2\sigma_\delta + \frac{1}{3}\{3\}\sigma_\alpha^3\sigma_\gamma^3 + 3\sigma_\alpha^2\sigma_\gamma^2\sigma_\delta^2\right), \quad (d = 3), \quad (3.86a)$$

$$B_3^{\mathrm{AHS}} = \left(\frac{\pi}{6}\right)^2\left(6M_1M_2M_3 + 3M_2^3 + M_3^2\right), \quad (d = 3), \quad (3.86b)$$

$$\widehat{B}_{\alpha\gamma\delta}^{\mathrm{AHS}} = \left(\frac{\pi^2}{60}\right)^2\left[\frac{5}{3}\{6\}\left(\sigma_\alpha^5\sigma_\gamma^4\sigma_\delta + 2\sigma_\alpha^5\sigma_\gamma^3\sigma_\delta^2\right)\right.$$

$$\left. + \frac{1}{3}\{3\}\left(\sigma_\alpha^5\sigma_\gamma^5 + 25\sigma_\alpha^4\sigma_\gamma^4\sigma_\delta^2 + 50\sigma_\alpha^4\sigma_\gamma^3\sigma_\delta^3\right)\right], \quad (d = 5), \quad (3.86c)$$

$$B_3^{\mathrm{AHS}} = \left(\frac{\pi^2}{60}\right)^2\left(10M_1M_4M_5 + 20M_2M_3M_5 + 25M_2M_4^2 + 50M_3^2M_4 + M_5^2\right),$$

$$(d = 5). \quad (3.86d)$$

Thus, from (3.69b) and (3.85) we see that, if $d =$ odd, the whole dependence of the second and third virial coefficients on composition (number of components, mole fractions, and sizes) is encapsulated by the first d moments. This implies that the EoS of two completely different AHS mixtures sharing the same values of the first d moments is exactly the same up to the level of the third virial coefficient, provided that $d =$ odd. Nevertheless, in the case $d =$ even this property is broken down after the second virial coefficient. In particular, from (3.82) one obtains

$$\widehat{B}_{\alpha\gamma\delta}^{\text{AHS}} = \left(\frac{\pi}{4}\right)^2 \frac{1}{3\pi} \{3\} \Big[(\sigma_\alpha + \sigma_\gamma)^2 (\sigma_\alpha + \sigma_\delta)^2 \cos^{-1} \frac{\sigma_\alpha(\sigma_\alpha + \sigma_\gamma + \sigma_\delta) - \sigma_\gamma \sigma_\delta}{(\sigma_\alpha + \sigma_\gamma)(\sigma_\alpha + \sigma_\delta)}$$
$$- (\sigma_\gamma + \sigma_\delta)^2 \sqrt{\sigma_\alpha \sigma_\gamma \sigma_\delta (\sigma_\alpha + \sigma_\gamma + \sigma_\delta)} \Big], \quad (d = 2). \tag{3.87}$$

The non-polynomial form of $\widehat{B}_{\alpha\gamma\delta}^{\text{AHS}}$ if $d =$ even prevents one from obtaining a compact expression for B_3^{AHS} in terms of a finite number of moments. On the other hand, a practical approximation of $\widehat{B}_{\alpha\gamma\delta}^{\text{AHS}}$ for $d = 2$ is [70]

$$\widehat{B}_{\alpha\gamma\delta}^{\text{AHS}} \simeq \left(\frac{\pi}{4}\right)^2 \frac{1}{3} \Big[\sigma_\alpha^2 \sigma_\gamma^2 + \sigma_\alpha^2 \sigma_\delta^2 + \sigma_\gamma^2 \sigma_\delta^2 + (b_3 - 1)\sigma_\alpha \sigma_\gamma \sigma_\delta (\sigma_\alpha + \sigma_\gamma + \sigma_\delta) \Big],$$
$$(d = 2), \tag{3.88}$$

what implies

$$B_3^{\text{AHS}} \simeq \left(\frac{\pi}{4}\right)^2 \Big[M_2^2 + (b_3 - 1)M_1^2 M_2 \Big], \quad (d = 2). \tag{3.89}$$

As can be seen from Fig. 3.13, the polynomial approximation (3.88) is practically indistinguishable from the exact expression (3.87).

3.8.2.3 Fourth Virial Coefficient

Obviously, the mathematical difficulties already found in the evaluation of the third virial coefficient of HS mixtures are significantly increased when going to the fourth virial coefficient. Not only is $\widehat{B}_{\alpha\gamma\delta\epsilon}$ made of three different classes of diagrams [see (3.37c) and (3.38)] but also each one of them represents a higher order integral.

The two first classes of diagrams can be expressed as [71]

$$\widehat{B}_{\alpha\gamma\delta\epsilon}^{(\text{I})} = -d2^{d-3} v_d \int_0^{\min(\sigma_{\alpha\gamma} + \sigma_{\alpha\delta}, \sigma_{\gamma\epsilon} + \sigma_{\delta\epsilon})} dr\, r^{d-1} \mathcal{V}_{\sigma_{\alpha\gamma},\sigma_{\alpha\delta}}(r) \mathcal{V}_{\sigma_{\gamma\epsilon},\sigma_{\delta\epsilon}}(r), \tag{3.90a}$$

$$\widehat{B}_{\alpha\gamma\delta\epsilon}^{(\text{II})} = d2^{d-3} v_d \int_0^{\min(\sigma_{\gamma\delta},\sigma_{\alpha\gamma} + \sigma_{\alpha\delta},\sigma_{\gamma\epsilon} + \sigma_{\delta\epsilon})} dr\, r^{d-1} \mathcal{V}_{\sigma_{\alpha\gamma},\sigma_{\alpha\delta}}(r) \mathcal{V}_{\sigma_{\gamma\epsilon},\sigma_{\delta\epsilon}}(r). \tag{3.90b}$$

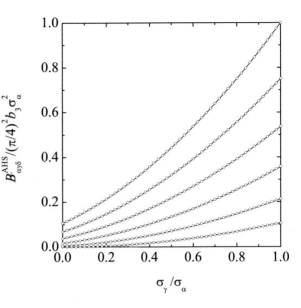

Fig. 3.13 Plot of the (reduced) two-dimensional composition-independent third virial coefficient $\widehat{B}^{\mathrm{AHS}}_{\alpha\gamma\delta}/(\pi/4)^2 b_3 \sigma_\alpha^2$ versus $\sigma_\gamma/\sigma_\alpha$ for, *from bottom to top*, $\sigma_\delta/\sigma_\alpha = 0, 0.2, 0.4, 0.6, 0.8,$ and 1. The *solid lines* and the *circles* represent the exact and approximate values, as given by (3.87) and (3.88), respectively

The upper limit in the integral of (3.90a) is a consequence of the property $\mathscr{V}_{a,b}(r) = 0$ if $r \geq a + b$ [see (3.80)]. In (3.90b) the Mayer function $f^{\mathrm{HS}}_{\gamma\delta}(r) = -\Theta(\sigma_{\gamma\delta} - r)$ imposes the additional constraint $r \leq \sigma_{\gamma\delta}$. The third diagram, $\widehat{B}^{(\mathrm{III})}_{\alpha\gamma\delta\epsilon}$, cannot be expressed as a single integral.

For concreteness, now we focus on three-dimensional ($d = 3$) AHS mixtures. Making use of (3.80) and (3.81a), it is possible to obtain

$$\widehat{B}^{(\mathrm{I})}_{\alpha\gamma\delta\epsilon} + \widehat{B}^{(\mathrm{II})}_{\alpha\gamma\delta\epsilon} + \widehat{B}^{(\mathrm{II})}_{\gamma\alpha\epsilon\delta}\bigg|_{\mathrm{AHS}} = \left(\frac{\pi}{6}\right)^3 \frac{1}{8}\bigg[\left(\sigma_\alpha^3\sigma_\gamma^3\sigma_\delta^3 + \sigma_\alpha^3\sigma_\gamma^3\sigma_\epsilon^3 + \sigma_\alpha^3\sigma_\delta^3\sigma_\epsilon^3\right.$$

$$+ \sigma_\gamma^3\sigma_\delta^3\sigma_\epsilon^3\bigg) + 3\sigma_\alpha\sigma_\gamma\sigma_\delta\sigma_\epsilon \left(\sigma_\alpha\sigma_\gamma\sigma_\delta + \sigma_\alpha\sigma_\gamma\sigma_\epsilon\right.$$

$$+ \sigma_\alpha\sigma_\delta\sigma_\epsilon + \sigma_\gamma\sigma_\delta\sigma_\epsilon\big)\left(\sigma_\alpha\sigma_\gamma + \sigma_\alpha\sigma_\delta + \sigma_\alpha\sigma_\epsilon\right.$$

$$+ \sigma_\gamma\sigma_\delta + \sigma_\gamma\sigma_\epsilon + \sigma_\delta\sigma_\epsilon\big)\bigg], \quad (d = 3) . \quad (3.91)$$

The above combination is invariant under index permutation and therefore [see (3.37c)] $\widehat{B}_{\alpha\gamma\delta\epsilon} = 3[\widehat{B}^{(\mathrm{I})}_{\alpha\gamma\delta\epsilon} + \widehat{B}^{(\mathrm{II})}_{\alpha\gamma\delta\epsilon} + \widehat{B}^{(\mathrm{II})}_{\gamma\alpha\epsilon\delta}] + \widehat{B}^{(\mathrm{III})}_{\alpha\gamma\delta\epsilon}$. Thus, only the complete star diagram $\widehat{B}^{(\mathrm{III})}_{\alpha\gamma\delta\epsilon}$ [see (3.38c)] remains to be determined. Unfortunately, the latter diagram can be analytically evaluated only in some limiting situations. First, if all the indices are equal,

$$\widehat{B}^{(\mathrm{III})}_{\alpha\alpha\alpha\alpha} = \left(\frac{\pi}{6}\right)^3 \sigma_\alpha^9 b^{(\mathrm{III})}_{4;0} , \quad (d = 3) , \quad (3.92)$$

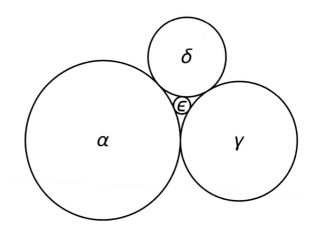

Fig. 3.14 Configuration where the three *bigger spheres* (α, γ, and δ) are tangent to each other and the *smallest sphere* (ϵ) has the maximum possible size to fit in the inner hole [see (3.94)]. In this particular example, $\sigma_\gamma/\sigma_\alpha = \frac{3}{4}$, $\sigma_\delta/\sigma_\alpha = \frac{1}{2}$, and $\sigma_\epsilon/\sigma_\alpha = \frac{3}{47}\left(6\sqrt{6}-13\right) \simeq 0.108$

where

$$b_{4;0}^{(III)} = -\left[\frac{4\,131\cos^{-1}(1/3)}{70\pi} - \frac{219\sqrt{2}}{35\pi} - \frac{356}{35}\right], \quad (d=3). \tag{3.93}$$

Next, if the smallest of the four spheres α, γ, δ, and ϵ (say sphere ϵ) fits in the inner hole created when the other three spheres are in a tangent configuration (see Fig. 3.14), the associated coefficient $\widehat{B}_{\alpha\gamma\delta\epsilon}^{(III)}$ can be obtained exactly [71]. This Apollonian condition is satisfied if

$$\sigma_\epsilon \leq \frac{\sigma_\alpha\sigma_\gamma\sigma_\delta}{\sigma_\alpha\sigma_\gamma + \sigma_\alpha\sigma_\delta + \sigma_\gamma\sigma_\delta + 2\sqrt{\sigma_\alpha\sigma_\gamma\sigma_\delta(\sigma_\alpha + \sigma_\gamma + \sigma_\delta)}}. \tag{3.94}$$

In that case, one has [71]

$$\begin{aligned}
\widehat{B}_{\alpha\gamma\delta\epsilon}^{(III)}\Big|_{AHS} &= -\left[\widehat{B}_{\alpha\gamma\delta\epsilon}^{(I)} + \widehat{B}_{\alpha\gamma\delta\epsilon}^{(II)} + \widehat{B}_{\gamma\alpha\epsilon\delta}^{(II)}\right]_{AHS} - \left(\frac{\pi}{6}\right)^3\frac{9\sigma_\epsilon^3}{280}\Big[70\sigma_\alpha^2\sigma_\gamma^2\sigma_\delta^2 \\
&\quad -35\sigma_\epsilon\sigma_\alpha\sigma_\gamma\sigma_\delta(\sigma_\alpha\sigma_\gamma + \sigma_\alpha\sigma_\delta + \sigma_\gamma\sigma_\delta) + 7\sigma_\epsilon^2(\sigma_\alpha^2\sigma_\gamma^2 + \sigma_\alpha^2\sigma_\delta^2 \\
&\quad +\sigma_\gamma^2\sigma_\delta^2 - 2\sigma_\alpha^2\sigma_\gamma\sigma_\delta - 2\sigma_\gamma^2\sigma_\alpha\sigma_\delta - 2\sigma_\delta^2\sigma_\alpha\sigma_\gamma) + 21\sigma_\epsilon^3(\sigma_\alpha + \sigma_\gamma) \\
&\quad \times(\sigma_\alpha + \sigma_\delta)(\sigma_\gamma + \sigma_\delta) + 12\sigma_\epsilon^4(\sigma_\alpha^2 + \sigma_\gamma^2 + \sigma_\delta^2 + 3\sigma_\alpha\sigma_\gamma + 3\sigma_\alpha\sigma_\delta \\
&\quad +3\sigma_\gamma\sigma_\delta) + 15\sigma_\epsilon^5(\sigma_\alpha + \sigma_\gamma + \sigma_\delta) + 5\sigma_\epsilon^6\Big], \quad (d=3). \tag{3.95}
\end{aligned}$$

In the particular case $\alpha = \gamma = \delta$, the condition (3.94) becomes $\sigma_\epsilon/\sigma_\alpha \leq q_0$, with $q_0 \equiv \frac{2}{3}\sqrt{3} - 1 \simeq 0.1547$, and (3.95) reduces to

$$\widehat{B}_{\alpha\alpha\alpha\epsilon}^{(III)}\Big|_{AHS} = \left(\frac{\pi}{6}\right)^3\sigma_\alpha^9 b_{3;1}^{(III)}(\sigma_\epsilon/\sigma_\alpha), \quad \sigma_\epsilon/\sigma_\alpha \leq q_0, \quad (d=3), \tag{3.96}$$

where

$$b_{3;1}^{(III)}(q) = -\frac{1}{8} - \frac{9q}{8} - \frac{9q^2}{2} - 6q^3 + \frac{27q^4}{8} + \frac{27q^5}{40} - \frac{27q^6}{5} - \frac{162q^7}{35} - \frac{81q^8}{56} - \frac{9q^9}{56}. \tag{3.97}$$

In the complementary situation $\sigma_\epsilon/\sigma_\alpha > q_0$, the exact expression for $\widehat{B}_{\alpha\alpha\alpha\epsilon}^{(III)}$ is [46, 72]

$$\widehat{B}_{\alpha\alpha\alpha\epsilon}^{(III)}\Big|_{AHS} = \left(\frac{\pi}{6}\right)^3 \sigma_\alpha^9 \left[b_{3;1}^{(III)}(\sigma_\epsilon/\sigma_\alpha) + \Delta b_{3;1}(\sigma_\epsilon/\sigma_\alpha)\right], \quad \sigma_\epsilon/\sigma_\alpha > q_0,$$

$$(d = 3), \quad (3.98)$$

where

$$\Delta b_{3;1}(q) = \frac{3\sqrt{3q^2 + 6q - 1}}{1\,120\pi}\left(745 + 228q - 210q^2 - 84q^3 + 279q^4 + 180q^5\right.$$

$$+ 30q^6\bigg) - \frac{243\left(8 + 6q + 3q^2\right)}{140\pi} \tan^{-1}\sqrt{3q^2 + 6q - 1} + \frac{27(1 + q)}{280\pi}$$

$$\times\left(144 - 36q + 90q^2 - 20q^3 - 85q^4 + 64q^5 + 104q^6 + 40q^7 + 5q^8\right)$$

$$\times \tan^{-1}\frac{\sqrt{3q^2 + 6q - 1}}{q + 1}. \tag{3.99}$$

It can be checked that, in the region $q \gtrsim q_0$,

$$\Delta b_{3;1}(q) \simeq \frac{7\,776 \times 3^{3/4}}{385\pi}(q - q_0)^{11/2}. \tag{3.100}$$

This means that $\widehat{B}_{\alpha\alpha\alpha\epsilon}^{(III)}$ and its first five derivatives with respect to σ_ϵ are continuous at $\sigma_\epsilon = q_0\sigma_\alpha$. On the other hand, the sixth derivative is finite when the limit is taken from the left and infinite if the limit is taken from the right.

The non-polynomial form of $\Delta b_{3;1}(q)$ implies that, in contrast to the second- and third-order composition-independent virial coefficients [see (3.69a) and (3.84)], the fourth-order coefficients fail to have a polynomial dependence on the sphere diameters, even if $d = $ odd. As a consequence, the full coefficient B_4 cannot be expressed in terms of a finite number of moments.

Apart from $\widehat{B}_{\alpha\alpha\alpha\alpha}^{(III)}$ [see (3.92)] and $\widehat{B}_{\alpha\alpha\alpha\epsilon}^{(III)}$ [see (3.96) and (3.98)], the diagram $\widehat{B}_{\alpha\gamma\delta\alpha}^{(III)}$ is not exactly known for general three-dimensional AHS mixtures, except if (3.94) is fulfilled. On the other hand, an extremely accurate semiempirical formula for $\widehat{B}_{\alpha\alpha\epsilon\epsilon}^{(III)}$ is available [46]. It is given by

$$\widehat{B}_{\alpha\alpha\epsilon\epsilon}^{(III)}\Big|_{AHS} = \left(\frac{\pi}{6}\right)^3 \sigma_\alpha^9 b_{2;2}^{(III)}(\sigma_\epsilon/\sigma_\alpha), \quad (d = 3), \tag{3.101}$$

Table 3.10 Numerical values of the coefficients in (3.102b)

j	A_j	C_j
0	1.490 955 765 990 51	1
1	−6.667 961 931 174 69	−5.449 505 086 375 50
2	4.150 337 817 416 92	7.741 899 937 434 46
3	7.167 099 184 147 82	−1.407 969 391 696 07
4	11.407 948 295 081 4	8.637 943 845 908 24

where

$$b_{2;2}^{(\mathrm{III})}(q) = -q^3 \frac{1 + \frac{15}{4}q + \frac{28}{5}q^2 + \frac{15}{4}q^3 + q^4}{\sqrt{q^2 + 1}} + \frac{2\sqrt{2}q^6}{(q^2 + 1)^{3/2}}\Re(q) , \qquad (3.102a)$$

$$\Re(q) = b_{4;0}^{(\mathrm{III})} + \frac{151\sqrt{2}}{20} + [1 - t(q)]\frac{\sum_{j=0}^{4} A_j \, t^j(q)}{\sum_{j=0}^{4} C_j \, t^j(q)} , \qquad t(q) \equiv \sqrt{\frac{2q}{q^2 + 1}} .$$

$$(3.102b)$$

Here, $b_{4;0}^{(\mathrm{III})}$ is given by (3.93) and the numerical values of the coefficients A_j and C_j are displayed in Table 3.10

All the above results for the composition-independent fourth virial coefficients of three-dimensional AHS mixtures are enough to fully characterize the coefficients of a binary mixture, as defined by (3.41a) and (3.42):

$$b_{4;0} = \frac{2\,707}{70} + \frac{219\sqrt{2}}{35\pi} - \frac{4\,131\cos^{-1}(1/3)}{70\pi} , \qquad (3.103a)$$

$$b_{3;1}(q) = \frac{1}{4} + \frac{9q}{4} + 9q^2 + \frac{21q^3}{4} + \frac{27q^4}{8} + \frac{27q^5}{40} - \frac{27q^6}{5} - \frac{162q^7}{35}$$

$$- \frac{81q^8}{56} - \frac{9q^9}{56} + \Theta(q - q_0)\Delta b_{3;1}(q) , \qquad (3.103b)$$

$$b_{2;2}(q) = 3q^3(1 + q)\left(1 + \frac{11q}{4} + q^2\right) - q^3\frac{1 + \frac{15}{4}q + \frac{28}{5}q^2 + \frac{15}{4}q^3 + q^4}{\sqrt{q^2 + 1}}$$

$$+ \frac{2\sqrt{2}q^6}{(q^2 + 1)^{3/2}}\Re(q) , \qquad (3.103c)$$

$$b_{1;3}(q) = q^9 b_{3;1}(1/q) \qquad (3.103d)$$

$$b_{0;4}(q) = q^9 b_{4;0} . \qquad (3.103e)$$

These coefficients are plotted in Fig. 3.15.

The computation of the fourth virial coefficient $\widehat{B}_{\alpha\gamma\delta\epsilon}^{(\mathrm{III})}$ for NAHS mixtures and of virial coefficients beyond fourth order for both AHS and NAHS mixtures requires numerical MC methods [38, 73–86].

Fig. 3.15 Plot of the rescaled
composition-independent
fourth virial coefficients for a
binary three-dimensional
AHS mixture

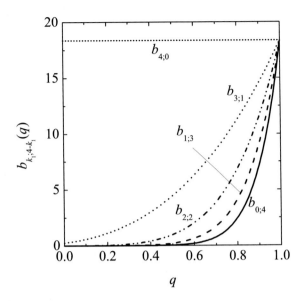

3.9 Simple Approximations for the Equation of State of Hard Disks and Spheres

In terms of the packing fraction η defined by (3.51), the virial series (3.8) for d-dimensional HS systems becomes

$$Z = 1 + 2^{d-1}\eta + b_3\eta^2 + b_4\eta^3 + \cdots , \qquad (3.104)$$

where the rescaled virial coefficients are defined by (3.46). The numerical values of b_2–b_{12} as obtained from Tables 3.7, 3.8 and 3.9 for hard-disk (HD) and HS fluids are displayed in Table 3.11.

Although incomplete, the knowledge of the first few virial coefficients is practically the only access to exact information about the EoS of the HS fluid. If the packing fraction η is low enough, the virial expansion truncated after a given order is an accurate representation of the exact EoS. However, this tool is not practical at moderate or high values of η. In those cases, instead of truncating the series, it is far more convenient to construct an *approximant* which, while keeping a number of exact virial coefficients, includes all orders in density [87]. The most popular class is made of Padé approximants [88], where the compressibility factor Z is approximated by the ratio of two polynomials. Obviously, as the number of retained exact virial coefficients increases so does the complexity of the approximant. In this section, however, we will deal with simpler, but yet accurate, approximations. They will be tested by comparison with MC and/or molecular dynamics (MD) computer simulations, the latter method having been pioneered by Berni J. Alder (see Fig. 3.16).

Table 3.11 Numerical values of the rescaled virial coefficients b_2–b_{12} for HDs ($d = 2$) and HSs ($d = 3$)

k	$d = 2$	$d = 3$
2	2	4
3	$3.128\,017\,75\cdots$	10
4	$4.257\,854\,46\cdots$	$18.364\,768\,38\cdots$
5	$5.336\,896\,6(6)$	$28.224\,376(15)$
6	$6.363\,027\,2(10)$	$39.815\,23(10)$
7	$7.352\,08(3)$	$53.342\,1(5)$
8	$8.318\,67(6)$	$68.529(3)$
9	$9.272\,4(3)$	$85.83(2)$
10	$10.215(3)$	$105.70(10)$
11	$11.195(19)$	$126.5(6)$
12	$12.00(10)$	$130(25)$

Fig. 3.16 Berni Julian Alder (b. 1925) (Photograph by Julie Russell, reproduced with permission)

3.9.1   Hard Disks ($d = 2$)

In the two-dimensional case, the virial series truncated after the third virial coefficient is

$$Z = 1 + 2\eta + b_3\eta^2 + \cdots, \quad \eta \equiv \frac{\pi}{4}n\sigma^2 , \qquad (3.105)$$

where

$$b_3 = 4\left(\frac{4}{3} - \sqrt{3}\pi\right) = 3.128\cdots \simeq \frac{25}{8} . \qquad (3.106)$$

In 1975, Henderson (see Fig. 3.17) proposed a simple EoS with two main ingredients: (1) consistency with the first three virial coefficients (with the replacement $b_3 \to \frac{25}{8}$ for simplicity) and (2) a double pole at $\eta = 1$. The resulting Henderson (H) EoS is [89]

$$\boxed{Z_{\mathrm{H}} = \frac{1 + \eta^2/8}{(1 - \eta)^2} .} \qquad (3.107)$$

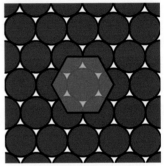

Despite the simplicity of this prescription, it provides fairly accurate values, although it tends to overestimate the compressibility factor for high densities of the stable fluid phase [90, 91]. A more accurate prescription was proposed by Luding (L) [92–94] by adding a correction term:

$$Z_L = \frac{1 + \eta^2/8}{(1-\eta)^2} - \frac{\eta^4}{64(1-\eta)^4} \, . \tag{3.108}$$

On the other hand, both (3.107) and (3.108) assume that the pressure is finite for any $\eta < 1$, whereas by geometrical reasons the maximum conceivable packing fraction is the close-packing value $\eta_{cp} = \frac{\sqrt{3}\pi}{6} \simeq 0.907$ (see Fig. 3.18).

Another simple approximation is based on (1) consistency with the first two virial coefficients and (2) a single pole at η_{cp}. Thus, the constraints are

$$Z = \begin{cases} 1 + 2\eta + \cdots \, , & \eta \ll 1 \, , \\ \infty \, , & \eta \to \eta_{cp} \, . \end{cases} \tag{3.109}$$

A simple expression satisfying those requirements was constructed by Santos, López de Haro, and Yuste (SHY) [95, 96]. It reads

$$Z_{SHY} = \frac{1}{1 - 2\eta + \frac{2\eta_{cp}-1}{\eta_{cp}^2}\eta^2} \, . \tag{3.110}$$

Fig. 3.19 Comparison between MC and MD computer-simulation values of the EoS of a HD fluid [92, 99] and the theoretical approximations (3.107), (3.110), and (3.111)

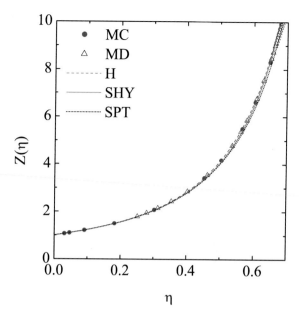

It is interesting to note that, if one *formally* makes the replacement $\eta_{cp} \rightarrow 1$, (3.110) becomes

$$Z_{SPT} = \frac{1}{(1-\eta)^2},\qquad (3.111)$$

which coincides with the EoS obtained from the so-called Scaled Particle Theory (SPT) [97, 98].

Figure 3.19 compares the predictions of (3.107), (3.110), and (3.111) against MC [99] and MD [92] computer simulations. Despite their simplicity, the three approximations, especially Z_H and Z_{SHY}, exhibit an excellent performance, even at packing fractions where the pressure is about ten times higher than the ideal-gas one. The curve representing (3.108) is not plotted in Fig. 3.19 because it tends to lie in between those representing Z_H and Z_{SHY}.

3.9.2 Hard Spheres (d = 3)

In the three-dimensional case, $\eta = (\pi/6)n\sigma^3$ and the second and third reduced virial coefficients are integer numbers: $b_2 = 4$ and $b_3 = 10$ (see Table 3.11). The fourth virial coefficient, however, is a transcendental number (see Table 3.8), namely $b_4 = 18.364\,768\,38\cdots$. If we round off this coefficient ($b_4 \simeq 18$), we realize that $b_4 - b_3 = (b_3 - b_2) + 2$. Interestingly, by continuing the rounding-off process

Fig. 3.20 Norman Frederick
Carnahan (b. 1942)
(Photograph courtesy of N.F.
Carnahan)

Fig. 3.21 Kenneth Earl
Starling (b. 1935)
(Photograph courtesy of N.F.
Carnahan)

($b_5 \simeq 28$, $b_6 \simeq 40$, see Table 3.11), the relationship $b_k - b_{k-1} = (b_{k-1} - b_{k-2}) + 2$ is seen to extend up to $k = 6$.

In the late sixties only the first six virial coefficients were accurately known and thus Carnahan (see Fig. 3.20) and Starling (see Fig. 3.21) proposed to extrapolate the relationship $b_k - b_{k-1} = (b_{k-1} - b_{k-2}) + 2$ to any $k \geq 2$, what is equivalent to the approximation [100]

$$b_k = k^2 + k - 2 .\qquad(3.112)$$

By summing the virial series within that approximation, they obtained the famous Carnahan–Starling (CS) EoS:

$$\boxed{Z_{\mathrm{CS}} = \frac{1 + \eta + \eta^2 - \eta^3}{(1 - \eta)^3} .}\qquad(3.113)$$

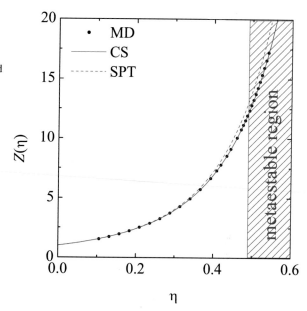

Fig. 3.22 Compressibility factor for three-dimensional HSs, as obtained from MD computer simulations [102], from the CS EoS (3.113), and from the SPT EoS (3.115)

The corresponding isothermal susceptibility and (excess) Helmholtz free energy per particle are

$$\chi_{T,CS} = \left[\frac{\partial (\eta Z_{CS})}{\partial \eta} \right]^{-1} = \frac{(1-\eta)^4}{1 + 4\eta + 4\eta^2 - 4\eta^3 + \eta^4}, \qquad (3.114a)$$

$$\beta a_{CS}^{ex} = \int_0^1 dt \, \frac{Z_{CS}(\eta t) - 1}{t} = \eta \frac{4 - 3\eta}{(1-\eta)^2}. \qquad (3.114b)$$

Equation (3.113) can also be seen as a small correction to the SPT EoS for HSs. The latter reads [97, 98, 101]

$$Z_{SPT} = \frac{1 + \eta + \eta^2}{(1-\eta)^3}. \qquad (3.115)$$

Figure 3.22 shows that, despite its simplicity, the CS equation exhibits an excellent performance over the whole fluid stable region and even in the metastable fluid region ($\eta \geq 0.492$ [103]), where the crystal is the stable phase. This is remarkable because the approximation $b_k = k^2 + k - 2$ (according to which $b_7 = 54$, $b_8 = 70$, $b_9 = 88$, $b_{10} = 108$, $b_{11} = 130$, $b_{12} = 154$) fails to capture the rounding-off of the virial coefficient b_k for $k \geq 7$ (see Table 3.11), the deviation increasing with k. The explanation of this paradox might partially lie in the fact that the CS recipe underestimates b_4 and b_5 but this is compensated by an overestimate of the higher virial coefficients. Apart from that, and analogously to Henderson's EoS (3.107), the CS EoS (3.113) artificially provides finite values even for packing fractions higher than the close-packing value $\eta_{cp} = \pi\sqrt{2}/6 \simeq 0.7405$.

There are other many empirical or semi-empirical proposals for the thermo-dynamic properties of HDs, HSs, or even hard hyperspheres [87, 104, 105], but most of them have more sophisticated forms than (3.107), (3.108), (3.110), (3.111), (3.113), or (3.115), and rely on the knowledge of a larger number of empirical virial coefficients.

3.9.3   Extension to Mixtures: Effective One-Component Fluid Approaches

3.9.3.1   Heuristic Approximations

As expected, the search for reliable EoS for HS mixtures is much more difficult than in the one-component case. To begin with, the (rescaled) virial coefficients are not pure numbers but functions of the size composition of the mixture.

Here we will focus on simple approaches consisting in extending a given *one-component* compressibility factor $Z_{oc}(\eta)$, such as (3.107), (3.108), (3.110), (3.111), (3.113), or (3.115), to construct a *multicomponent* compressibility factor $Z(\eta)$, where in the latter case the total packing fraction is defined as

$$\eta \equiv nv_d \sum_\alpha x_\alpha \sigma_\alpha^d .\tag{3.116}$$

The problem can be stated as follows. Given a HS fluid mixture with a certain size composition at a packing fraction η, can we find an *effective* one-component HS fluid such that the EoS of the former system can be mapped onto that of the latter? A specially simple recipe is provided by the so-called one-fluid van der Waals (see Fig. 3.12) theory [106, 107], according to which

$$Z_{vdW}(\eta) = Z_{oc}(\eta_{eff}) , \qquad \eta_{eff} \equiv \eta \frac{\sum_{\alpha\gamma} x_\alpha x_\gamma \sigma_{\alpha\gamma}^d}{M_d} ,\tag{3.117}$$

where we recall the definition (3.70) of the size distribution moments. As can be seen from (3.36a) and (3.68), this approximation is consistent with the second virial coefficient of the mixture.

Equation (3.117) is in principle applicable to both AHS and NAHS mixtures, but here we will restrict ourselves to the additive case. In an alternative proposal by Santos, Yuste, and López de Haro (SYH) [108–110], the *excess* quantity $Z(\eta) - 1$ is expressed as a linear combination of $Z_{oc}(\eta) - 1$ and $(1 - \eta)^{-1} - 1 = \eta/(1 - \eta)$, with coefficients such that the second and third virial coefficients of the mixture are exactly reproduced:

$$Z_{SYH}(\eta) = 1 + \frac{\hat{b}_3 - \hat{b}_2}{b_3 - b_2} [Z_{oc}(\eta) - 1] + \frac{b_3\hat{b}_2 - b_2\hat{b}_3}{b_3 - b_2} \frac{\eta}{1 - \eta} .\tag{3.118}$$

Here, $\hat{b}_k \equiv B_k/(v_d M_d)^{k-1}$ are the rescaled virial coefficients of the HS mixture. Note that $\hat{b}_k \to b_k$ in the one-component limit. At the level of the free energy, the approximation (3.118) can be written as

$$\beta a_{\mathrm{SYH}}^{\mathrm{ex}}(\eta) = \frac{\hat{b}_3 - \hat{b}_2}{b_3 - b_2} \beta a_{\mathrm{oc}}^{\mathrm{ex}}(\eta) - \frac{b_3 \hat{b}_2 - b_2 \hat{b}_3}{b_3 - b_2} \ln(1 - \eta) , \qquad (3.119)$$

where $a^{\mathrm{ex}} \equiv F^{\mathrm{ex}}/N$ is the excess Helmholtz free energy per particle, $a_{\mathrm{oc}}^{\mathrm{ex}}$ is the corresponding one-component quantity, and use has been made of the thermodynamic relation (1.36b).

In a slightly different proposal, Barrio and Solana (BS) [111, 112] assumed that the ratio of excess compressibility factors, $[Z(\eta)-1]/[Z_{\mathrm{oc}}(\eta)-1]$, is a linear function of η, with coefficients adjusted to reproduce the second and third virial coefficients:

$$\boxed{Z_{\mathrm{BS}}(\eta) = 1 + [Z_{\mathrm{oc}}(\eta) - 1] \left(\frac{\hat{b}_2}{b_2} - \frac{b_3 \hat{b}_2 - b_2 \hat{b}_3}{b_2^2} \eta \right) .} \qquad (3.120)$$

The associated free energy is

$$\beta a_{\mathrm{BS}}^{\mathrm{ex}}(\eta) = \beta a_{\mathrm{oc}}^{\mathrm{ex}}(\eta) \left(\frac{\hat{b}_2}{b_2} - \frac{b_3 \hat{b}_2 - b_2 \hat{b}_3}{b_2^2} \eta \right) + \frac{b_3 \hat{b}_2 - b_2 \hat{b}_3}{b_2^2} \eta \int_0^1 \mathrm{d}t \, \beta a_{\mathrm{oc}}^{\mathrm{ex}}(\eta t) . \qquad (3.121)$$

In the case of AHS fluids, $\hat{b}_2$ can be expressed in terms of the first d moments of the size distribution [see (3.69b)]. In particular,

$$\hat{b}_2 = 1 + m_2^{-1} , \quad (d = 2) , \qquad (3.122a)$$

$$\hat{b}_2 = 1 + 3\frac{m_2}{m_3} , \quad (d = 3) , \qquad (3.122b)$$

where

$$m_k \equiv \frac{M_k}{M_1^k} \qquad (3.123)$$

are reduced moments. Nevertheless, the third coefficient $\hat{b}_3$ is a function of the first d moments only if $d = $ odd. In particular, (3.86b) implies that

$$\hat{b}_3 = 1 + 6\frac{m_2}{m_3} + 3\frac{m_2^3}{m_3^2} , \quad (d = 3) . \qquad (3.124)$$

On the other hand, as seen from Fig. 3.13, an excellent approximation in the case $d = 2$ is provided by (3.88) and (3.89), i.e.,

$$\hat{b}_3 \simeq 1 + (b_3 - 1)m_2^{-1} , \quad (d = 2) . \qquad (3.125)$$

Taking all of this into account, (3.118) and (3.120) yield

$$Z_{\text{SYH}}(\eta) = 1 + m_2^{-1} [Z_{\text{oc}}(\eta) - 1] + (1 - m_2^{-1}) \frac{\eta}{1 - \eta} , \quad (d = 2) , \tag{3.126a}$$

$$Z_{\text{SYH}}(\eta) = 1 + \frac{1}{2} \left(\frac{m_2}{m_3} + \frac{m_2^3}{m_3^2} \right) [Z_{\text{oc}}(\eta) - 1] + \left(1 + \frac{m_2}{m_3} - 2\frac{m_2^3}{m_3^2} \right) \frac{\eta}{1 - \eta} , \quad (d = 3) , \tag{3.126b}$$

$$Z_{\text{BS}}(\eta) = 1 + [Z_{\text{oc}}(\eta) - 1] \left[\frac{1 + m_2^{-1}}{2} - \frac{b_3 - 2}{4} (1 - m_2^{-1})\eta \right] , \quad (d = 2) , \tag{3.127a}$$

$$Z_{\text{BS}}(\eta) = 1 + [Z_{\text{oc}}(\eta) - 1] \left[\frac{1 + 3m_2/m_3}{4} - \frac{3}{8} \left(1 + \frac{m_2}{m_3} - 2\frac{m_2^3}{m_3^2} \right) \eta \right] , \quad (d = 3) , \tag{3.127b}$$

respectively. If the EoS (3.107) is used for $Z_{\text{oc}}(\eta)$ in (3.126a), the Jenkins–Mancini generalization to mixtures [113] is readily obtained.

Expressions similar to (3.126) and (3.127) hold for the excess Helmholtz free energy per particle, which is then expressed in terms of η and the first d moments. In general, an excess AHS free energy $\beta a^{\text{ex}}(n; \{x_v\})$ is said to have a *truncatable* structure if it depends on the size distribution $\{x_v\}$ only through a finite number of moments [114–116].

In the heuristic proposals (3.118) and (3.120), the effective one-component fluid associated with a given mixture has the same packing fraction as that of the mixture, i.e., $\eta_{\text{eff}} = \eta$, but this is not a necessary condition, as exemplified by the vdW prescription (3.117). In order to construct a proposal where $\eta_{\text{eff}} = \eta$ is not imposed a priori [117], let us resort to a couple of consistency conditions.

3.9.3.2   Two Consistency Conditions

We first consider an AHS mixture characterized by a total number density $n = N/V$ and a set of mole fractions $\{x_1, x_2, \ldots\}$. According to (3.116), the packing fraction is $\eta = nv_d M_d$. Let us assume that, without modifying the volume V, we add $N_0 = x_0 N$ extra particles of diameter σ_0, so that the augmented system has a number density $n' = (N + N_0)/V = n(1 + x_0)$, a set of mole fractions $\{x_0', x_1', x_2', \ldots\}$, where $x_v' = N_v/(N + N_0) = x_v/(1 + x_0)$, and a packing fraction $\eta' = \eta + nx_0 v_d \sigma_0^d$. Now, if the extra particles have zero diameter ($\sigma_0 \to 0$), it can be proved [118] that

$$\lim_{\sigma_0 \to 0} \beta a^{\text{ex}} (\eta; \{x_0', x_1', x_2', \ldots\}) = \frac{\beta a^{\text{ex}}(\eta; \{x_1, x_2, \ldots\})}{1 + x_0} - \frac{x_0}{1 + x_0} \ln(1 - \eta) , \tag{3.128a}$$

$$\lim_{\sigma_0 \to 0} Z\left(\eta; \{x_0', x_1', x_2', \ldots\}\right) - 1 = \frac{Z(\eta; \{x_1, x_2, \ldots\}) - 1}{1 + x_0} + \frac{x_0}{1 + x_0}\frac{\eta}{1 - \eta}.$$

(3.128b)

Equations (3.128) hold for arbitrary $x_0 > 0$. The scaling relation (3.128b) implies

$$\lim_{\sigma_0 \to 0} \hat{b}_k\left(\{x_0', x_1', x_2', \ldots\}\right) = 1 + \frac{\hat{b}_k(\{x_1, x_2, \ldots\}) - 1}{1 + x_0}.$$

(3.129)

This can be used to prove that the approximation (3.118) is fully consistent with the exact condition (3.128b), regardless of the choice for $Z_{\text{oc}}(\eta)$, while the approximation (3.120) is not.

Now we turn to another more stringent condition. Instead of taking the limit $\sigma_0 \to 0$ for an arbitrary number N_0 of extra particles, we assume that $N_0 \ll N$ (i.e., $x_0 \to 0$) and $\sigma_0 \to \infty$, in such a way that $x_0 \sigma_0^d / \eta \to 0$ (i.e., the extra "big" particles occupy a negligible volume). In that case [119–122],

$$\beta p(\eta; \{x_1, x_2, \ldots\}) = \lim_{x_0 \to 0, \sigma_0 \to \infty} \frac{\beta \mu_0^{\text{ex}}(\eta'; \{x_0', x_1', x_2', \ldots\})}{v_d \sigma_0^d}.$$

(3.130)

This condition is related to the reversible work needed to create a cavity large enough to accommodate a particle of infinite diameter. Using (1.36b) and (1.36c), one can rewrite (3.130) as

$$1 + \eta \frac{\partial \beta a^{\text{ex}}(\eta; \{x_1, x_2, \ldots\})}{\partial \eta} = \lim_{x_0 \to 0, \sigma_0 \to \infty} \frac{1}{v_d \sigma_0^d}\frac{\partial \beta a^{\text{ex}}(\{n_0, n_1, n_2, \ldots\})}{\partial n_0},$$

(3.131)

where on the right hand side the change of independent variables $(\eta'; \{x_0', x_1', x_2', \ldots\}) \to (\{n_0, n_1, n_2, \ldots\})$ has been carried out.

The exact conditions (3.128a) and (3.131) complement each other since the former accounts for the limit where one of the species is made of point particles, whereas the latter accounts for the opposite limit where a few particles have a very large size.

3.9.3.3 Consistent Approximations

Now we restrict ourselves to (approximate) free energies with a truncatable structure involving the first d moments, i.e.,

$$\beta a^{\text{ex}}(\eta; \{x_\nu\}) \to \beta a^{\text{ex}}(\eta; \{m_2, \ldots, m_d\}),$$

(3.132)

and enforce (3.128a) and (3.131). First, noting that in the augmented system of (3.128a) the reduced moments are $m'_k = (1 + x_0)^{k-1} m_k$, we get

$$\beta a^{\text{ex}}(\eta; \{\lambda m_2, \dots, \lambda^{d-1} m_d\}) + \ln(1 - \eta) = \lambda^{-1} [\beta a^{\text{ex}}(\eta; \{m_2, \dots, m_d\})$$
$$+ \ln(1 - \eta)] , \qquad (3.133)$$

where $\lambda = 1 + x_0$ is arbitrary. This scaling property yields

$$\beta a^{\text{ex}}(\eta; \{x_\nu\}) = m_2^{-1} \mathscr{A}(\eta; \{\varpi_3, \dots, \varpi_d\}) - \ln(1 - \eta) , \qquad \varpi_k \equiv \frac{m_k}{m_2^{k-1}} , \qquad (3.134)$$

where $\mathscr{A}$ is so far an unknown function of $d - 1$ variables. In the two-dimensional case ($d = 2$), $\mathscr{A}$ is a function of η only and, therefore, it is fixed by the one-component free energy, namely $\mathscr{A}(\eta) = \beta a_{\text{oc}}^{\text{ex}}(\eta) + \ln(1 - \eta)$, so that

$$\beta a_{\text{SYH}}^{\text{ex}}(\eta) = m_2^{-1} [\beta a_{\text{oc}}^{\text{ex}}(\eta) + \ln(1 - \eta)] - \ln(1 - \eta) , \quad (d = 2) . \qquad (3.135)$$

This is fully equivalent to (3.126a) (hence the label SYH), this time obtained by implementing the consistency condition (3.128a) on the truncatability ansatz (3.132) for $d = 2$.

Thus far, we have not used the more stringent consistency condition (3.131) yet. In order to apply it on (3.134) (for general d), we will need the mathematical properties

$$\frac{1}{v_d \sigma_0^d} \frac{\partial \eta'}{\partial n_0} = 1 , \qquad (3.136a)$$

$$\lim_{x_0 \to 0, \sigma_0 \to \infty} \frac{1}{v_d \sigma_0^d} \frac{\partial \varpi'_k}{\partial n_0} = \frac{\varpi_d}{\eta} \delta_{k,d} , \quad (k \geq 3) , \qquad (3.136b)$$

$$\lim_{x_0 \to 0, \sigma_0 \to \infty} \frac{1}{v_d \sigma_0^d} \frac{\partial m'_2}{\partial n_0} = \frac{m_2}{\eta} \delta_{d,2} . \qquad (3.136c)$$

In the case $d = 2$, insertion of (3.134) into (3.131) gives the linear ordinary differential equation

$$\frac{\mathrm{d}\mathscr{A}(\eta)}{\mathrm{d}\eta} = \frac{\mathscr{A}(\eta)}{\eta(1 - \eta)} , \quad (d = 2) , \qquad (3.137)$$

whose solution is

$$\mathscr{A}(\eta) = \frac{\eta}{1 - \eta} , \quad (d = 2) . \qquad (3.138)$$

This completes the determination of the free energy and its associated compressibility factor:

$$\beta a_{\text{SPT}}^{\text{ex}}(\eta) = m_2^{-1}\frac{\eta}{1-\eta} - \ln(1-\eta), \quad (d=2), \tag{3.139a}$$

$$Z_{\text{SPT}}(\eta) = \frac{1}{1-\eta} + m_2^{-1}\frac{\eta}{(1-\eta)^2}, \quad (d=2). \tag{3.139b}$$

This is not but the SPT for HD mixtures [101]. In fact, (3.139b) can be reobtained from (3.126a) by injecting the SPT compressibility factor (3.111) for $Z_{\text{oc}}(\eta)$. Thus, we conclude that the only truncatable free energy density in two dimensions consistent with the exact requirements (3.128a) and (3.131) is the SPT one.

On the other hand, if $d \geq 3$, combination of (3.134) and (3.131) yields a linear partial differential equation:

$$\eta(1-\eta)\frac{\partial \mathscr{A}}{\partial \eta} + \varpi_d\frac{\partial \mathscr{A}}{\partial \varpi_d} = 0, \quad (d \geq 3). \tag{3.140}$$

Its solution is

$$\mathscr{A}(\eta; \{\varpi_3, \dots, \varpi_d\}) = \mathscr{A}_0\left(\frac{\eta}{\varpi_d(1-\eta)}; \{\varpi_3, \dots, \varpi_{d-1}\}\right), \tag{3.141}$$

where $\mathscr{A}_0$ is a function of $d-2$ variables that remains arbitrary, except for the one-component constraint

$$\mathscr{A}_0\left(\frac{\eta}{1-\eta}; \underbrace{\{1, \dots, 1\}}_{d-3}\right) = \beta a_{\text{oc}}^{\text{ex}}(\eta) + \ln(1-\eta). \tag{3.142}$$

In the physically important three-dimensional case, $\mathscr{A}_0$ is fully determined from $\beta a_{\text{oc}}^{\text{ex}}$, so that combining (3.134), (3.141), and (3.142) we obtain [117]

$$\beta a_{\text{consist}}^{\text{ex}}(\eta) + \ln(1-\eta) = \frac{\varsigma_p}{\varsigma_\eta}\left[\beta a_{\text{oc}}^{\text{ex}}(\eta_{\text{eff}}) + \ln(1-\eta_{\text{eff}})\right], \tag{3.143a}$$

$$\eta Z_{\text{consist}}(\eta) - \frac{\eta}{1-\eta} = \varsigma_p\left[\eta_{\text{eff}}Z_{\text{oc}}(\eta_{\text{eff}}) - \frac{\eta_{\text{eff}}}{1-\eta_{\text{eff}}}\right], \tag{3.143b}$$

where the label "consist" stands for "consistent" [in the sense of (3.128a) and (3.131)]. In (3.143), η_{eff} and η are related by

$$\frac{\eta_{\text{eff}}}{1-\eta_{\text{eff}}} = \varsigma_\eta^{-1}\frac{\eta}{1-\eta} \Rightarrow \eta_{\text{eff}} = \left[1 + \varsigma_\eta\frac{1-\eta}{\eta}\right]^{-1}, \tag{3.144}$$

where we have called

$$\varsigma_\eta \equiv \frac{m_3}{m_2^2}, \quad \varsigma_p \equiv \frac{m_3}{m_2^3}, \quad (d = 3). \tag{3.145}$$

Note that the inequalities $m_3 \geq m_2^2 \geq m_2 \geq 1$ [123] imply $\varsigma_p \geq \varsigma_\eta \geq 1$. Equation (3.143b) can be equivalently written as

$$Z_{\text{consist}}(\eta) = \frac{\varsigma_p/\varsigma_\eta}{1 - (1 - \varsigma_\eta^{-1})\eta} Z_{\text{oc}}(\eta_{\text{eff}}) - \frac{\varsigma_p/\varsigma_\eta - 1}{1 - \eta}. \tag{3.146}$$

Interestingly, using the SPT one-component EoS (3.115), one can obtain

$$Z_{\text{SPT}}(\eta) = \frac{1}{1 - \eta} + \frac{m_2}{m_3} \frac{3\eta}{(1 - \eta)^2} + \frac{m_2^3}{m_3^2} \frac{3\eta^2}{(1 - \eta)^3}. \quad (d = 3). \tag{3.147}$$

As might be expected, this is precisely the SPT expression for three-dimensional AHS mixtures [97, 101, 124–126].

To sum up, the enforcement of the conditions (3.128a) and (3.131) on truncatable free energies leads to the SPT EoS (3.139) for $d = 2$ and to the simple one-component↔multicomponent mapping (3.143) for $d = 3$. For higher dimensionalities ($d \geq 4$), however, (3.142) is not sufficient to fix the function $\mathscr{A}_0$, so that $d - 3$ additional conditions would in principle be needed.

3.9.3.4 Surplus Free Energy and Pressure

Equations (3.143) and (3.144) have an appealing physical interpretation [127], even for arbitrary d. First, note that the ratio $\eta/(1 - \eta)$ represents a *rescaled* packing fraction, i.e., the ratio between the volume Nv_dM_d occupied by the spheres and the *void* volume $V - Nv_dM_d$. According to (3.144), the effective one-component fluid associated with a given mixture has a rescaled packing fraction that is ς_η times smaller than that of the mixture. Next, we can realize that $\eta Z(\eta) - \eta/(1 - \eta)$ represents a (reduced) *modified* excess pressure with respect to a *modified* ideal-gas value corresponding to the void volume $V - Nv_dM_d$, namely

$$\Delta p^* \equiv \beta v_d M_d \left[p - \frac{Nk_BT}{V - Nv_dM_d} \right] = \eta Z(\eta) - \frac{\eta}{1 - \eta}. \tag{3.148}$$

To avoid confusion with the conventional excess pressure $p^{\text{ex}} = p - Nk_BT/V$, we will refer to the quantity Δp^* as the *surplus* pressure. Analogously, we can define the surplus Helmholtz free energy per particle $\Delta a^* \equiv \beta a^{\text{ex}}(\eta) + \ln(1 - \eta)$ as the difference between the Helmholtz free energy per particle (in units of k_BT) and the ideal-gas value corresponding to N particles occupying the void volume $V - Nv_dM_d$. In terms of those quantities, the approximation (3.143) establishes

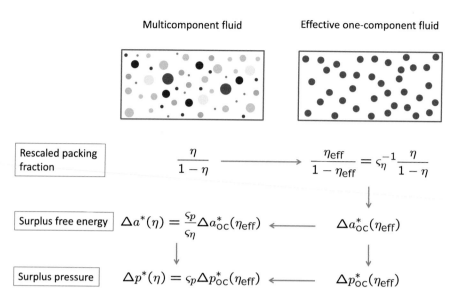

Fig. 3.23 Schematical view of the one-component↔multicomponent mapping represented by (3.143) and (3.144)

that the surplus free energy Δa^* and pressure Δp^* of the multicomponent fluid are just proportional to their respective one-component quantities Δa_{oc}^* and Δp_{oc}^*, as schematically depicted in Fig. 3.23. The surplus free energy of the mixture is never smaller than that of the effective one-component fluid, but the surplus pressure Δp^* can be larger than, equal to, or smaller than Δp_{oc}^* since $\varsigma_p - 1$ has not a definite sign.

The free energy density obtained from (3.143a) has been extended to inhomogeneous systems [117, 128, 129] in the context of the fundamental-measure density functional theory pioneered by Rosenfeld (see Fig. 3.24) [120, 130].

The physical interpretation of (3.143) and (3.144) suggests their generalization to any dimensionality $d \neq 3$. In that case, the parameters ς_η and ς_p are no longer given by (3.145) but can be determined, in analogy with (3.118), by requiring consistency with the second and third virial coefficients. This leads to

$$\varsigma_\eta = \frac{\hat{b}_2 - 1}{b_2 - 1}\frac{b_3 - 2b_2 + 1}{\hat{b}_3 - 2\hat{b}_2 + 1}, \quad \varsigma_p = \left(\frac{\hat{b}_2 - 1}{b_2 - 1}\right)^3 \left(\frac{b_3 - 2b_2 + 1}{\hat{b}_3 - 2\hat{b}_2 + 1}\right)^2. \quad (3.149)$$

Since the third virial coefficient for $d = $ odd can be expressed in terms of moments of the size distribution [see (3.85)], it turns out that (3.149) implies a free energy with a truncatable structure if $d = $ odd. In particular, in the three-dimensional case, use of (3.122b) and (3.124) shows that (3.149) reduces to (3.145). As an extra

Fig. 3.24 Yaakov (Yasha)
Rosenfeld (1948–2002)
(Photograph from Wikimedia
Commons, https://upload.
wikimedia.org/wikipedia/en/
2/23/Yasha_Rosenfeld.jpg)

example, consider five-dimensional systems, in which case (3.69b) and (3.86d) yield

$$\hat{b}_2 = 1 + 5\frac{m_4 + 2m_2 m_3}{m_5}, \quad (d = 5), \tag{3.150a}$$

$$\hat{b}_3 = 1 + 10\frac{m_4 + 2m_2 m_3}{m_5} + 25m_4 \frac{m_2 m_4 + 2m_3^2}{m_5^2}, \quad (d = 5), \tag{3.150b}$$

respectively. Consequently,

$$\varsigma_\eta = \frac{m_5}{m_4}\frac{m_4 + 2m_2 m_3}{m_2 m_4 + 2m_3^2} = \frac{\varpi_5}{\varpi_4}\frac{\varpi_4 + 2\varpi_3}{\varpi_4 + 2\varpi_3^2}, \quad (d = 5), \tag{3.151a}$$

$$\varsigma_p = \frac{m_5}{3m_4^2}\frac{(m_4 + 2m_2 m_3)^3}{(m_2 m_4 + 2m_3^2)^2} = m_2^{-1}\frac{\varpi_5}{3\varpi_4^2}\frac{(\varpi_4 + 2\varpi_3)^3}{(\varpi_4 + 2\varpi_3^2)^2}, \quad (d = 5). \tag{3.151b}$$

From (3.143a) and (3.144) it is easy to check that the function $\mathscr{A}$ defined by (3.134) has the structure (3.141), meaning that the requirement (3.131) is indeed fulfilled. Therefore, for $d = $ odd ≥ 5, the prescription (3.143a) and (3.144), combined with (3.149), avoids the need of $d - 3$ extra conditions to fully determine the function $\mathscr{A}_0$ in (3.141).

In the case of $d = $ even, however, the use of (3.149) in (3.143a) and (3.144) gives a non-truncatable free energy, unless an approximation of the third virial coefficient in terms of moments is introduced [70]. For instance, in the case of HD mixtures, the accurate approximation (3.125) implies $\varsigma_\eta \simeq 1$ and $\varsigma_p \simeq m_2^{-1}$, so that (3.143b) becomes equivalent to (3.126a).

Table 3.12 summarizes the different one-component↔multicomponent mappings described in this section for generic d. The last row must be complemented with (3.144) and (3.149).

3.9 Simple Approximations for the Equation of State of Hard Disks and Spheres

Table 3.12 Summary of different approximations for thermodynamic properties of AHS mixtures in terms of one-component HS properties

Label	$\beta a^{ex}(\eta)$	$Z(\eta)$
vdW	$\beta a_{oc}^{ex}(\eta \hat{b}_2/b_2)$	$Z_{oc}(\eta \hat{b}_2/b_2)$
SYH	$\dfrac{\hat{b}_3 - \hat{b}_2}{b_3 - b_2}\beta a_{oc}^{ex}(\eta) - \dfrac{b_3\hat{b}_2 - b_2\hat{b}_3}{b_3 - b_2}\ln(1-\eta)$	$1 + \dfrac{\hat{b}_3 - \hat{b}_2}{b_3 - b_2}[Z_{oc}(\eta) - 1]$ $+ \dfrac{b_3\hat{b}_2 - b_2\hat{b}_3}{b_3 - b_2}\dfrac{\eta}{1-\eta}$
BS	$\beta a_{oc}^{ex}(\eta)\left(\dfrac{\hat{b}_2}{b_2} - \dfrac{b_3\hat{b}_2 - b_2\hat{b}_3}{b_2^2}\eta\right)$ $+ \dfrac{b_3\hat{b}_2 - b_2\hat{b}_3}{b_2^2}\eta\displaystyle\int_0^1 dt\, \beta a_{oc}^{ex}(\eta t)$	$1 + [Z_{oc}(\eta) - 1]\left(\dfrac{\hat{b}_2}{b_2} - \dfrac{b_3\hat{b}_2 - b_2\hat{b}_3}{b_2^2}\eta\right)$
consist	$\dfrac{\varsigma_p}{\varsigma_\eta}\left[\beta a_{oc}^{ex}(\eta_{eff}) + \ln(1-\eta_{eff})\right] - \ln(1-\eta)$	$\dfrac{\varsigma_p/\varsigma_\eta}{1-(1-\varsigma_\eta^{-1})\eta}Z_{oc}(\eta_{eff}) - \dfrac{\varsigma_p/\varsigma_\eta - 1}{1-\eta}$

3.9.3.5 Polydisperse Systems

Although along this section we have assumed a discrete distribution of sizes $\{x_\nu\}$, the results are readily generalized to a *continuous* distribution $x(\sigma)$ (polydisperse mixture). In the latter case, $x(\sigma)d\sigma$ is the fraction of spheres having a diameter comprised between σ and $\sigma + d\sigma$. For instance, in a top-hat distribution,

$$x(\sigma) = \frac{1}{\sigma_{max} - \sigma_{min}}\begin{cases}0, & 0 < \sigma < \sigma_{min},\\ 1, & \sigma_{min} < \sigma < \sigma_{max},\\ 0, & \sigma > \sigma_{max}.\end{cases} \tag{3.152}$$

In a continuous size distribution, the definition (3.70) for the moment M_k is obviously replaced by

$$M_k = \int_0^\infty d\sigma\, x(\sigma)\sigma^k . \tag{3.153}$$

For instance, in the case of the distribution (3.152), $M_k = (\sigma_{max}^{k+1} - \sigma_{min}^{k+1})/[(k+1)(\sigma_{max} - \sigma_{min})]$. The knowledge of M_k (and hence of m_k) is enough to obtain the rescaled virial coefficients $\hat{b}_2$ for all d and $\hat{b}_3$ for $d = $ odd. In general, combination of (3.36b) and (3.78) yields

$$\hat{b}_3 = \frac{2^{d-1}}{v_d M_d^2}\int_0^\infty d\sigma_1 \int_0^\infty d\sigma_2 \int_0^\infty d\sigma_3\, x(\sigma_1)x(\sigma_2)x(\sigma_3)\sigma_{12}^d \mathscr{V}_{\sigma_{13},\sigma_{23}}(\sigma_{12}) . \tag{3.154}$$

This triple integration must be used to strictly obtain $\hat{b}_3$ for $d = $ even, unless a moment approximation, like in (3.125), is employed.

Fig. 3.25 Compressibility factor for (*a*) a HD binary mixture with $x_1 = x_2 = \frac{1}{2}$, $\sigma_2/\sigma_1 = \frac{5}{7}$, and (*b*) a top-hat HD polydisperse mixture with $\sigma_{max}/\sigma_{min} = 100$. The *symbols* correspond to computer simulations [131], while the *solid lines* correspond to (3.126a) complemented by (3.107) for $Z_{oc}(\eta)$. Use of (3.127a) gives practically indistinguishable results

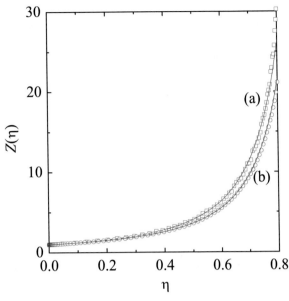

Fig. 3.26 Compressibility factor for (*a*) a HS binary mixture with $x_1 = x_2 = \frac{1}{2}$, $\sigma_2/\sigma_1 = \frac{10}{13}$, and (*b*) a top-hat HS polydisperse mixture with $\sigma_{max}/\sigma_{min} = 100$. The *symbols* correspond to computer simulations [127], while the *solid lines* correspond to (3.126b) complemented by (3.113) for $Z_{oc}(\eta)$. Use of (3.127b) or (3.146) gives practically indistinguishable results

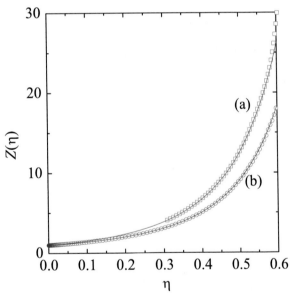

3.9.3.6   Comparison with Computer Simulations

As representative examples, let us consider an equimolar ($x_1 = x_2 = \frac{1}{2}$) binary mixture with a size ratio $\sigma_2/\sigma_1 = \frac{5}{7}$ ($d = 2$) and $\sigma_2/\sigma_1 = \frac{10}{13}$ ($d = 3$) and a top-hat continuous distribution with $\sigma_{max}/\sigma_{min} = 100$ ($d = 2$ and $d = 3$). Computer simulations of those mixtures [127, 131] are compared with theoretical predictions in Figs. 3.25 and 3.26 for $d = 2$ and $d = 3$, respectively.

An excellent general agreement between simulation and theory can be observed, although it tends to worsen at high packing fractions for the binary mixtures. We also note that the influence of multicomposition is more important in the three-dimensional case than in the two-dimensional one. Moreover, comparison with Figs. 3.19 and 3.22 shows that higher packing fractions in the metastable region can be reached in a mixture as compared to the one-component case. This is due to the fact that the presence of different sizes tends to frustrate crystallization [132].

Before closing this chapter, it is worth saying that, despite the weight given here to HS systems, some important topics such as inhomogeneous fluids and density functional theory [122, 133–137], metastable glassy states [138–140], and perturbation theories [3, 4] have been left out. The interested reader is referred to the relevant literature for further details on those topics.

Exercises

3.1 What relationship exists between the Mayer functions for the PS and HS potentials?

3.2 As an extension of both the PS and the SW potentials one can introduce a penetrable square-well (PSW) potential. How would you define it? How many free parameters (energy and length scales) would it include? Make a schematic graph of the associated Mayer function. Compare your definition with that of Fantoni et al. [141].

3.3 The PSW model can violate Ruelle's stability criterion [142, 143]. Why? Hint: Consult Fantoni et al. [141].

3.4 Download and install the Wolfram CDF Player (http://www.wolfram.com/cdf-player) in your computer. Play with the Demonstration of Pajuelo and Santos [144] to explore how the scattering angle depends on the impact parameter in (a) the HS, (b) the PS, (c) the SW, and (d) the PSW interactions.

3.5 Derive (3.12).

3.6 Check (3.15).

3.7 Derive (3.18) and (3.19).

3.8 Derive (3.33).

3.9 Check the correctness of the numerical coefficient associated with each diagram of Table 3.4.

3.10 Suppose that the SHS limit from the SW potential is taken as $e^{\beta \varepsilon} \sim (\sigma'/\sigma - 1)^{-k}$ with $k > 0$, $k \neq 1$, instead of as in (3.4). How would the corresponding Mayer function look like? Would the associated second virial coefficient be different from that of HSs? Would it be finite?

3.11 Obtain the second virial coefficient for the PSW fluid defined in Exercise 3.2.

3.12 Derive (3.53).

3.13 Download and install the Wolfram CDF Player (http://www.wolfram.com/cdf-player) in your computer. Play with the Demonstration of Santos [50] to explore how the Boyle temperature and the maximum value of the second virial coefficient for the generalized LJ potential change with the parameter s.

3.14 Download and install the Wolfram CDF Player (http://www.wolfram.com/cdf-player) in your computer. Play with the Demonstration of Blinder [145] to explore how the LJ interaction potential, the Boyle temperature, and the second virial coefficient (all in real units) differ as the gas changes.

3.15 Derive (3.55) and (3.56).

3.16 Derive (3.61) and (3.62).

3.17 Assuming that Argon can be modeled by means of the LJ potential (3.52) ($s = 6$) with $\sigma = 3.405$ Å and $\varepsilon/k_B = 119.8$ K, plot the second virial coefficient (see Table 3.6) $B_2(T)$ (in cm^3/mol) for the temperature range $T = 80$–1100 K. Compare with the empirical equation (5), Fig. 5, and Table 7 of Stewart and Jacobsen [146].

3.18 Derive (3.66).

3.19 Check that (3.63) indeed leads to Table 3.7. Extend Table 3.7 to higher values of d. Compare the obtained numerical values with those from the asymptotic formula (3.66).

3.20 Check that (3.65) indeed leads to the *exact* values of Table 3.7 for $d =$ odd.

3.21 Check that (3.65) with the approximation $j_{max}^{(d)} \to 7$ leads to the *numerical* values of Table 3.7 for $d =$ even.

3.22 Derive (3.74) and (3.75).

3.23 With the help of a computational software program, check that both sides of (3.76) yield the same result for $d = 1$–12.

3.24 With the help of a computational software program, check that (3.75) and (3.78) yield the same numerical results for different choices of $\sigma_{\alpha\gamma}$, $\sigma_{\alpha\delta}$, and $\sigma_{\gamma\delta}$ in the cases $d = 2$ [you need to use (3.82)], $d = 3$ [you need to use (3.81a)], and $d = 5$ [you need to use (3.81b)].

3.25 Particularize the second line of (3.83) to $d = 3$ and 5 and check that the obtained polynomials are consistent with (3.81).

3.26 Make $a = b$ in (3.82) and expand $\bar{\mathcal{V}}_{a,a}(r)$ in powers of r. Check that the result coincides with (3.83) particularized to $d = 2$.

3.27 Derive (3.86), and (3.87) by complementing (3.78) with (3.81) and (3.82), respectively.

3.28 Assuming, without loss of generality, that $\sigma_\alpha \geq \sigma_\gamma \geq \sigma_\delta \geq \sigma_\epsilon$, derive (3.91) from (3.90).

3.29 Derive (3.96) from (3.95).

3.30 Check that (3.98) reduces to (3.92) if $\sigma_\epsilon = \sigma_\alpha$.

3.31 With the help of a computational software program, check (3.100).

3.32 Check from (3.103c) that the symmetry condition $b_{2;2}(q) = q^9 b_{2;2}(1/q)$ is satisfied.

3.33 Check that (3.103) is consistent with (3.47).

3.34 Download and install the Wolfram CDF Player (http://www.wolfram.com/cdf-player) in your computer. Play with the Demonstration of Santos [147] to explore how the second, third, and fourth virial coefficients of a binary three-dimensional AHS mixture depend on the composition and size ratio of the mixture.

3.35 Check the numerical values of Table 3.11.

3.36 Prove from Fig. 3.18 that $\eta_{cp} = \pi \sqrt{2}/6$ at close packing.

3.37 Obtain the first twelve virial coefficients for HDs obtained from (3.107), (3.108), (3.110), and (3.111), and compare the results with the exact values of Table 3.11.

3.38 Check that the EoS (3.110) presents an additional mathematical pole at $\eta = \eta_{cp}/(2\eta_{cp} - 1) \simeq 1.1144$.

3.39 Check that a HD EoS alternative to (3.110) but yet complying with the requirements (3.109) is

$$Z = \frac{1 + (2 - 1/\eta_{cp})\eta}{1 - \eta/\eta_{cp}}.$$

Does this EoS behave reasonably well? Obtain its first few virial coefficients.

3.40 Derive (3.113) by inserting the approximation (3.112) into the virial series (3.104).

3.41 Derive (3.114).

3.42 Obtain the first twelve virial coefficients from the SPT EoS (3.115) and compare with the results of Table 3.11.

3.43 Check that (3.118) and (3.120) are consistent with the first three virial coefficients, i.e., $Z(\eta) = 1 + \hat{b}_2\eta + \hat{b}_3\eta^2 + \cdots$.

3.44 Derive (3.119) and (3.121).

3.45 Prove that

$$\lim_{\sigma_0 \to 0} \mathscr{Q}_{\{N_0,N_1,N_2,\ldots\}} = (1 - \eta)^{N_0} \mathscr{Q}_{\{N_1,N_2,\ldots\}} \,,$$

where $\mathscr{Q}_{\{N_\nu\}}$ is the configuration integral of an AHS mixture [see (4.66) in Chap. 4 for its general definition]. From the above relation, derive (3.128).

3.46 Check (3.131).

3.47 Prove that (3.118) is consistent with (3.128b). Is also (3.120) consistent with (3.128b)?

3.48 Prove (3.136).

3.49 Check (3.137) and (3.140).

3.50 Check that (3.141) satisfies (3.140).

3.51 Check (3.143). Hint: Upon deriving (3.143b) from (3.143a), take into account that $\partial \eta_{\mathrm{eff}}/\partial \eta = \varsigma_\eta (\eta_{\mathrm{eff}}/\eta)^2$.

3.52 Check (3.146) and (3.147).

3.53 Insert the SPT one-component EoS (3.115) into (3.126b). Does the result coincide with the SPT EoS for mixtures (3.147)?

3.54 Check that insertion of (3.149) into (3.146) gives $Z(\eta) = 1 + \hat{b}_2 \eta + \hat{b}_3 \eta^2 + \mathcal{O}(\eta^4)$.

3.55 Using (3.143a), (3.144), and (3.151) for $d = 5$, prove that the function $\mathscr{A}$ defined by (3.134) has the structure (3.141).

References

1. J.E. Mayer, M. Goeppert Mayer, *Statistical Mechanics* (Wiley, New York, 1940)
2. A. Mulero (ed.), *Theory and Simulation of Hard-Sphere Fluids and Related Systems*. Lecture Notes in Physics, vol. 753 (Springer, Berlin, 2008)
3. J.A. Barker, D. Henderson, Rev. Mod. Phys. **48**, 587 (1976)
4. J.R. Solana, *Perturbation Theories for the Thermodynamic Properties of Fluids and Solids* (CRC Press, Boca Raton, 2013)
5. W. Schirmacher, *Theory of Liquids and Other Disordered Media. A Short Introduction*. Lecture Notes in Physics, vol. 887 (Springer, Cham, 2014)
6. R. Piazza, *Soft Matter. The Stuff that Dreams Are Made of* (Springer, Dordrecht, 2011)
7. C.N. Likos, Phys. Rep. **348**, 267 (2001)
8. P.N. Pusey, E. Zaccarelli, C. Valeriani, E. Sanz, W.C.K. Poon, M.E. Cates, Philos. Trans. R. Soc. A **367**, 4993 (2009)
9. C. Marquest, T.A. Witten, J. Phys. Fr. **50**, 1267 (1989)
10. W. Klein, H. Gould, R.A. Ramos, I. Clejan, A.I. Mel'cuk, Physica A **205**, 738 (1994)
11. C.N. Likos, M. Watzlawek, H. Löwen, Phys. Rev. E **58**, 3135 (1998)
12. M. Schmidt, J. Phys. Condens. Matter **11**, 10163 (1999)

13. M.J. Fernaud, E. Lomba, L.L. Lee, J. Chem. Phys. **112**, 810 (2000)
14. Y. Rosenfeld, M. Schmidt, M. Watzlawek, H. Löwen, Phys. Rev. E **62**, 5006 (2000)
15. C.N. Likos, A. Lang, M. Watzlawek, H. Löwen, Phys. Rev. E **63**, 031206 (2001)
16. M. Schmidt, M. Fuchs, J. Chem. Phys. **117**, 6308 (2002)
17. S.C. Kim, S.H. Suh, J. Chem. Phys. **117**, 9880 (2002)
18. L. Acedo, A. Santos, Phys. Lett. A **323**, 427 (2004). Erratum: **376**, 2274–2275 (2012)
19. F.H. Stillinger, D.K. Stillinger, Physica A **244**, 358 (1997)
20. C.N. Likos, H. Löwen, M. Watzlawek, B. Abbas, O. Jucknischke, J. Allgaier, D. Richter, Phys. Rev. Lett. **80**, 4450 (1998)
21. P.C. Hemmer, G. Stell, Phys. Rev. Lett. **24**, 1284 (1970)
22. Z. Yan, S.V. Buldyrev, N. Giovambattista, H.E. Stanley, Phys. Rev. Lett. **95**, 130604 (2005)
23. R.J. Baxter, J. Chem. Phys. **49**, 2770 (1968)
24. G. Stell, J. Stat. Phys. **63**, 1203 (1991)
25. B. Borštnik, C.G. Jesudason, G. Stell, J. Chem. Phys. **106**, 9762 (1997)
26. S.H. Chen, J. Rouch, F. Sciortino, P. Tartaglia, J. Phys. Condens. Matter **6**, 109855 (1994)
27. H. Verduin, J.K.G. Dhont, J. Colloid Interface Sci. **172**, 425 (1995)
28. D. Rosenbaum, P.C. Zamora, C.F. Zukoski, Phys. Rev. Lett. **76**, 150 (1996)
29. D. Pontoni, S. Finet, T. Narayanan, A.R. Rennie, J. Chem. Phys. **119**, 6157 (2003)
30. S. Buzzaccaro, R. Rusconi, R. Piazza, Phys. Rev. Lett. **99**, 098301 (2007)
31. R. Piazza, V. Peyre, V. Degiorgio, Phys. Rev. E **58**, R2733 (1998)
32. M.G. Noro, D. Frenkel, J. Chem. Phys. **113**, 2941 (2000)
33. M.A.G. Maestre, R. Fantoni, A. Giacometti, A. Santos, J. Chem. Phys. **138**, 094904 (2013)
34. M. Thiesen, Ann. Phys. **260**, 467 (1885)
35. H. Kamerlingh Onnes, Commun. Phys. Lab. Univ. Leiden **71**, 3 (1901). Reprinted in Expression of the equation of state of gases and liquids by means of series, in *Through Measurement to Knowledge*. Boston Studies in the Philosophy and History of Science, vol. 124 (Springer, Netherlands, 1991), pp. 146–163
36. E.G.D. Cohen, Einstein and Boltzmann: Determinism and Probability or The Virial Expansion Revisited (2013), http://arxiv.org/abs/1302.2084
37. L.E. Reichl, *A Modern Course in Statistical Physics*, 1st edn. (University of Texas Press, Austin, 1980)
38. E. Enciso, N.G. Almarza, M.A. González, F.J. Bermejo, Phys. Rev. E **57**, 4486 (1998)
39. A.J. Schultz, D.A. Kofke, Phys. Rev. E **90**, 023301 (2014)
40. The On-Line Encyclopedia of Integer Sequences (OEIS) (1996), http://oeis.org/A002218, http://oeis.org/A013922
41. R.J. Wheatley, Phys. Rev. Lett. **110**, 200601 (2013)
42. C. Zhang, B.M. Pettitt, Mol. Phys. **112**, 1427 (2014)
43. R.J. Wheatley, Mol. Phys. **93**, 965 (1998)
44. R.J. Wheatley, J. Chem. Phys. **111**, 5455 (1999)
45. C. Barrio, J.R. Solana, in *Theory and Simulation of Hard-Sphere Fluids and Related Systems*, ed. by A. Mulero. Lecture Notes in Physics, vol. 753 (Springer, Berlin, 2008), pp. 133–182
46. S. Labík, J. Kolafa, Phys. Rev. E **80**, 051122 (2009). Erratum: **84**, 069901 (2011)
47. M. Abramowitz, I.A. Stegun (eds.), *Handbook of Mathematical Functions* (Dover, New York, 1972)
48. F.W.J. Olver, D.W. Lozier, R.F. Boisvert, C.W. Clark (eds.), *NIST Handbook of Mathematical Functions* (Cambridge University Press, New York, 2010)
49. A. Santos, in *5th Warsaw School of Statistical Physics*, ed. by B. Cichocki, M. Napiórkowski, J. Piasecki (Warsaw University Press, Warsaw, 2014). http://arxiv.org/abs/1310.5578
50. A. Santos, Second Virial Coefficients for the Lennard-Jones (2n-n) Potential. Wolfram Demonstrations Project (2012), http://demonstrations.wolfram.com/SecondVirialCoefficientsForTheLennardJones2nNPotential/
51. A. Erdélyi, *Asymptotic Expansions* (Dover, New York, 1956)
52. N. Clisby, B. McCoy, J. Stat. Phys. **114**, 1361 (2004)
53. I. Lyberg, J. Stat. Phys. **119**, 747 (2005)

54. K.W. Kratky, J. Stat. Phys. **27**, 533 (1982)
55. N. Clisby, B.M. McCoy, J. Stat. Phys. **114**, 1343 (2004)
56. S. Labík, J. Kolafa, A. Malijevský, Phys. Rev. E **71**, 021105 (2005)
57. N. Clisby, B.M. McCoy, Pramana **64**, 775 (2005)
58. N. Clisby, B.M. McCoy, J. Stat. Phys. **122**, 15 (2006)
59. M. Luban, A. Baram, J. Chem. Phys. **76**, 3233 (1982)
60. M. Baus, J.L. Colot, Phys. Rev. A **36**, 3912 (1987)
61. R.D. Rohrmann, A. Santos, Phys. Rev. E **76**, 051202 (2007)
62. J.H. Nairn, J.E. Kilpatrick, Am. J. Phys. **40**, 503 (1972)
63. B.R.A. Nijboer, L. van Hove, Phys. Rev. **85**, 777 (1952)
64. R.D. Rohrmann, M. Robles, M. López de Haro, A. Santos, J. Chem. Phys. **129**, 014510 (2008)
65. H.L. Frisch, N. Rivier, D. Wyler, Phys. Rev. Lett. **54**, 2061 (1985)
66. H.O. Carmesin, H. Frisch, J. Percus, J. Stat. Phys. **63**, 791 (1991)
67. R.D. Rohrmann, A. Santos, Phys. Rev. E **83**, 011201 (2011)
68. S. Torquato, F.H. Stillinger, Phys. Rev. E **68**, 041113 (2003)
69. S. Torquato, F.H. Stillinger, Exp. Math. **15**, 307 (2006)
70. A. Santos, S.B. Yuste, M. López de Haro, Mol. Phys. **99**, 1959 (2001)
71. R. Blaak, Mol. Phys. **95**, 695 (1998)
72. I. Urrutia, Phys. Rev. E **84**, 062101 (2011)
73. F. Saija, G. Fiumara, P.V. Giaquinta, Mol. Phys. **87**, 991 (1996). Erratum: **92**, 1089 (1997)
74. F. Saija, G. Fiumara, P.V. Giaquinta, Mol. Phys. **89**, 1181 (1996)
75. F. Saija, G. Fiumara, P.V. Giaquinta, Mol. Phys. **90**, 679 (1997)
76. E. Enciso, N.G. Almarza, D.S. Calzas, M.A. González, Mol. Phys. **92**, 173 (1997)
77. R.J. Wheatley, Mol. Phys. **93**, 665 (1998)
78. R.J. Wheatley, F. Saija, P.V. Giaquinta, Mol. Phys. **94**, 877 (1998)
79. F. Saija, G. Fiumara, P.V. Giaquinta, J. Chem. Phys. **108**, 9098 (1998)
80. R.J. Wheatley, Mol. Phys. **96**, 1805 (1999)
81. E. Enciso, N.G. Almarza, M.A. González, F.J. Bermejo, Mol. Phys. **100**, 1941 (2002)
82. A.Y. Vlasov, A.J. Masters, Fluid Phase Equilib. **212**, 183 (2003)
83. G. Pellicane, C. Caccamo, P.V. Giaquinta, F. Saija, J. Phys. Chem. B **111**, 4503 (2007)
84. M. López de Haro, A. Malijevský, S. Labík, Collect. Czech. Chem. Commun. **75**, 359 (2010)
85. F. Saija, Phys. Chem. Chem. Phys. **13**, 11885 (2011)
86. F. Saija, A. Santos, S.B. Yuste, M. López de Haro, J. Chem. Phys. **136**, 184505 (2012)
87. A. Mulero, C.A. Galán, M.I. Parra, F. Cuadros, in *Theory and Simulation of Hard-Sphere Fluids and Related Systems*, ed. by A. Mulero. Lecture Notes in Physics, vol. 753 (Springer, Berlin, 2008), pp. 37–109
88. C.M. Bender, S.A. Orszag, *Advanced Mathematical Methods for Scientists and Engineers* (McGraw-Hill, Auckland, 1987)
89. D. Henderson, Mol. Phys. **30**, 971 (1975)
90. L. Verlet, D. Levesque, Mol. Phys. **46**, 969 (1982)
91. A. Mulero, F. Cuadros, C. Galán, J. Chem. Phys. **107**, 6887 (1997)
92. S. Luding, Phys. Rev. E **63**, 042201 (2001)
93. S. Luding, O. Strauß, in *Granular Gases*, ed. by T. Pöschel, S. Luding. Lecture Notes in Physics, vol. 564 (Springer, Berlin, 2001), pp. 389–409
94. S. Luding, A. Santos, J. Chem. Phys. **121**, 8458 (2004)
95. A. Santos, M. López de Haro, S.B. Yuste, J. Chem. Phys. **103**, 4622 (1995)
96. M. López de Haro, A. Santos, S.B. Yuste, Eur. J. Phys. **19**, 281 (1998)
97. H. Reiss, H.L. Frisch, J.L. Lebowitz, J. Chem. Phys. **31**, 369 (1959)
98. E. Helfand, H.L. Frisch, J.L. Lebowitz, J. Chem. Phys. **34**, 1037 (1961)
99. J.J. Erpenbeck, M.J. Luban, Phys. Rev. A **32**, 2920 (1985)
100. N.F. Carnahan, K.E. Starling, J. Chem. Phys. **51**, 635 (1969)
101. J.L. Lebowitz, E. Helfand, E. Praestgaard, J. Chem. Phys. **43**, 774 (1965)
102. J. Kolafa, S. Labík, A. Malijevský, Phys. Chem. Chem. Phys. **6**, 2335 (2004). See also http://www.vscht.cz/fch/software/hsmd/

103. L.A. Fernández, V. Martín-Mayor, B. Seoane, P. Verrocchio, Phys. Rev. Lett. **108**, 165701 (2012)
104. A. Santos, M. López de Haro, J. Chem. Phys. **130**, 214104 (2009)
105. M. Luban, J.P.J. Michels, Phys. Rev. A **41**, 6796 (1990)
106. D. Henderson, P.J. Leonard, Proc. Natl. Acad. Sci. U. S. A. **67**, 1818 (1970)
107. D. Henderson, P.J. Leonard, Proc. Natl. Acad. Sci. U. S. A. **68**, 2354 (1971)
108. M. López de Haro, S.B. Yuste, A. Santos, in *Theory and Simulation of Hard-Sphere Fluids and Related Systems*, ed. by A. Mulero. Lecture Notes in Physics, vol. 753 (Springer, Berlin, 2008), pp. 183–245
109. A. Santos, S.B. Yuste, M. López de Haro, Mol. Phys. **96**, 1 (1999)
110. A. Santos, M. López de Haro, S.B. Yuste, J. Chem. Phys. **122**, 024514 (2005)
111. C. Barrio, J.R. Solana, Mol. Phys. **97**, 797 (1999)
112. C. Barrio, J.R. Solana, Phys. Rev. E **63**, 011201 (2001)
113. J.T. Jenkins, F. Mancini, J. Appl. Mech. **54**, 27 (1987)
114. J.A. Gualtieri, J.M. Kincaid, G. Morrison, J. Chem. Phys. **77**, 521 (1982)
115. P. Sollich, P.B. Warren, M.E. Cates, Adv. Chem. Phys. **116**, 265 (2001)
116. P. Sollich, J. Phys. Condens. Matter **14**, R79 (2002)
117. A. Santos, Phys. Rev. E **86**, 040102(R) (2012)
118. A. Santos, J. Chem. Phys. **136**, 136102 (2012)
119. H. Reiss, H.L. Frisch, E. Helfand, J.L. Lebowitz, J. Chem. Phys. **32**, 119 (1960)
120. Y. Rosenfeld, Phys. Rev. Lett. **63**, 980 (1989)
121. R. Roth, R. Evans, A. Lang, G. Kahl, J. Phys. Condens. Matter **14**, 12063 (2002)
122. R. Roth, J. Phys. Condens. Matter **22**, 063102 (2010)
123. V. Ogarko, S. Luding, J. Chem. Phys. **136**, 124508 (2012)
124. M. Mandell, H. Reiss, J. Stat. Phys. **13**, 113 (1975)
125. Y. Rosenfeld, J. Chem. Phys. **89**, 4272 (1988)
126. M. Heying, D. Corti, J. Phys. Chem. B **108**, 19756 (2004)
127. A. Santos, S.B. Yuste, M. López de Haro, G. Odriozola, V. Ogarko, Phys. Rev. E **89**, 040302(R) (2014)
128. J.F. Lutsko, Phys. Rev. E **87**, 014103 (2013)
129. H. Hansen-Goos, M. Mortazavifar, M. Oettel, R. Roth, Phys. Rev. E **91**, 052121 (2015)
130. Y. Rosenfeld, M. Schmidt, H. Löwen, P. Tarazona, Phys. Rev. E **55**, 4245 (1997)
131. A. Santos, S.B. Yuste, M. López de Haro, V. Ogarko, Equation of state of polydisperse hard-disk mixtures (2016, in preparation)
132. P.N. Pusey, J. Phys. France **48**, 709 (1987)
133. R. Evans, in *Fundamentals of Inhomogeneous Fluids*, ed. by D. Henderson (Dekker, New York, 1992)
134. P. Tarazona, J.A. Cuesta, Y. Martínez-Ratón, in *Theory and Simulation of Hard-Sphere Fluids and Related Systems*, ed. by A. Mulero. Lecture Notes in Physics, vol. 753 (Springer, Berlin, 2008), pp. 247–341
135. R. Evans, in *3rd Warsaw School of Statistical Physics*, ed. by B. Cichocki, M. Napiórkowski, J. Piasecki (Warsaw University Press, Warsaw, 2010), pp. 43–85. http://agenda.albanova.se/getFile.py/access?contribId=260&resId=251&materialId=250&confId=2509
136. H. Löwen, in *3rd Warsaw School of Statistical Physics*, ed. by B. Cichocki, M. Napiórkowski, J. Piasecki (Warsaw University Press, Warsaw, 2010), pp. 87–121. http://www2.thphy.uni-duesseldorf.de/~hlowen/doc/ra/ra0025.pdf
137. J.F. Lutsko, Adv. Chem. Phys. **144**, 1 (2010)
138. G. Parisi, F. Zamponi, Rev. Mod. Phys. **82**, 789 (2010)
139. L. Berthier, G. Biroli, Rev. Mod. Phys. **3**, 587 (2011)
140. J. Kurchan, in *4th Warsaw School of Statistical Physics*, ed. by B. Cichocki, M. Napiórkowski, J. Piasecki (Warsaw University Press, Warsaw, 2012), pp. 131–167. https://www.icts.res.in/media/uploads/Program/Files/kurchan.pdf
141. R. Fantoni, A. Malijevský, A. Santos, A. Giacometti, Mol. Phys. **109**, 2723 (2011)
142. M.E. Fisher, D. Ruelle, J. Math. Phys. **7**, 260 (1966)

143. D. Ruelle, *Statistical Mechanics: Rigorous Results* (World Scientific, Singapore, 1999)
144. P. Pajuelo, A. Santos, Classical Scattering with a Penetrable Square-Well Potential, Wolfram Demonstrations Project (2011), http://demonstrations.wolfram.com/ClassicalScatteringWithAPenetrableSquareWellPotential
145. S.M. Blinder, Second Virial Coefficients Using the Lennard-Jones Potential, Wolfram Demonstrations Project (2010), http://demonstrations.wolfram.com/SecondVirialCoefficientsUsingTheLennardJonesPotential/
146. R.B. Stewart, R.T. Jacobsen, J. Phys. Chem. Ref. Data **18**, 639 (1989). http://www.nist.gov/data/PDFfiles/jpcrd363.pdf
147. A. Santos, Virial Coefficients for a Hard-Sphere Mixture, Wolfram Demonstrations Project (2014), http://demonstrations.wolfram.com/VirialCoefficientsForAHardSphereMixture/

Chapter 4
Spatial Correlation Functions and Thermodynamic Routes

This chapter introduces the reduced (or marginal) distribution functions describing groups of s particles. The fundamental one is the pair configurational distribution function, from which the radial distribution function $g(r)$ is defined as a key quantity in the statistical-mechanical description of liquids. Most of the chapter is devoted to the derivation of thermodynamic quantities in terms of integrals involving $g(r)$. Apart from the conventional compressibility, energy, and virial routes, the less known chemical-potential and free-energy routes are worked out.

4.1 Reduced Distribution Functions

The N-body probability distribution function $\rho_N(\mathbf{x}^N)$ contains all the (equilibrium or nonequilibrium) statistical-mechanical information about the system. On the other hand, partial information embedded in *marginal* few-body distributions are usually enough for the most relevant quantities. Moreover, it is much simpler to introduce useful approximations at the level of the marginal distributions than at the N-body level.

Let us introduce the s-body *reduced distribution function* $f_s(\mathbf{x}^s)$ so that $f_s(\mathbf{x}^s)d\mathbf{x}^s$ is the (average) number of groups of s particles such that one particle lies inside a volume $d\mathbf{x}_1$ around the (one-body) phase-space point $\mathbf{x}_1$, other particle lies inside a volume $d\mathbf{x}_2$ around the (one-body) phase-space point $\mathbf{x}_2$, ... and so on (see Fig. 4.1 for $s = 3$). More explicitly,

$$
f_s(\mathbf{x}^s) = \sum_{i_1 \neq i_2 \neq \cdots \neq i_s} \int d\mathbf{x}'^N \, \delta(\mathbf{x}'_{i_1} - \mathbf{x}_1) \cdots \delta(\mathbf{x}'_{i_s} - \mathbf{x}_s) \rho_N(\mathbf{x}'^N)
$$

$$
= \frac{N!}{(N-s)!} \int d\mathbf{x}_{s+1} \int d\mathbf{x}_{s+2} \cdots \int d\mathbf{x}_N \, \rho_N(\mathbf{x}^N) . \tag{4.1}
$$

© Springer International Publishing Switzerland 2016
A. Santos, *A Concise Course on the Theory of Classical Liquids*,
Lecture Notes in Physics 923, DOI 10.1007/978-3-319-29668-5_4

Fig. 4.1 Sketch of the
one-body phase space. The
horizontal axis represents the
d position coordinates, while
the *vertical axis* represents
the *d* momentum
components. Three points ($\mathbf{x}_1$,
$\mathbf{x}_2$, and $\mathbf{x}_3$) are represented.
Compare with Fig. 2.1

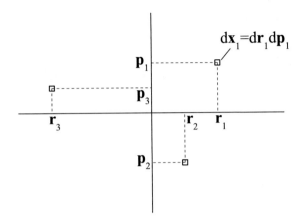

In (4.1) it is implicitly assumed that the total number N of particles is fixed. If
that is not the case (for instance, in the grand canonical ensemble), the adequate
definition is

$$f_s(\mathbf{x}^s) = s! \rho_s(\mathbf{x}^s) + \sum_{N=s+1}^{\infty} \frac{N!}{(N-s)!} \int d\mathbf{x}_{s+1} \int d\mathbf{x}_{s+2} \cdots \int d\mathbf{x}_N \, \rho_N(\mathbf{x}^N), \qquad (4.2)$$

where the first term on the right-hand side is the contribution corresponding to
$N = s$.

In equilibrium situations, momenta are uncorrelated and thus the relevant
information is contained in the *s*-body *configurational* distribution functions

$$n_s(\mathbf{r}^s) = \int d\mathbf{p}^s \, f_s(\mathbf{x}^s) . \qquad (4.3)$$

Obviously, the normalization conditions of $f_s(\mathbf{x}^s)$ and $n_s(\mathbf{r}^s)$ are

$$\int d\mathbf{x}^s \, f_s(\mathbf{x}^s) = \int d\mathbf{r}^s \, n_s(\mathbf{r}^s) = \left\langle \frac{N!}{(N-s)!} \right\rangle , \qquad (4.4)$$

where the angular brackets are unnecessary if N is fixed.

The relevance of n_s arises especially when one is interested in evaluating the
average of a dynamical variable that can be expressed in terms of *s*-body functions,
namely

$$A(\mathbf{r}^N) = \frac{1}{s!} \sum_{i_1 \neq i_2 \neq \cdots \neq i_s} A_s(\mathbf{r}_{i_1}, \mathbf{r}_{i_2}, \ldots, \mathbf{r}_{i_s}) . \qquad (4.5)$$

In that case, it is easy to obtain

$$\langle A \rangle \equiv \sum_{N=s}^{\infty} \int d\mathbf{x}^N \, A(\mathbf{r}^N) \rho_N(\mathbf{x}^N) = \frac{1}{s!} \int d\mathbf{r}^s \, A_s(\mathbf{r}^s) n_s(\mathbf{r}^s) . \qquad (4.6)$$

Needless to say, the summation $\sum_{N=s}^{\infty}$ is only needed if the number of particles in the system is not fixed

The quantities (4.1) and (4.3) can be defined both out of and in equilibrium. In the latter case, however, we can benefit from the (formal) knowledge of ρ_N. In particular, in the canonical [see (2.17) and (2.64)] and grand canonical [see (2.31) and (2.67)] ensembles one has

$$n_s(\mathbf{r}^s) = \frac{1}{V^N \mathcal{Q}_N} \frac{N!}{(N-s)!} \int d\mathbf{r}_{s+1} \cdots \int d\mathbf{r}_N \, e^{-\beta \Phi_N(\mathbf{r}^N)} \,, \tag{4.7a}$$

$$n_s(\mathbf{r}^s) = \frac{1}{\Xi} \sum_{N=s}^{\infty} \frac{\hat{z}^N}{(N-s)!} \int d\mathbf{r}_{s+1} \cdots \int d\mathbf{r}_N \, e^{-\beta \Phi_N(\mathbf{r}^N)} \,, \tag{4.7b}$$

respectively. We recall that the rescaled fugacity $\hat{z}$ is defined in (2.68). Note that, because of the translational invariance property (2.62), $n_1 = n$, where $n \equiv \langle N \rangle / V$ is the number density. More in general, the functions n_s are translationally invariant, namely

$$n_s(\mathbf{r}_1 + \mathbf{a}, \mathbf{r}_2 + \mathbf{a}, \ldots, \mathbf{r}_s + \mathbf{a}) = n_s(\mathbf{r}_1, \mathbf{r}_2, \ldots, \mathbf{r}_s) \,. \tag{4.8}$$

The grand-canonical expression (4.7b) allows us to establish an important relationship between n_s, $\partial n_s / \partial z$, and an integral of n_{s+1}. First, note that

$$z \frac{\partial n_s(\mathbf{r}^s)}{\partial z} = -\langle N \rangle \, n_s(\mathbf{r}^s) + \frac{1}{\Xi} \sum_{N=s}^{\infty} \frac{N \hat{z}^N}{(N-s)!} \int d\mathbf{r}_{s+1} \cdots \int d\mathbf{r}_N \, e^{-\beta \Phi_N(\mathbf{r}^N)} \,, \tag{4.9}$$

where use has been made of the thermodynamic relation $\langle N \rangle = z \partial \ln \Xi / \partial z$ [see (2.36b)]. Next, replacing the factor N by $s + N - s$ inside the summation of the second term, we obtain

$$\left(z \frac{\partial}{\partial z} - s \right) n_s(\mathbf{r}^s) = \int d\mathbf{r}_{s+1} \left[n_{s+1}(\mathbf{r}^{s+1}) - n n_s(\mathbf{r}^s) \right] \,. \tag{4.10}$$

Noting the chain of identities

$$z \frac{\partial}{\partial z} = k_B T \frac{\partial}{\partial \mu} = k_B T \frac{\partial p}{\partial \mu} \frac{\partial n}{\partial p} \frac{\partial}{\partial n} = \chi_T n \frac{\partial}{\partial n} \,, \tag{4.11}$$

where in the last step use has been made of (1.22), (1.28), and (1.34), we can rewrite (4.10) as

$$\boxed{\left(\chi_T n \frac{\partial}{\partial n} - s \right) n_s(\mathbf{r}^s) = \int d\mathbf{r}_{s+1} \left[n_{s+1}(\mathbf{r}^{s+1}) - n n_s(\mathbf{r}^s) \right] \,.} \tag{4.12}$$

The exact relation (4.12) was derived by Baxter (see Fig. 3.4) by a different method
[1]. It can also be derived from the canonical representation (4.7a).

4.2   Correlation Functions

In the absence of interactions ($\Phi_N = 0$),

$$n_s^{\text{id}} = \frac{1}{V^s} \left\langle \frac{N!}{(N-s)!} \right\rangle \approx n^s , \tag{4.13}$$

where the thermodynamic limit has been applied (assuming $s \ll \langle N \rangle$) in the second
step. Again, the angular brackets involving functions of N are not needed in the
canonical ensemble.

In general, the existence of interactions ($\Phi_N \neq 0$) creates spatial correlations and,
consequently, $n_s \neq n^s$ (except, of course, in the trivial case $s = 1$). This suggest the
introduction of the *correlation* functions g_s by

$$n_s(\mathbf{r}^s) = n^s g_s(\mathbf{r}^s) . \tag{4.14}$$

Thus, according to (4.7a),

$$g_s(\mathbf{r}^s) = \frac{V^{-(N-s)}}{\mathcal{Q}_N} \int d\mathbf{r}_{s+1} \cdots \int d\mathbf{r}_N e^{-\beta\Phi_N(\mathbf{r}^N)} \tag{4.15}$$

in the canonical ensemble.

An interesting normalization relation holds in the grand canonical ensemble.
Inserting (4.14) into (4.4) we get

$$V^{-s} \int d\mathbf{r}^s g_s(\mathbf{r}^s) = \frac{1}{\langle N \rangle^s} \left\langle \frac{N!}{(N-s)!} \right\rangle . \tag{4.16}$$

In analogy with (3.18), we can define the *cluster* correlation functions $h_s(\mathbf{r}^s)$
by [2]

$$g_1(1) = h_1(1) = 1 , \tag{4.17a}$$

$$g_2(1,2) = h_1(1)h_1(2) + h_2(1,2) , \tag{4.17b}$$

$$g_3(1,2,3) = h_1(1)h_1(2)h_1(3) + \{3\}h_1(1)h_2(2,3) + h_3(1,2,3) , \tag{4.17c}$$

$$g_4(1,2,3,4) = h_1(1)h_1(2)h_1(3)h_1(4) + \{6\}h_1(1)h_1(2)h_2(3,4)$$
$$+ \{3\}h_2(1,2)h_2(3,4) + \{4\}h_1(1)h_3(2,3,4) + h_4(1,2,3,4) , \tag{4.17d}$$

and so on, where we recall that each numerical factor enclosed by curly braces represents the number of terms equivalent (except for particle labeling) to the indicated canonical term. More in general, the functions $\{h_s\}$ and $\{g_s\}$ are *formally* related by

$$1 + \sum_{s=1}^{\infty} \frac{\epsilon^s}{s!} g_s \leftrightarrow \exp\left[\sum_{s=1}^{\infty} \frac{\epsilon^s}{s!} h_s\right], \tag{4.18}$$

where ϵ is a formal expansion parameter. Expressed in terms of the cluster correlation functions, the hierarchy (4.12) becomes [1]

$$\left(\chi_T n \frac{\partial}{\partial n} - s\right) n^s h_s(\mathbf{r}^s) = n^{s+1} \int d\mathbf{r}_{s+1} \, h_{s+1}(\mathbf{r}^{s+1}) . \tag{4.19}$$

4.3 Radial Distribution Function

If the potential energy function $\Phi_N(\mathbf{r}^N)$ is assumed to be pairwise additive [see (3.1)], the basic correlation function is g_2. Now, taking into account the translational invariance property (4.8), one has $g_2(\mathbf{r}_1, \mathbf{r}_2) = g(\mathbf{r}_1 - \mathbf{r}_2)$. Moreover, a fluid is rotationally invariant, so that (assuming central forces), $g(\mathbf{r}_1 - \mathbf{r}_2) = g(r_{12})$, where $r_{12} \equiv |\mathbf{r}_1 - \mathbf{r}_2|$ is the distance between the points $\mathbf{r}_1$ and $\mathbf{r}_2$. In such a case, the function $g(r)$ is called *radial distribution function* (RDF) and will play a very important role henceforth. According to (4.15),

$$g(r_{12}) = \frac{V^{-(N-2)}}{\mathscr{Q}_N} \int d\mathbf{r}_3 \cdots \int d\mathbf{r}_N \, e^{-\beta \Phi_N(\mathbf{r}^N)} \tag{4.20}$$

in the canonical ensemble. Also, the grand-canonical normalization condition (4.16) becomes

$$V^{-1} \int d\mathbf{r} \, g(r) = \frac{\langle N(N-1) \rangle}{\langle N \rangle^2} = \frac{\langle N^2 \rangle}{\langle N \rangle^2} - \frac{1}{\langle N \rangle} . \tag{4.21}$$

In the thermodynamic limit ($\langle N \rangle \to \infty$ and $V \to \infty$ with $n = $ const), we know that $\langle N^2 \rangle / \langle N \rangle^2 \to 1$ [see (2.40)] (except near the critical point, where κ_T diverges). This implies that $V^{-1} \int d\mathbf{r} \, g(r) \approx 1$, meaning that $g(r) \approx 1$ for *macroscopic* distances r, which are those dominating the value of the integral. In other words, $\int d\mathbf{r} \, [g(r) - 1] \ll V$.

Apart from the formal definition provided by (4.14) and (4.20), it is important to have a more intuitive physical interpretation of $g(r)$. Two simple equivalent

Fig. 4.2 Schematic view of
how $g(r)$ is determined. The
central (*red*) particle is the
reference one, while the dark
(*blue*) peripheral particles are
those whose centers are at a
distance between r and
$r + dr$. The average number
of those particles, divided by
$n4\pi r^2 dr$ (in three
dimensions) or by $n2\pi r dr$ (in
two dimensions), gives $g(r)$
(Image from Wikimedia
Commons, http://commons.
wikimedia.org/wiki/File:Rdf_
schematic.jpg)

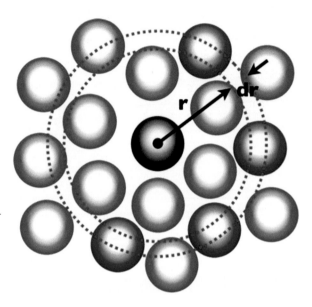

interpretations are:

- $g(r)$ is the probability density of finding a particle at a distance r away
  from a given reference particle, *relative* to the probability density for an
  ideal gas.
- If a given reference particle is taken to be at the origin, then the *local*
  average density at a distance r from that particle is $ng(r)$.

Figure 4.2 illustrates the meaning of $g(r)$ and how this quantity can be measured
in MC or MD computer simulations [3, 4].

The typical shape of the RDF for (three-dimensional) fluids of particles interacting via the LJ and HS potentials (see Table 3.1) is displayed in Figs. 4.3 and 4.4,
respectively. As we see in Fig. 4.3, $g(r)$ is practically zero in the region $0 \leq r \lesssim \sigma$
(due to the strongly repulsive force exerted by the reference particle at those
distances), presents a peak at $r \approx \sigma$, oscillates thereafter, and eventually tends to
unity for longer distances. Those features are enhanced as the density increases.
A similar behavior is observed in Fig. 4.4, except that now $g(r)$ presents a jump
discontinuity at $r = \sigma$ as a consequence of the absolute impenetrability of two
particles separated a distance smaller that σ. In general, we observe that the RDF
captures an interesting structure exhibited by liquids.

It is useful to define some functions related to the RDF $g(r)$. The first one is
simply the *pair* cluster correlation function $h_2(\mathbf{r}_1, \mathbf{r}_2) = h(r_{12})$ [see (4.17b)], i.e.,

$$\boxed{h(r) = g(r) - 1,}$$ (4.22)

Fig. 4.3 Plot of $g(r)$ for a LJ fluid simulated with MD at a reduced temperature $T^* \equiv k_B T/\varepsilon = 2.0$ and reduced number densities $n^* \equiv n\sigma^3 = 0.2, 0.4, 0.6,$ and 0.7 [5]

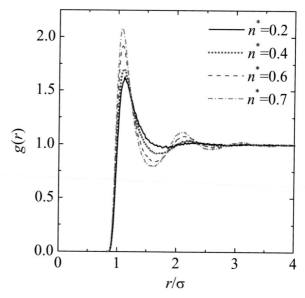

Fig. 4.4 Plot of $g(r)$ for a HS fluid simulated with MD at reduced number densities $n^* \equiv n\sigma^3 = 0.2, 0.4, 0.6,$ and 0.9 [6]

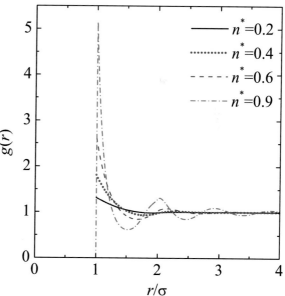

usually denoted as the *total correlation function*. Its Fourier transform

$$\tilde{h}(k) \equiv \int d\mathbf{r} \, e^{-i\mathbf{k}\cdot\mathbf{r}} h(r) \tag{4.23}$$

is directly connected to the (static) *structure factor*:

$$\boxed{\widetilde{S}(k) = 1 + n\tilde{h}(k) \,.} \tag{4.24}$$

Fig. 4.5 Cavity function in the overlapping region $r < \sigma$ for a HS fluid at three different densities, as obtained from MC simulations [7]

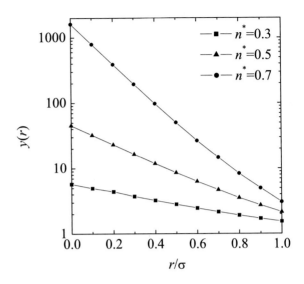

The structure factor is a very important quantity because it is experimentally accessible by elastic scattering of radiation (x-rays or neutrons) by the fluid [8, 9]. Thus, while $g(r)$ can be measured directly in simulations (either MC or MD) [3, 4, 10], it can be obtained indirectly in experiments from a numerical inverse Fourier transform of $\widetilde{S}(k) - 1$.

Another important quantity closely related to the RDF is the so-called *cavity function*

$$y(r) \equiv g(r)e^{\beta\phi(r)} .$$
(4.25)

This is a much more regular function than the RDF $g(r)$. As we will see in Sect. 6.5, it is continuous even if the interaction potential is discontinuous or diverges. In the HS case, for instance, while $g(r) = 0$ if $r < \sigma$ (see Fig. 4.4), $y(r)$ is well defined in that region, as illustrated by Fig. 4.5.

4.4 Ornstein–Zernike Relation and the Direct Correlation Function

The total correlation function (4.22) owes its name to the fact that it measures the degree of spatial correlation between two particles separated a distance r due not only to their *direct* interaction but also *indirectly* through other intermediate or "messenger" particles. In fact, the range of $h(r)$ is usually much larger than that of the potential $\phi(r)$ itself, as illustrated by Figs. 4.3 and 4.4. In fluids with a vapor–liquid phase transition, $h(r)$ decays algebraically at the *critical point*, so

Fig. 4.6 Leonard Salomon
Ornstein (1880–1941)
(Photograph reproduced with
permission from AIP Emilio
Segrè Visual Archives, W. F.
Meggers Collection, https://
photos.aip.org/history-
programs/niels-bohr-library/
photos/ornstein-leonard-b1)

Fig. 4.7 Frits Zernike
(1888–1966)
(Photograph from Wikimedia
Commons, http://en.
wikipedia.org/wiki/File:
Zernike.jpg)

that the integral $\tilde{h}(0) = \int d\mathbf{r}\, h(r)$ diverges and so does the isothermal compressibility κ_T [see (4.29) in Sect. 4.5], a phenomenon known as *critical opalescence* [8, 9].

It is then important to disentangle from $h(r)$ its direct and indirect contributions. This aim was addressed in 1914 by the Dutch physicists L.S. Ornstein (see Fig. 4.6) and F. Zernike (see Fig. 4.7). They defined the *direct* correlation function (DCF) $c(r)$ by the integral relation

$$h(r_{12}) = c(r_{12}) + n \int d\mathbf{r}_3\, c(r_{13}) h(r_{32}) \,. \tag{4.26}$$

The idea behind the Ornstein–Zernike (OZ) relation (4.26) is sketched in Fig. 4.8: the total correlation function h_{12} between particles 1 and 2 can be decomposed into

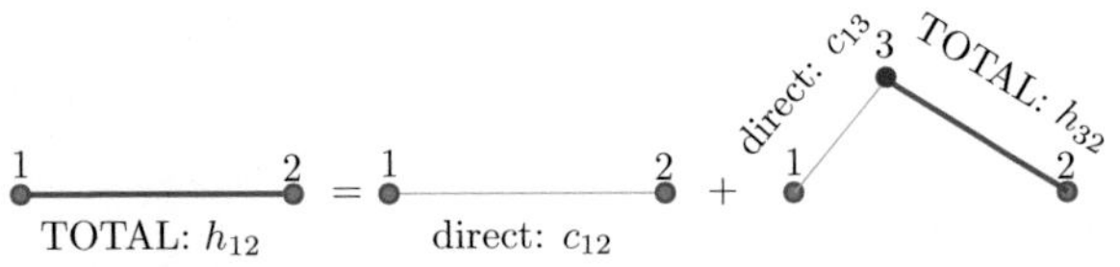

Fig. 4.8 Sketch of the meaning of the OZ relation (4.26)

the DCF c_{12} plus the indirect part, the latter being mediated by a messenger particle 3 that is directly correlated to 1 and totally correlated to 2.

Thanks to the convolution structure of the indirect part, the OZ relation (4.26) becomes $\tilde{h}(k) = \tilde{c}(k) + n\tilde{c}(k)\tilde{h}(k)$ in Fourier space or, equivalently,

$$\tilde{h}(k) = \frac{\tilde{c}(k)}{1 - n\tilde{c}(k)} \,, \quad \tilde{c}(k) = \frac{\tilde{h}(k)}{1 + n\tilde{h}(k)} \,. \tag{4.27}$$

Thus, the relationship (4.24) can also be written as

$$\widetilde{S}(k) = \frac{1}{1 - n\tilde{c}(k)} \,. \tag{4.28}$$

4.5 Thermodynamics from the Radial Distribution Function

As shown by (2.12), (2.21), (2.35), and (2.50) (see also Table 2.1), the knowledge of any of the ensemble partition functions allows one to obtain the full thermodynamic information about the system. But now imagine that instead of the partition function (for instance, the canonical one), we are given (from experimental measurements, computer simulations, or a certain theory) the RDF $g(r)$. Can we have access to thermodynamics directly from $g(r)$? As we will see in this section, the answer is affirmative in the case of pairwise interactions.

4.5.1 Compressibility Route

The most straightforward route to thermodynamics from $g(r)$ is provided by choosing the grand canonical ensemble and simply combining (2.40) and (4.21) to obtain

$$\boxed{\chi_T \equiv n k_B T \kappa_T = k_B T \left(\frac{\partial n}{\partial p}\right)_T = 1 + n \int d\mathbf{r}\, h(r) = \widetilde{S}(0) \,,} \tag{4.29}$$

where in the last step use has been made of (4.24). Therefore, the zero wavenumber limit of the structure factor is directly related to the isothermal compressibility. Equation (4.29) is usually known as the *compressibility* EoS or the *compressibility route* to thermodynamics.

Using (4.28), the compressibility route to the EoS (4.29) can be rewritten as

$$\chi_T = \frac{1}{1 - n\tilde{c}(0)} \, . \tag{4.30}$$

Therefore, even if $\tilde{h}(0) \rightarrow \infty$ (at the critical point), $\tilde{c}(0) \rightarrow n^{-1} =$ finite, thus showing that $c(r)$ is much shorter ranged than $h(r)$, as expected.

It must be noticed that (4.29) applies regardless of the specific form of the potential energy function $\Phi_N(\mathbf{r}^N)$ (whether pairwise additive or not) and can be recovered as the first equation ($s = 1$) of the hierarchy (4.19).

4.5.2   Energy Route

From now on we assume that the interaction is *pairwise additive*, as given by (3.1). This implies that Φ_N is a dynamical variable of the form (4.5) with $s = 2$. As a consequence, we can apply the property (4.6) to the average potential energy:

$$\langle E \rangle^{\text{ex}} = \langle \Phi_N(\mathbf{r}^N) \rangle = \frac{1}{2} \int d\mathbf{r}_1 \int d\mathbf{r}_2 \, n_2(\mathbf{r}_1, \mathbf{r}_2) \phi(r_{12}) \, . \tag{4.31}$$

Adding the ideal-gas term $\langle E \rangle^{\text{id}}$ (see Table 2.2) and taking into account (4.14), we finally obtain

$$\langle E \rangle = N \left[\frac{d}{2} k_B T + \frac{n}{2} \int d\mathbf{r} \, \phi(r) g(r) \right] , \tag{4.32}$$

where we have used the general property $\int d\mathbf{r}_1 \int d\mathbf{r}_2 \, \mathscr{F}(r_{12}) = V \int d\mathbf{r} \, \mathscr{F}(r)$, $\mathscr{F}(r)$ being an arbitrary function.

Equation (4.32) defines the *energy route* to thermodynamics. It can be equivalently written in terms of the cavity function (4.25) as

$$\langle E \rangle = N \left[\frac{d}{2} k_B T + \frac{n}{2} \int d\mathbf{r} \, \phi(r) e^{-\beta \phi(r)} y(r) \right] . \tag{4.33}$$

4.5.3 Virial Route

Now we consider the pressure, which is the quantity more directly related to the EoS. In the canonical ensemble, the excess pressure is proportional to $\partial \ln \mathscr{Q}_N / \partial V$ [see (2.65c)] and thus it is not the average of a dynamical variable of type (4.6). To make things worse, the volume V appears in the configuration integral [see (2.64)] both explicitly and *implicitly* through the integration limits. Let us make this more evident by writing

$$\mathscr{Q}_N(V) = V^{-N} \int_{V^N} d\mathbf{r}^N \, e^{-\beta \Phi_N(\mathbf{r}^N)} \,. \tag{4.34}$$

To get rid of this difficulty, we imagine now that the system is a sphere of volume V and the origin of coordinates is chosen at the center of the sphere. If the whole system is blown up by a factor λ [8], the volume changes from V to $\lambda^d V$ and the configuration integral changes from $\mathscr{Q}_N(V)$ to $\mathscr{Q}_N(\lambda^d V)$ with

$$\mathscr{Q}_N(\lambda^d V) = (\lambda^d V)^{-N} \int_{(\lambda^d V)^N} d\mathbf{r}^N \, e^{-\beta \Phi_N(\mathbf{r}^N)} = V^{-N} \int_{V^N} d\mathbf{r}'^N \, e^{-\beta \Phi_N(\lambda^N \mathbf{r}'^N)} \,, \tag{4.35}$$

where in the last step the change $\mathbf{r}_i \to \mathbf{r}_i' = \mathbf{r}_i/\lambda$ has been performed. We see that $\mathscr{Q}_N(\lambda^d V)$ depends on λ explicitly through the argument of the interaction potential. Next, taking into account the identity

$$\frac{\partial \ln \mathscr{Q}_N(\lambda^d V)}{\partial V} = \frac{\lambda}{Vd} \frac{\partial \ln \mathscr{Q}_N(\lambda^d V)}{\partial \lambda} \,, \tag{4.36}$$

we can write

$$\frac{\partial \ln \mathscr{Q}_N(V)}{\partial V} = \frac{1}{Vd} \left. \frac{\partial \ln \mathscr{Q}_N(\lambda^d V)}{\partial \lambda} \right|_{\lambda=1} \,, \tag{4.37}$$

so that

$$
\begin{aligned}
\left. \frac{\partial \ln \mathscr{Q}_N(\lambda^d V)}{\partial \lambda} \right|_{\lambda=1} &= -\beta \left\langle \left. \frac{\partial \Phi_N(\lambda^N \mathbf{r}^N)}{\partial \lambda} \right|_{\lambda=1} \right\rangle \\
&= -\frac{\beta}{2} \int d\mathbf{r}_1 \int d\mathbf{r}_2 \, n_2(\mathbf{r}_1, \mathbf{r}_2) \left. \frac{\partial \phi(\lambda r_{12})}{\partial \lambda} \right|_{\lambda=1} \\
&= -\frac{\beta}{2} n^2 V \int d\mathbf{r} \, g(r) \left. \frac{\partial \phi(\lambda r)}{\partial \lambda} \right|_{\lambda=1} \,.
\end{aligned}
\tag{4.38}
$$

In the second equality use has been made of (4.6) with $s = 2$. Finally, a mathematical property similar to (4.37) is

$$\left.\frac{\partial \phi(\lambda r)}{\partial \lambda}\right|_{\lambda=1} = r\frac{d\phi(r)}{dr} .$$ (4.39)

Inserting (4.39) into (4.38), and using (4.37), we obtain the sought result:

$$\boxed{Z \equiv \frac{p}{nk_BT} = 1 - \frac{n\beta}{2d}\int dr\, r\frac{d\phi(r)}{dr}g(r) .}$$ (4.40)

This is known as the pressure route or *virial route* to the EoS, where we recall that Z is the *compressibility factor* [see (1.35)]. Expressed in terms of the cavity function (4.25), the virial route becomes

$$Z \equiv \frac{p}{nk_BT} = 1 + \frac{n}{2d}\int dr\, y(r) r\frac{\partial f(r)}{\partial r} ,$$ (4.41)

where we recall that the Mayer function is defined in (3.3).

4.5.4 Chemical-Potential Route

A look at (2.65b) and (2.65c) shows that we have already succeeded in expressing the first two derivatives of $\ln \mathcal{Q}_N$ in terms of integrals involving the RDF. The third derivative yields the chemical potential and is much more delicate. First, noting that N is actually a discrete variable, we can rewrite (2.65d) as

$$\beta\mu^{\text{ex}} = -\frac{\partial \ln \mathcal{Q}_N}{\partial N} \to -\ln\frac{\mathcal{Q}_{N+1}(\beta, V)}{\mathcal{Q}_N(\beta, V)} .$$ (4.42)

Thus, the (excess) chemical potential is related to the response of the system to the addition of one more particle without changing either temperature or volume.

The N-body potential energy is expressed by (3.1). Now we add an *extra* particle (labeled as $i = 0$), so that the $(N + 1)$-body potential energy becomes

$$\Phi_{N+1}(\mathbf{r}^{N+1}) = \sum_{i=1}^{N-1}\sum_{j=i+1}^{N}\phi(r_{ij}) + \sum_{j=1}^{N}\phi(r_{0j}) .$$ (4.43)

Making use of (2.64), (2.66), and the translational invariance property (2.62), it turns out that (4.42) can be rewritten as

$$\beta\mu^{\text{ex}} = -\ln\left\langle e^{-\beta\sum_{j=1}^{N}\phi(r_{0j})}\right\rangle .$$ (4.44)

Fig. 4.9 Benjamin Widom
(b. 1927)
(Photograph reproduced with
permission from Cornell
Chronicle, http://www.news.
cornell.edu/stories/2005/11/
molecular-physics-journal-
pays-tribute-benjamin-
widom)

This represents the insertion method introduced by Widom (see Fig. 4.9) in 1963
[11].

Our goal now is to derive an expression for the chemical potential in terms of
the RDF. The trick consists in inserting the extra particle (the "solute") little by
little through a *charging process* [8, 12–18]. We do so by introducing a *coupling
parameter* ξ such that its value $0 \leq \xi \leq 1$ controls the strength of the interaction of
particle $i = 0$ to the rest of particles (the "solvent"):

$$\phi^{(\xi)}(r) = \begin{cases} 0, & \xi = 0, \\ \phi(r), & \xi = 1. \end{cases} \tag{4.45}$$

The associated total potential energy and configuration integral are

$$\Phi_{N+1}^{(\xi)}(\mathbf{r}^{N+1}) = \Phi_N(\mathbf{r}^N) + \sum_{j=1}^{N} \phi^{(\xi)}(r_{0j}), \tag{4.46a}$$

$$\mathcal{Q}_{N+1}^{(\xi)}(\beta, V) = V^{-(N+1)} \int d\mathbf{r}^{N+1} e^{-\beta \Phi_{N+1}^{(\xi)}(\mathbf{r}^{N+1})}. \tag{4.46b}$$

Thus, assuming that $\mathcal{Q}_{N+1}^{(\xi)}$ is a smooth function of ξ, (4.42) becomes

$$\beta\mu^{\text{ex}} = -\int_0^1 d\xi \, \frac{\partial \ln \mathcal{Q}_{N+1}^{(\xi)}(\beta, V)}{\partial \xi}. \tag{4.47}$$

Since the dependence of $\mathcal{Q}_{N+1}^{(\xi)}$ on ξ takes place through the extra summation in
(4.46a) and all the solvent particles are assumed to be identical, then

$$\frac{\partial \ln \mathcal{Q}_{N+1}^{(\xi)}}{\partial \xi} = -\frac{n\beta V^{-N}}{\mathcal{Q}_{N+1}^{(\xi)}} \int d\mathbf{r}^{N+1} e^{-\beta \Phi_{N+1}^{(\xi)}(\mathbf{r}^{N+1})} \frac{\partial \phi^{(\xi)}(r_{01})}{\partial \xi}. \tag{4.48}$$

Now we realize that, similarly to (4.20), the solute–solvent RDF is defined as

$$g_{01}^{(\xi)}(r_{01}) = \frac{V^{-(N-1)}}{\mathscr{Q}_{N+1}^{(\xi)}} \int d\mathbf{r}_2 \cdots \int d\mathbf{r}_N \, e^{-\beta \Phi_{N+1}^{(\xi)}(\mathbf{r}^{N+1})} . \tag{4.49}$$

This allows us to rewrite (4.48) in the form

$$\frac{\partial \ln \mathscr{Q}_{N+1}^{(\xi)}}{\partial \xi} = -\frac{n\beta}{V} \int d\mathbf{r}_0 \int d\mathbf{r}_1 \, \frac{\partial \phi^{(\xi)}(r_{01})}{\partial \xi} g_{01}^{(\xi)}(r_{01}) . \tag{4.50}$$

Finally, after taking into account that $\mu^{\mathrm{id}} = k_B T \ln(n\Lambda^d)$ (see Table 2.2), (4.47) yields

$$\boxed{\mu = k_B T \ln(n\Lambda^d) + n \int_0^1 d\xi \int d\mathbf{r} \, \frac{\partial \phi^{(\xi)}(r)}{\partial \xi} g_{01}^{(\xi)}(r) ,} \tag{4.51}$$

or, equivalently,

$$\beta\mu = \ln(n\Lambda^d) - n \int_0^1 d\xi \int d\mathbf{r} \, \frac{\partial e^{-\beta\phi^{(\xi)}(r)}}{\partial \xi} y_{01}^{(\xi)}(r) , \tag{4.52}$$

where the cavity function is $y_{01}^{(\xi)}(r) = g_{01}^{(\xi)}(r)e^{\beta\phi^{(\xi)}(r)}$.

In contrast to the other three conventional routes [see (4.29), (4.32), and (4.40)], the *chemical-potential route* (4.51) requires the knowledge of the solute–solvent correlation functions for all the values $0 \le \xi \le 1$ of the coupling parameter ξ.

4.5.5   A Master Route: The Free Energy

The formulas relating the isothermal compressibility, the internal energy, the pressure, and the chemical potential to the RDF are summarized in Table 4.1. The thermodynamic relation between the quantities in the second column and the free energy are given in the third column [see (1.36)].

In principle, the excess free energy can be obtained by integration from any of those routes. The corresponding expressions are displayed in the first four rows of Table 4.2, where it has been taken into account that the fluid behaves as an ideal gas in the limits of zero density and/or infinite temperature. The latter limit assumes that the interaction potential is finite at nonzero distances, what includes the LJ potential but discards the HS, SW, SS, and SHS potentials of Table 3.1. In the latter cases, one needs to add the HS excess free energy per particle to the energy-route expression of Table 4.2.

The question we now may ask is, can we derive a more direct relationship between the free energy and the RDF? To address this question we proceed in a

Table 4.1 Summary of the main thermodynamic routes in a one-component system

Route	Quantity	Thermodynamic relation	Expression
Compressibility	χ_T	$\left(\dfrac{\partial}{\partial n} n^2 \dfrac{\partial(\beta a)}{\partial n} \right)_\beta^{-1}$	$1 + n \displaystyle\int \mathrm{d}\mathbf{r}\, h(r)$
Energy	u^{ex}	$\left(\dfrac{\partial(\beta a^{\mathrm{ex}})}{\partial \beta} \right)_n$	$\dfrac{n}{2} \displaystyle\int \mathrm{d}\mathbf{r}\, \phi(r) g(r)$
Virial	$Z - 1$	$n \left(\dfrac{\partial(\beta a^{\mathrm{ex}})}{\partial n} \right)_\beta$	$-\dfrac{n\beta}{2d} \displaystyle\int \mathrm{d}\mathbf{r}\, r \dfrac{\mathrm{d}\phi(r)}{\mathrm{d}r} g(r)$
Chemical potential	μ^{ex}	$\left(\dfrac{\partial(n a^{\mathrm{ex}})}{\partial n} \right)_\beta$	$n \displaystyle\int_0^1 \mathrm{d}\xi \int \mathrm{d}\mathbf{r}\, \dfrac{\partial \phi^{(\xi)}(r)}{\partial \xi} g_{01}^{(\xi)}(r)$

Here, $a^{\mathrm{ex}} = F^{\mathrm{ex}}/N$ is the excess Helmholtz free energy per particle and $u^{\mathrm{ex}} = \langle E \rangle^{\mathrm{ex}}/N$ is the excess internal energy per particle

Table 4.2 Expressions of the excess Helmholtz free energy per particle (in units of the thermal energy) as a function of n and β from different thermodynamic routes

Route	Expression of βa^{ex}
Compressibility	$-\displaystyle\int_0^n \dfrac{\mathrm{d}n'}{n'^2} \int_0^{n'} \mathrm{d}n'' \dfrac{n'' \int \mathrm{d}\mathbf{r}\, h(r; n'', \beta)}{1 + n'' \int \mathrm{d}\mathbf{r}\, h(r; n'', \beta)}$
Energy	$\dfrac{n}{2} \displaystyle\int_0^\beta \mathrm{d}\beta' \int \mathrm{d}\mathbf{r}\, \phi(r) g(r; n, \beta')$
Virial	$-\dfrac{\beta}{2d} \displaystyle\int_0^n \mathrm{d}n' \int \mathrm{d}\mathbf{r}\, r \dfrac{\mathrm{d}\phi(r)}{\mathrm{d}r} g(r; n', \beta)$
Chemical potential	$\dfrac{\beta}{n} \displaystyle\int_0^n \mathrm{d}n'\, n' \int_0^1 \mathrm{d}\xi \int \mathrm{d}\mathbf{r}\, \dfrac{\partial \phi^{(\xi)}(r)}{\partial \xi} g_{01}^{(\xi)}(r; n', \beta)$
Free energy	$\dfrac{n\beta}{2} \displaystyle\int_0^1 \mathrm{d}\xi \int \mathrm{d}\mathbf{r}\, \dfrac{\partial \phi^{(\xi)}(r)}{\partial \xi} g^{(\xi)}(r; n, \beta)$

way similar to the one followed for the derivation of the chemical-potential route, except that now the charging process affects *all* the particles of the system and not only a "solute" particle. Thus, the N-body potential energy function is

$$\Phi_N^{(\xi)}(\mathbf{r}^N) = \sum_{1 \leq i < j \leq N} \phi^{(\xi)}(r_{ij}) , \qquad (4.53)$$

where $\phi^{(\xi)}(r)$ is still defined by (4.45). The associated configuration integral and RDF are

$$\mathscr{Q}_N^{(\xi)}(\beta, V) = V^{-N} \int \mathrm{d}\mathbf{r}^N\, \mathrm{e}^{-\beta \Phi_N^{(\xi)}(\mathbf{r}^N)} , \qquad (4.54a)$$

$$g^{(\xi)}(r_{12}) = \frac{V^{-(N-2)}}{\mathscr{Q}_N^{(\xi)}} \int d\mathbf{r}_3 \cdots \int d\mathbf{r}_N \, e^{-\beta \Phi_N^{(\xi)}(\mathbf{r}^N)} . \qquad (4.54b)$$

Then, the free energy of the true system can be obtained as

$$F^{\text{ex}} = -k_B T \ln \mathscr{Q}_N = -k_B T \int_0^1 d\xi \, \frac{\partial \ln \mathscr{Q}_N^{(\xi)}}{\partial \xi} . \qquad (4.55)$$

Finally, by using (4.53) and (4.54) one easily obtains

$$a^{\text{ex}} = \frac{F^{\text{ex}}}{N} = \frac{n}{2} \int_0^1 d\xi \int d\mathbf{r} \, \frac{\partial \phi^{(\xi)}(r)}{\partial \xi} g^{(\xi)}(r) . \qquad (4.56)$$

In terms of the cavity function,

$$\beta a^{\text{ex}} = -\frac{n}{2} \int_0^1 d\xi \int d\mathbf{r} \, \frac{\partial e^{-\beta \phi^{(\xi)}(r)}}{\partial \xi} y^{(\xi)}(r) . \qquad (4.57)$$

Equation (4.56) or, equivalently, (4.57) can be considered as an alternative "free-energy" route and is included in Table 4.2 for comparison with the other four routes. In all the cases we observe that the knowledge of the RDF at the state point (n, β) of interest (and with the full interaction potential) is not enough to obtain the excess free energy at that very point. This situation is sketched in Fig. 4.10. In the virial and compressibility routes one needs to know the RDF at all smaller densities ($n' < n$), while the knowledge of the RDF at higher temperatures ($\beta' < \beta$) is required in the energy route. In the free-energy route the density and temperature do not change

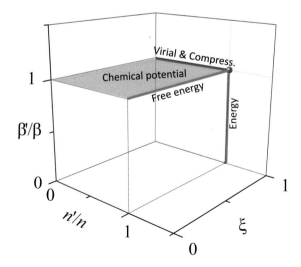

Fig. 4.10 Sketch of the points in the (n', β', ξ) space where the RDF needs to be known in order to evaluate the free energy according to different thermodynamic routes. The *bullet* at the point $(n'/n, \beta'/\beta, \xi) = (1, 1, 1)$ represents the state of interest

but, in contrast, the pertinent piece of information is the RDF for intermediate systems where the particle interactions are progressively switched on, the ξ-protocol followed to go from the ideal gas to the full interacting system being arbitrary. This protocol-dependent charging process (but only for one solute particle) is also present in the chemical-potential route, where, in addition, systems with $n' < n$ need to be considered.

The free-energy route (4.56) can be seen as a *master route* in the sense that it includes the energy and virial routes as particular cases related to two specific choices for the protocol $\phi^{(\xi)}(r)$.

As a first specific choice, let us take

$$\phi^{(\xi)}(r) = \xi\phi(r) \, , \tag{4.58}$$

where it is assumed that $\phi(r) = $ finite for all $r > 0$. This protocol is equivalent to an energy rescaling. As a consequence, on physical grounds,

$$g^{(\xi)}(r; n, \beta) = g(r; n, \beta\xi) \, . \tag{4.59}$$

Consequently, the free-energy route (4.56) reduces to

$$a^{\text{ex}} = \frac{n}{2} \int_0^1 d\xi \int d\mathbf{r}\, \phi(r) g(r; n, \beta\xi) \, . \tag{4.60}$$

Comparison with the second row of Table 4.2 shows that (4.60) coincides indeed with the energy-route expression.

The second choice consists in the protocol

$$\phi^{(\xi)}(r) = \phi(r/\xi) \, , \tag{4.61}$$

where it is assumed that $\lim_{r\to\infty} \phi(r) = 0$. This is equivalent to a distance rescaling or blowing up of the system. Therefore, one must have

$$g^{(\xi)}(r; n, \beta) = g(r/\xi; n\xi^d, \beta) \, . \tag{4.62}$$

Moreover,

$$\frac{\partial \phi^{(\xi)}(r)}{\partial \xi} = -\frac{r}{\xi^2} \frac{d\phi(r/\xi)}{d(r/\xi)} \, . \tag{4.63}$$

Thus, (4.56) becomes

$$a^{\text{ex}} = -\frac{n}{2} \int_0^1 d\xi \int d\mathbf{r}\, \frac{r}{\xi^2} \frac{d\phi(r/\xi)}{d(r/\xi)} g(r/\xi; n\xi^d, \beta)$$

$$= -\frac{n}{2} \int_0^1 d\xi \, \xi^{d-1} \int d\mathbf{r} \, r \frac{d\phi(r)}{dr} g(r; n\xi^d, \beta) \, . \tag{4.64}$$

Finally, a change of variables $\xi \to n\xi^d$ makes (4.64) coincide with the virial-route expression in Table 4.2.

4.6 Extension to Mixtures

As mentioned before [see, for instance, (1.2), (1.3), and (3.6)], the main quantities in a mixture or multicomponent system are

- Number of particles of species α: N_α .
- Total number of particles: $N = \sum_\alpha N_\alpha$.
- Mole fraction of species α: $x_\alpha = \dfrac{N_\alpha}{N}$, $\sum_\alpha x_\alpha = 1$.
- Number density of species α: $n_\alpha = \dfrac{N_\alpha}{V}$.
- Total number density: $n = \dfrac{N}{V} = \sum_\alpha n_\alpha$.

Assuming again pairwise additivity and denoting by $\phi_{\alpha\gamma}(r)$ the interaction potential between a particle of species α and a particle of species γ [see (3.7) for the HS case], the total potential energy can be written as

$$\Phi_{\{N_\nu\}}(\mathbf{r}^N) = \sum_{i=1}^{N-1} \sum_{j=i+1}^{N} \phi_{\vartheta_i \vartheta_j}(r_{ij}) = \frac{1}{2} \sum_{i \neq j} \phi_{\vartheta_i \vartheta_j}(r_{ij}) \, , \tag{4.65}$$

where the label ϑ_i denotes the species to which particle i belongs. The corresponding configuration integral is

$$\mathcal{Q}_{\{N_\nu\}}(\beta, V) = V^{-N} \int d\mathbf{r}^N \, e^{-\beta \Phi_{\{N_\nu\}}(\mathbf{r}^N)} \, . \tag{4.66}$$

In analogy with the one-component case, the *pair* configurational distribution function $n_{\alpha\gamma}$ is defined as

$$n_{\alpha\gamma}(\mathbf{r}_\alpha, \mathbf{r}_\gamma) = \left\langle \sum_{i \neq j} \delta_{\vartheta_i, \alpha} \delta_{\vartheta_j, \gamma} \delta(\mathbf{r}_i - \mathbf{r}_\alpha) \delta(\mathbf{r}_j - \mathbf{r}_\gamma) \right\rangle \, , \tag{4.67}$$

where the Kronecker and Dirac deltas select those particles of species α and γ sitting at $\mathbf{r}_\alpha$ and $\mathbf{r}_\gamma$, respectively. Its normalization condition is then

$$\int d\mathbf{r}_\alpha \int d\mathbf{r}_\gamma \, n_{\alpha\gamma}(\mathbf{r}_\alpha, \mathbf{r}_\gamma) = \langle N_\alpha (N_\gamma - \delta_{\alpha\gamma}) \rangle . \tag{4.68}$$

In particular, the canonical-ensemble expression is

$$n_{\alpha\gamma}(\mathbf{r}_\alpha, \mathbf{r}_\gamma) = \frac{N_\alpha (N_\gamma - \delta_{\alpha\gamma})}{V^N \mathcal{Q}_{\{N_\nu\}}} \int d\mathbf{r}^N \, e^{-\beta \Phi_{\{N_\nu\}}(\mathbf{r}^N)} \delta(\mathbf{r}_1 - \mathbf{r}_\alpha)\delta(\mathbf{r}_2 - \mathbf{r}_\gamma) , \tag{4.69}$$

where, without loss of generality, particles $i = 1$ and $j = 2$ have been assumed to belong to species α and γ, respectively. The RDF for the pair $\alpha\gamma$, $g_{\alpha\gamma}(r)$, is defined by

$$n_{\alpha\gamma}(\mathbf{r}_\alpha, \mathbf{r}_\gamma) = n_\alpha n_\gamma g_{\alpha\gamma}(r_{\alpha\gamma}) . \tag{4.70}$$

Inserting (4.70) into (4.69), we obtain

$$g_{\alpha\gamma}(r_{12}) = \frac{V^{-(N-2)}}{\mathcal{Q}_{\{N_\nu\}}} \int d\mathbf{r}_3 \cdots \int d\mathbf{r}_N \, e^{-\beta \Phi_{\{N_\nu\}}(\mathbf{r}^N)} . \tag{4.71}$$

In the grand canonical ensemble, the normalization condition (4.68) implies

$$V^{-1} \int d\mathbf{r} \, g_{\alpha\gamma}(r) = \frac{\langle N_\alpha N_\gamma \rangle}{\langle N_\alpha \rangle \langle N_\gamma \rangle} - \frac{\delta_{\alpha\gamma}}{\langle N_\alpha \rangle} . \tag{4.72}$$

Equations (4.69)–(4.72) are the multicomponent equivalents of (4.7a), (4.14), (4.20), and (4.21), respectively.

The physical meaning of $g_{\alpha\gamma}(r)$ is still captured by Fig. 4.2, except that now the central particle must belong to species α and, out of all the spheres at a distance between r and $r + dr$, only those belonging to species γ are considered. Since $g_{\alpha\gamma}(r) = g_{\gamma\alpha}(r)$, the same result is obtained if the central particle belongs to γ and only α particles between r and $r + dr$ are taken. Stated in simple terms,

- $g_{\alpha\gamma}(r)$ is the probability density of finding a particle of species γ at a distance r away from a given reference particle of species α, *relative* to the probability density for an ideal gas.
- If a given reference particle of species α is taken to be at the origin, then the *local* average density of species γ at a distance r from that particle is $n_\gamma g_{\alpha\gamma}(r)$.

Fig. 4.11 John Gamble Kirkwood (1907–1959) (Photograph by C. T. Alburtus, Yale University News Bureau, courtesy of AIP Emilio Segrè Visual Archives, https://photos.aip.org/history-programs/niels-bohr-library/photos/kirkwood-john-a2)

As in the one-component case, it is convenient to introduce the total correlation function and the cavity function as

$$h_{\alpha\gamma}(r) = g_{\alpha\gamma}(r) - 1 \, , \tag{4.73a}$$

$$y_{\alpha\gamma}(r) = g_{\alpha\gamma}(r)e^{\beta\phi_{\alpha\gamma}(r)} \, . \tag{4.73b}$$

Also, the generalization to mixtures of the (static) structure factor (4.24) is [9]

$$\widetilde{S}_{\alpha\gamma}(k) = x_\alpha \delta_{\alpha\gamma} + nx_\alpha x_\gamma \tilde{h}_{\alpha\gamma}(k) \, , \tag{4.74}$$

where $\tilde{h}_{\alpha\gamma}(k)$ is the Fourier transform of $h_{\alpha\gamma}(r)$ [see (4.23)]. The zero wavenumber limit of $\tilde{h}_{\alpha\gamma}(k)$, i.e., $\tilde{h}_{\alpha\gamma}(0) = \int d\mathbf{r}\, h_{\alpha\gamma}(r)$, is called a Kirkwood–Buff integral [19, 20] (see Fig. 4.11) and plays a central role in relating the structure of a solution to its thermodynamic properties [20]. For instance, combination of (2.74b), (4.72), and (4.74) yields

$$\boxed{\widetilde{S}_{\alpha\gamma}(0) = \frac{k_B T}{\langle N \rangle} \left(\frac{\partial \langle N_\alpha \rangle}{\partial \mu_\gamma} \right)_{\beta, V, \{\mu_\nu \neq \gamma\}} \, .} \tag{4.75}$$

The OZ relation (4.26) can be easily extended to mixtures. In consistency with the physical idea sketched in Fig. 4.8, one now has

$$\boxed{h_{\alpha\gamma}(r_{12}) = c_{\alpha\gamma}(r_{12}) + n \sum_\nu x_\nu \int d\mathbf{r}_3\, c_{\alpha\nu}(r_{13})h_{\nu\gamma}(r_{32}) \, .} \tag{4.76}$$

In Fourier space,

$$\check{c}(k) = \check{h}(k) \cdot \left[\mathsf{I} + \check{h}(k) \right]^{-1} = \mathsf{I} - \left[\mathsf{I} + \check{h}(k) \right]^{-1} \, , \tag{4.77}$$

where $\check{c}(k)$ and $\check{h}(k)$ are matrices with elements $\check{c}_{\alpha\gamma}(k) \equiv n\sqrt{x_\alpha x_\gamma}\,\tilde{c}_{\alpha\gamma}(k)$ and $\check{h}_{\alpha\gamma}(k) \equiv n\sqrt{x_\alpha x_\gamma}\,\tilde{h}_{\alpha\gamma}(k)$, respectively, and I is the identity matrix.

4.6.1 Thermodynamic Routes

The generalization of the compressibility route (4.29) to mixtures is not trivial [14, 21]. First we note that, as a purely mathematical property, the inverse of a matrix of the form $(\partial a_\alpha/\partial b_\gamma)_{\{b_\nu \neq \gamma\}}$ is the matrix $(\partial b_\alpha/\partial a_\gamma)_{\{a_\nu \neq \gamma\}}$ [14]. Therefore, (4.75) can be rewritten as

$$\left(\frac{\partial \mu_\alpha}{\partial N_\gamma}\right)_{\beta,V,\{N_\nu \neq \gamma\}} = \frac{k_B T}{N}\left[\widetilde{\mathsf{S}}^{-1}(0)\right]_{\alpha\gamma}, \tag{4.78}$$

where the matrix $\widetilde{\mathsf{S}}(k)$ is defined by the elements $\widetilde{S}_{\alpha\gamma}(k)$ and we have dropped the angular brackets in N and N_γ because the left-hand side of (4.78) is understood in the canonical ensemble rather than in the grand canonical one. Then, from the thermodynamic relation (1.33) we finally obtain

$$\chi_T^{-1} = \sum_{\alpha,\gamma} x_\alpha x_\gamma \left[\widetilde{\mathsf{S}}^{-1}(0)\right]_{\alpha\gamma}. \tag{4.79}$$

In particular, in the case of a *binary* mixture,

$$\begin{aligned}\chi_T^{-1} &= \frac{x_1^2 \widetilde{S}_{22}(0) + x_2^2 \widetilde{S}_{11}(0) - 2x_1 x_2 \widetilde{S}_{12}(0)}{\widetilde{S}_{11}(0)\widetilde{S}_{22}(0) - \widetilde{S}_{12}^2(0)} \\ &= \frac{1 + n x_1 x_2 \left[\tilde{h}_{11}(0) + \tilde{h}_{22}(0) - 2\tilde{h}_{12}(0)\right]}{\left[1 + n x_1 \tilde{h}_{11}(0)\right]\left[1 + n x_2 \tilde{h}_{22}(0)\right] - n^2 x_1 x_2 \left[\tilde{h}_{12}(0)\right]^2}.\end{aligned} \tag{4.80}$$

The first line of (4.80) can also be rewritten as

$$\chi_T = \widetilde{S}_{nn}(0) - \frac{\left[\widetilde{S}_{nc}(0)\right]^2}{\widetilde{S}_{cc}(0)}, \tag{4.81}$$

where [21, 22]

$$\widetilde{S}_{nn}(k) \equiv \widetilde{S}_{11}(k) + \widetilde{S}_{22}(k) + 2\widetilde{S}_{12}(k), \tag{4.82a}$$

$$\widetilde{S}_{nc}(k) \equiv x_2 \widetilde{S}_{11}(k) - x_1 \widetilde{S}_{22}(k) + (x_2 - x_1)\widetilde{S}_{12}(k), \tag{4.82b}$$

$$\widetilde{S}_{cc}(k) \equiv x_1^2 \widetilde{S}_{22}(k) + x_2^2 \widetilde{S}_{11}(k) - 2x_1 x_2 \widetilde{S}_{12}(k). \tag{4.82c}$$

The quantities $\widetilde{S}_{nn}(0)$, $\widetilde{S}_{nc}(0)$, and $\widetilde{S}_{cc}(0)$ measure density–density, density–concentration, and concentration–concentration fluctuations, respectively [21, 22].

Noting that (4.74) can be rewritten as $\widetilde{S}_{\alpha\gamma}(k)/\sqrt{x_\alpha x_\gamma} = \delta_{\alpha\gamma} + \check{h}_{\alpha\gamma}(k)$, one finds that (4.79) is equivalent to

$$\chi_T^{-1} = \sum_{\alpha,\gamma} \sqrt{x_\alpha x_\gamma} \left[1 + \check{h}(0)\right]_{\alpha\gamma}^{-1} = \sum_{\alpha,\gamma} \sqrt{x_\alpha x_\gamma} \left[1 - \check{c}(0)\right]_{\alpha\gamma} = 1 - n \sum_{\alpha,\gamma} x_\alpha x_\gamma \tilde{c}_{\alpha\gamma}(0) ,$$

(4.83)

where (4.77) has been used in the second step. Equation (4.83) is the counterpart for mixtures of (4.30).

The extensions to mixtures of the energy route (4.32), the virial route (4.40), the chemical-potential route (4.51), and the free-energy route (4.56) are relatively straightforward. The resulting expressions are summarized in Table 4.3 (compare with Table 4.1 and the last row of Table 4.2). Note that, in analogy with (4.41), the virial route can alternatively be written as

$$Z = 1 + \frac{n}{2d} \sum_{\alpha,\gamma} x_\alpha x_\gamma \int d\mathbf{r}\, y_{\alpha\gamma}(r) r \frac{\partial f_{\alpha\gamma}(r)}{\partial r} .$$

(4.84)

In the case of the chemical-potential route [18], the evaluation of the excess chemical potential associated with species v, μ_v^{ex}, assumes that the *solute* particle

Table 4.3 Summary of the main thermodynamic routes in a mixture

Route	Quantity	Thermodynamic relation	Expression
Compressibility	χ_T^{-1}	$\left(\frac{\partial}{\partial n} n^2 \frac{\partial(\beta a)}{\partial n}\right)_{\beta,\{x_\alpha\}}$	$\sum_{\alpha,\gamma} x_\alpha x_\gamma \left(\widetilde{\mathsf{S}}^{-1}\right)_{\alpha\gamma}$
Energy	u^{ex}	$\left(\frac{\partial(\beta a^{ex})}{\partial\beta}\right)_{n,\{x_\alpha\}}$	$\frac{n}{2}\sum_{\alpha,\gamma} x_\alpha x_\gamma \int d\mathbf{r}\,\phi_{\alpha\gamma}(r)g_{\alpha\gamma}(r)$
Virial	$Z-1$	$n\left(\frac{\partial(\beta a^{ex})}{\partial n}\right)_{\beta,\{x_\alpha\}}$	$-\frac{\beta n}{2d}\sum_{\alpha,\gamma} x_\alpha x_\gamma \int d\mathbf{r}\, r\frac{d\phi_{\alpha\gamma}(r)}{dr}g_{\alpha\gamma}(r)$
Chemical potential	μ_v^{ex}	$\left(\frac{\partial(n a^{ex})}{\partial n_v}\right)_{\beta,\{n_{\alpha\neq v}\}}$	$n\sum_\alpha x_\alpha \int_0^1 d\xi \int d\mathbf{r}\,\frac{\partial\phi_{0\alpha}^{(\xi)}(r)}{\partial\xi}g_{0\alpha}^{(\xi)}(r)$
	$\sum_v x_v \mu_v^{ex}$	$\left(\frac{\partial(n a^{ex})}{\partial n}\right)_{\beta,\{x_\alpha\}}$	$n\sum_{v,\alpha} x_v x_\alpha \int_0^1 d\xi \int d\mathbf{r}\,\frac{\partial\phi_{0\alpha}^{(\xi)}(r)}{\partial\xi}g_{0\alpha}^{(\xi)}(r)$
Free energy	a^{ex}	a^{ex}	$\frac{n}{2}\sum_{\alpha,\gamma} x_\alpha x_\gamma \int_0^1 d\xi \int d\mathbf{r}\,\frac{\partial\phi_{\alpha\gamma}^{(\xi)}(r)}{\partial\xi}g_{\alpha\gamma}^{(\xi)}(r)$

$i = 0$ is coupled to a particle of species α via an interaction potential $\phi_{0\alpha}^{(\xi)}(r)$ such that

$$\phi_{0\alpha}^{(\xi)}(r) = \begin{cases} 0, & \xi = 0, \\ \phi_{v\alpha}(r), & \xi = 1, \end{cases} \tag{4.85}$$

so that it becomes a particle of species v at the end of the charging process. The quantity $g_{0\alpha}^{(\xi)}(r)$ is the associated solute–solvent RDF. In contrast, in the free-energy route, *all* the particles of all the species are involved in switching the interactions on.

The third column of Table 4.3 gives the thermodynamic relation between each quantity and the free energy per particle. In the case of the single chemical potential μ_v^{ex}, the free energy per particle is seen as a function of temperature and *partial* number densities. On the other hand, thanks to (1.13) and the identity $G = -V^2 \partial (F/V)/\partial V$, in the thermodynamic relation between $\sum_v x_v \mu_v^{ex}$ and a^{ex}, the latter quantity is seen as a function of temperature, *total* number density, and mole fractions. Therefore, the free energy can be derived by integrating over n (compressibility, virial, and chemical-potential routes) or over β (energy route), analogously to the one-component case (see Table 4.2).

4.6.2   Hard Spheres

Let us now particularize the above expressions to multicomponent HS fluids [23], in which case the interaction potential function is given by the form (3.7a) for any pair of species. As a consequence,

$$\frac{\partial f_{\alpha\gamma}(r)}{\partial r} = \delta \left(r - \sigma_{\alpha\gamma}\right) . \tag{4.86}$$

The compressibility route (4.79) does not include the interaction potential explicitly and so it is not simplified in the HS case. As for the energy route, the integral $\int d\mathbf{r}\, \phi_{\alpha\gamma}(r) g_{\alpha\gamma}(r)$ vanishes because $\phi_{\alpha\gamma}(r) e^{-\beta\phi_{\alpha\gamma}(r)} \to 0$ both for $r < \sigma_{\alpha\gamma}$ and $r > \sigma_{\alpha\gamma}$, while $y_{\alpha\gamma}(r)$ is finite even in the region $r < \sigma_{\alpha\gamma}$ (see Fig. 4.5). Therefore,

$$\langle E \rangle = N \frac{d}{2} k_B T . \tag{4.87}$$

This is not but the ideal-gas internal energy! This is an expected result since the HS potential is only different from zero when two particles overlap and those configurations are forbidden by the pair Boltzmann factor $e^{-\beta\phi_{\alpha\gamma}(r)}$.

The virial route is highly simplified for HSs. First, it is convenient to change to spherical coordinates and take into account that the total d-dimensional solid angle (area of a d-dimensional sphere of unit radius) is $\int d\hat{\mathbf{r}} = d2^d v_d$, where the volume of a d-dimensional sphere of unit diameter is defined by (3.43). Next, using the property (4.86) in (4.84), we obtain

$$Z \equiv \frac{p}{nk_BT} = 1 + 2^{d-1}nv_d \sum_{\alpha,\gamma} x_\alpha x_\gamma \sigma_{\alpha\gamma}^d y_{\alpha\gamma}(\sigma_{\alpha\gamma}) \,. \tag{4.88}$$

The same method works for the chemical-potential route with the choice

$$e^{-\beta\phi_{0\alpha}^{(\xi)}(r)} = \Theta\left(r - \sigma_{0\alpha}^{(\xi)}\right) \,, \tag{4.89}$$

where $\sigma_{0\alpha}^{(0)} = 0$ and $\sigma_{0\alpha}^{(1)} = \sigma_{\nu\alpha}$ [see (4.85)]. Changing the integration variable from ξ to $\sigma_{\nu\alpha}^{(\xi)}$, one gets

$$\beta\mu_\nu = \ln\left(nx_\nu\Lambda_\nu^d\right) + d2^d nv_d \sum_\alpha x_\alpha \int_0^{\sigma_{\nu\alpha}} d\sigma_{0\alpha} \, \sigma_{0\alpha}^{d-1} y_{0\alpha}(\sigma_{0\alpha}) \,, \tag{4.90}$$

where use has been made of (2.75) and the notation has been simplified as $\sigma_{\nu\alpha}^{(\xi)} \to \sigma_{0\alpha}$ and $y_{\nu\alpha}^{(\xi)} \to y_{0\alpha}$. If $\sigma_{\alpha\gamma} \geq \frac{1}{2}(\sigma_\alpha + \sigma_\gamma)$ (positive or zero *nonadditivity*, see p. 39), it can be proved [18] that

$$d2^d nv_d \sum_\alpha x_\alpha \int_0^{\frac{1}{2}\sigma_\alpha} d\sigma_{0\alpha} \, \sigma_{0\alpha}^{d-1} y_{0\alpha}(\sigma_{0\alpha}) = -\ln(1 - \eta) \,, \tag{4.91}$$

where the total packing fraction is defined in (3.116). In that case, (4.90) can be rewritten as

$$\beta\mu_\nu = \ln\frac{nx_\nu\Lambda_\nu^d}{1 - \eta} + d2^d nv_d \sum_\alpha x_\alpha \int_{\frac{1}{2}\sigma}^{\sigma_{\nu\alpha}} d\sigma_{0\alpha} \, \sigma_{0\alpha}^{d-1} y_{0\alpha}(\sigma_{0\alpha}) \,. \tag{4.92}$$

Finally, in the case of the free-energy route, it seems natural to choose the intermediate potentials $\phi_{\alpha\gamma}(r)$ as maintaining a HS form, i.e.,

$$e^{-\beta\phi_{\alpha\gamma}^{(\xi)}(r)} = \Theta\left(r - \sigma_{\alpha\gamma}^{(\xi)}\right) \,, \tag{4.93}$$

where $\sigma_{\alpha\gamma}^{(0)} = 0$ and $\sigma_{\alpha\gamma}^{(1)} = \sigma_{\alpha\gamma}$. In such a case, the free-energy route becomes

$$\beta a^{\text{ex}} = 2^{d-1}dnv_d \sum_{\alpha,\gamma} x_\alpha x_\gamma \int_0^1 d\xi \left[\sigma_{\alpha\gamma}^{(\xi)}\right]^{d-1} \frac{\partial\sigma_{\alpha\gamma}^{(\xi)}}{\partial\xi} y_{\alpha\gamma}^{(\xi)}\left(\sigma_{\alpha\gamma}^{(\xi)}\right) \,. \tag{4.94}$$

Here the protocol $\sigma_{\alpha\gamma}^{(\xi)}$ is arbitrary. In the special case of a simple rescaling, i.e., $\sigma_{\alpha\gamma}^{(\xi)} = \xi\sigma_{\alpha\gamma}$, one has $y_{\alpha\gamma}^{(\xi)}\left(\sigma_{\alpha\gamma}^{(\xi)};n\right) = y_{\alpha\gamma}\left(\sigma_{\alpha\gamma};n\xi^d\right)$ [see (4.62)]. With that specific choice, (4.94) reduces to

$$\beta a^{\text{ex}} = 2^{d-1}v_d \sum_{\alpha,\gamma} x_\alpha x_\gamma \sigma_{\alpha\gamma}^d \int_0^n \mathrm{d}n' \, y_{\alpha\gamma}\left(\sigma_{\alpha\gamma};n'\right) , \qquad (4.95)$$

where the change of variables $\xi \to n' = n\xi^d$ has been carried out. As expected, (4.95) is fully equivalent to the virial route (4.88).

4.7 The Thermodynamic Inconsistency Problem

Going back to the case of an arbitrary interaction potential, we have seen that the knowledge of the RDF allows one to obtain the free energy from at least five different routes. This is shown explicitly by Table 4.2 for one-component systems and sketched in Fig. 4.12. The important question is, would one obtain consistent results?

Since all the thermodynamic routes are derived from formally exact statistical-mechanical formulas, it is obvious that the use of the *exact* RDF $g(r)$ must lead to the same exact free energy $F(T, V, N)$, regardless of the route followed. On the other hand, if an *approximate* $g(r)$ is used, one must be aware that (in general) a different approximate $F(T, V, N)$ is obtained from each separate route. This is known as the *thermodynamic (in)consistency problem*. Which route is more accurate, i.e., which route is more effective in concealing the deficiencies of an approximate $g(r)$, may depend on the approximation, the potential, and the thermodynamic state.

Fig. 4.12 Schematic view of the thermodynamic inconsistency problem

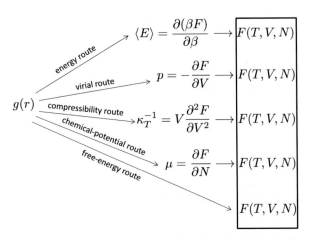

Same result?

Exercises

4.1 Derive (4.6) and (4.7).

4.2 Derive (4.9) and (4.10).

4.3 Use (4.18) to derive (4.17).

4.4 Inserting (4.17) into (4.12), check (4.19) for $s = 1, 2$, and 3.

4.5 Prove the equivalence between (4.42) and (4.44).

4.6 Making use of Table 1.1 and (1.28), check the correctness of the third column of Table 4.1.

4.7 Consider the protocol

$$\phi^{(\xi)}(r) = a(\xi)\phi(r/\xi) \,,$$

where $a(\xi)$ is any function subject to the constraints $\lim_{\xi \to 0} a(\xi)\phi(r/\xi) = 0$ and $\lim_{\xi \to 1} a(\xi) = 1$. Prove that, in such a case, the free-energy route (4.56) becomes

$$a^{\text{ex}} = \frac{n}{2} \int_0^1 d\xi \, \xi^{d-1} a(\xi) \int d\mathbf{r} \left[\xi \frac{d \ln a(\xi)}{d\xi} \phi(r) - \frac{d\phi(r)}{dr} \right] g\left(r; n\xi^d, \beta a(\xi)\right) \,.$$

4.8 Derive (4.69).

4.9 Derive (4.75).

4.10 Obtain (4.77) from (4.76).

4.11 Derive (4.80) from (4.79).

4.12 If all the species in a mixture are mechanically equivalent, then $g_{\alpha\gamma}(r) = g(r)$ for all pairs. Check in that case that (4.80) reduces to (4.29).

4.13 Taking into account (4.82), prove the equivalence between (4.80) and (4.81).

4.14 Starting from Table 4.3, construct a table similar to (4.2), but for mixtures.

4.15 Derive (4.94).

References

1. R.J. Baxter, J. Chem. Phys. **41**, 553 (1964)
2. G.E. Uhlenbeck, P.C. Hemmer, M. Kac, J. Math. Phys. **4**, 229 (1963)
3. M.P. Allen, D.J. Tildesley, *Computer Simulation of Liquids* (Clarendon Press, Oxford, 1987)
4. D. Frenkel, B. Smit, *Understanding Molecular Simulation: From Algorithms to Applications*, 2nd edn. (Academic, San Diego, 2002)

5. Simulations performed by the author with the Glotzilla library (2015), http://glotzerlab.engin. umich.edu/home/
6. J. Kolafa, S. Labík, A. Malijevský, Mol. Phys. **100**, 2629 (2002). See also http://www.vscht. cz/fch/software/hsmd/
7. S. Labík, A. Malijevský, Mol. Phys. **53**, 381 (1984)
8. R. Balescu, *Equilibrium and Nonequilibrium Statistical Mechanics* (Wiley, New York, 1974)
9. J.P. Hansen, I.R. McDonald, *Theory of Simple Liquids*, 3rd edn. (Academic, London, 2006)
10. D. Heyes, *The Liquid State: Applications of Molecular Simulations* (Wiley, Chichester, 1998)
11. B. Widom, J. Chem. Phys. **39**, 2808 (1963)
12. L.E. Reichl, *A Modern Course in Statistical Physics*, 1st edn. (University of Texas Press, Austin, 1980)
13. L. Onsager, Chem. Rev. **13**, 73 (1933)
14. T.L. Hill, *Statistical Mechanics* (McGraw-Hill, New York, 1956)
15. H. Reiss, H.L. Frisch, J.L. Lebowitz, J. Chem. Phys. **31**, 369 (1959)
16. M. Mandell, H. Reiss, J. Stat. Phys. **13**, 113 (1975)
17. A. Santos, Phys. Rev. Lett. **109**, 120601 (2012)
18. A. Santos, R.D. Rohrmann, Phys. Rev. E **87**, 052138 (2013)
19. J.G. Kirkwood, F.P. Buff, J. Chem. Phys. **19**, 774 (1951)
20. A. Ben-Naim, *Molecular Theory of Solutions* (Oxford University Press, Oxford, 2006)
21. A.B. Bhatia, D.E. Thornton, Phys. Rev. B **8**, 3004 (1970)
22. N.W. Ashcroft, D.C. Langreth, Phys. Rev. **156**, 685 (1967)
23. J.S. Rowlinson, F. Swinton, *Liquids and Liquid Mixtures* (Butterworth, London, 1982)

Chapter 5
One-Dimensional Systems: Exact Solution for Nearest-Neighbor Interactions

One-dimensional systems with interactions restricted to nearest neighbors lend themselves to a full exact statistical-mechanical solution, what has undoubtful pedagogical and illustrative values. It is first noted in this chapter that the pair correlation function in Laplace space can be expressed in terms of the nearest-neighbor distribution function. The latter quantity is subsequently obtained in the isothermal–isobaric ensemble. As explicit examples, the square-well, square-shoulder, sticky-hard-rod, and nonadditive hard-rod fluids are worked out in detail.

5.1 Nearest-Neighbor and Pair Correlation Functions

As is apparent from (4.20), the evaluation of $g(r)$ is in general a formidable task, comparable to that of the evaluation of the configuration integral itself. However, in the case of one-dimensional systems ($d = 1$) of particles which only interact with their nearest neighbors, the problem can be exactly solved [1–5]. Exact solutions are also possible for some one-dimensional systems with interactions extending beyond nearest neighbors, as happens in an isolated self-gravitating system [6], but those other cases will not be considered in this chapter.

Let us consider a one-dimensional system of N particles in a box of length L (so the number density is $n = N/L$) subject to an interaction potential $\phi(r)$ such that

1. $\lim_{r \to 0} \phi(r) = \infty$. This implies that the *order* of the particles in the line does not change.
2. $\lim_{r \to \infty} \phi(r) = 0$. The interaction has a *finite* range.
3. Each particle interacts *only* with its two nearest neighbors.

© Springer International Publishing Switzerland 2016
A. Santos, *A Concise Course on the Theory of Classical Liquids*,
Lecture Notes in Physics 923, DOI 10.1007/978-3-319-29668-5_5

The total potential energy is then

$$\Phi_N(\mathbf{r}^N) = \sum_{i=1}^{N-1} \phi(x_{i+1} - x_i) \ . \tag{5.1}$$

Given a particle at a certain position, let $p^{(1)}(r)dr$ be the *conditional* probability of finding its (right) *nearest neighbor* at a distance between r and $r + dr$ (see Fig. 5.1). More in general, we can define $p^{(\ell)}(r)dr$ as the conditional probability of finding its (right) ℓth neighbor ($1 \leq \ell \leq N - 1$) at a distance between r and $r + dr$ (see Fig. 5.2). Since the ℓth neighbor must be somewhere, the normalization condition is

$$\int_0^\infty dr \, p^{(\ell)}(r) = 1 \ . \tag{5.2}$$

In making the upper limit equal to infinity, we are implicitly assuming the thermodynamic limit ($L \to \infty$, $N \to \infty$, $n = $ const). Moreover, periodic boundary conditions are supposed to be applied when needed.

As illustrated by Fig. 5.3, the following recurrence relation holds

$$p^{(\ell)}(r) = \int_0^r dr' \, p^{(1)}(r')p^{(\ell-1)}(r - r') \ . \tag{5.3}$$

The convolution structure of the integral invites one to introduce the Laplace transform

$$\widehat{P}^{(\ell)}(s) \equiv \int_0^\infty dr \, e^{-rs} p^{(\ell)}(r) \ , \tag{5.4}$$

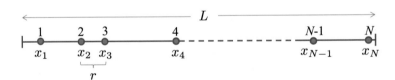

Fig. 5.1 Two nearest-neighbor particles separated a distance r

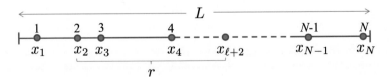

Fig. 5.2 Two ℓth-order neighbors separated a distance r

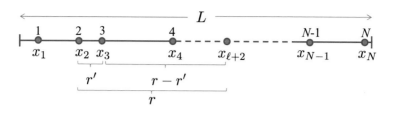

Fig. 5.3 Illustration of the convolution property (5.3)

so that (5.3) becomes

$$\widehat{P}^{(\ell)}(s) = \widehat{P}^{(1)}(s)\widehat{P}^{(\ell-1)}(s) \Rightarrow \widehat{P}^{(\ell)}(s) = \left[\widehat{P}^{(1)}(s)\right]^{\ell}. \tag{5.5}$$

The normalization condition (5.2) is equivalent to

$$\widehat{P}^{(\ell)}(0) = 1. \tag{5.6}$$

Now, given a reference particle at a certain position, the physical meaning of the RDF (see p. 102) implies that $ng(r)dr$ is the total number of particles at a distance between r and $r+dr$, regardless of whether those particles correspond to the nearest neighbor, the next-nearest neighbor, ... of the reference particle. Thus,

$$ng(r) = \sum_{\ell=1}^{N-1} p^{(\ell)}(r) \xrightarrow{N \to \infty} \sum_{\ell=1}^{\infty} p^{(\ell)}(r). \tag{5.7}$$

Introducing the Laplace transform

$$\widehat{G}(s) \equiv \int_0^\infty dr \, e^{-rs} g(r), \tag{5.8}$$

and using (5.5), we have

$$\widehat{G}(s) = \frac{1}{n} \sum_{\ell=1}^{\infty} \left[\widehat{P}^{(1)}(s)\right]^{\ell} = \frac{1}{n} \frac{\widehat{P}^{(1)}(s)}{1 - \widehat{P}^{(1)}(s)}. \tag{5.9}$$

Thus, the determination of the RDF $g(r)$ reduces to the determination of the nearest-neighbor distribution function $p^{(1)}(r)$.

To proceed, we take advantage of the ensemble equivalence in the thermodynamic limit and use the isothermal–isobaric ensemble.

5.2   Nearest-Neighbor Distribution: Isothermal–Isobaric Ensemble

The isothermal–isobaric ensemble is described by (2.45) (see also Table 2.1). The important point is that the N-body probability distribution function in configuration space is proportional to $e^{-\beta p V - \beta \Phi_N(\mathbf{r}^N)}$ and the evaluation of any physical quantity implies integrating over the volume and over the particle coordinates. Therefore, in this ensemble the one-dimensional nearest-neighbor probability distribution function is

$$p^{(1)}(r) \propto \int_r^\infty dL\, e^{-\beta p L} \int_{x_2}^L dx_3 \int_{x_3}^L dx_4 \cdots \int_{x_{N-1}}^L dx_N\, e^{-\beta \Phi_N(\mathbf{r}^N)} \,, \tag{5.10}$$

where we have identified the volume V with the length L and have taken the particles $i = 1$ (at $x_1 = 0$) and $i = 2$ (at $x_2 = r$) as the canonical nearest-neighbor pair (see Fig. 5.4). Next, using (5.1) and applying periodic boundary conditions,

$$p^{(1)}(r) \propto e^{-\beta \phi(r)} \int_r^\infty dL\, e^{-\beta p L} \int_0^{L-r} dr_3\, e^{-\beta \phi(r_3)} \int_0^{L-r-r_3} dr_4\, e^{-\beta \phi(r_4)}$$

$$\times \cdots \int_0^{L-r-r_3-\cdots-r_{N-1}} dr_N\, e^{-\beta \phi(r_N)} e^{-\beta \phi(r_{N+1})} \,, \tag{5.11}$$

where a change of variables $x_i \to r_i = x_i - x_{i-1}$ ($i = 3, \ldots, N$) has been carried out and $r_{N+1} = L - r - r_3 - r_4 - \cdots - r_N$. Finally, the change of variable $L \to L' = L - r$ implies that a factor $e^{-\beta p r}$ comes out of the integrals, the latter being independent of r. In summary,

$$p^{(1)}(r) = K e^{-\beta \phi(r)} e^{-\beta p r} \,, \tag{5.12}$$

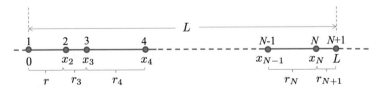

Fig. 5.4 Illustration of the evaluation of $p^{(1)}(r)$ in the isothermal–isobaric ensemble

where the proportionality constant K will be determined by normalization. The Laplace transform of (5.12) is

$$\widehat{P}^{(1)}(s) = K\widehat{\Omega}(s + \beta p) , \qquad (5.13)$$

where

$$\boxed{\widehat{\Omega}(s) \equiv \int_0^\infty dr\, e^{-rs} e^{-\beta\phi(r)}} \qquad (5.14)$$

is the Laplace transform of the pair Boltzmann factor $e^{-\beta\phi(r)}$. The normalization condition (5.6) yields

$$K = \frac{1}{\widehat{\Omega}(\beta p)} . \qquad (5.15)$$

5.3 Exact Radial Distribution Function and Thermodynamic Quantities

Insertion of (5.13) and (5.15) into (5.9) gives the exact RDF (in Laplace space):

$$\boxed{\widehat{G}(s) = \frac{1}{n} \frac{\widehat{\Omega}(s + \beta p)}{\widehat{\Omega}(\beta p) - \widehat{\Omega}(s + \beta p)} .} \qquad (5.16)$$

To fully close the problem, it remains to determine the EoS. i.e., a relation between the pressure p, the density n, and the temperature T. To do that, we apply the consistency condition

$$\lim_{r\to\infty} g(r) = 1 \Rightarrow \lim_{s\to 0} s\widehat{G}(s) = 1 . \qquad (5.17)$$

Expanding $\widehat{\Omega}(s + \beta p)$ in powers of s and imposing (5.17), we obtain

$$\boxed{n(p, T) = -\frac{\widehat{\Omega}(\beta p)}{\widehat{\Omega}'(\beta p)} ,} \qquad (5.18)$$

where

$$\widehat{\Omega}'(s) \equiv \frac{\partial \widehat{\Omega}(s)}{\partial s} = -\int_0^\infty dr\, e^{-rs} r\, e^{-\beta\phi(r)} . \qquad (5.19)$$

As discussed in Chap. 1 (see Table 1.1), the right thermodynamic potential in the isothermal–isobaric ensemble is the Gibbs free energy $G(N, p, T)$. Using the thermodynamic relation (1.16b) it is easy to obtain from (5.18) the explicit result

$$G(T, p, N) = Nk_BT \ln \frac{\Lambda(\beta)}{\widehat{\Omega}(\beta p; \beta)} , \qquad (5.20)$$

where the notation $\widehat{\Omega}(s) \to \widehat{\Omega}(s; \beta)$ gas been employed to emphasize the parametric dependence of the Laplace transform (5.14) on β. Upon deriving (5.20), the integration constant has been determined by noting that $G^{\text{id}} = Nk_BT \ln(\beta p \Lambda)$ (see Table 2.2) and

$$\lim_{s \to 0} s\widehat{\Omega}(s) = \lim_{s \to 0} \int_0^\infty dx\, e^{-x} e^{-\beta\phi(x/s)} = 1 . \qquad (5.21)$$

Using standard thermodynamic relations (see Table 1.1), one gets

$$\frac{\langle E \rangle}{Nk_BT} = \frac{1}{2} + \frac{\beta \widehat{\Upsilon}(\beta p; \beta)}{\widehat{\Omega}(\beta p; \beta)} , \qquad (5.22a)$$

$$\frac{F}{Nk_BT} = \ln \frac{\Lambda(\beta)}{\widehat{\Omega}(\beta p; \beta)} + \frac{\beta p \widehat{\Omega}'(\beta p; \beta)}{\widehat{\Omega}(\beta p; \beta)} , \qquad (5.22b)$$

$$\frac{S}{Nk_B} = \frac{1}{2} - \ln \frac{\Lambda(\beta)}{\widehat{\Omega}(\beta p; \beta)} - \frac{\beta p \widehat{\Omega}'(\beta p; \beta) - \beta \widehat{\Upsilon}(\beta p; \beta)}{\widehat{\Omega}(\beta p; \beta)} , \qquad (5.22c)$$

where

$$\widehat{\Upsilon}(s) \equiv \int_0^\infty dr\, e^{-rs} \phi(r) e^{-\beta\phi(r)} \qquad (5.23)$$

is the Laplace transform of $\phi(r)e^{-\beta\phi(r)}$.

As a simple test, let us check that the EoS (5.18) is consistent with the compressibility route (4.29). First, according to (5.18), the isothermal susceptibility is

$$\chi_T = \left(\frac{\partial n}{\partial \beta p} \right)_\beta = -1 + \frac{\widehat{\Omega}(\beta p)\widehat{\Omega}''(\beta p)}{\left[\widehat{\Omega}'(\beta p)\right]^2} , \qquad (5.24)$$

where

$$\widehat{\Omega}''(s) \equiv \frac{\partial \widehat{\Omega}'(s)}{\partial s} = \int_0^\infty dr\, e^{-rs} r^2 e^{-\beta\phi(r)} . \qquad (5.25)$$

Alternatively, the Laplace transform

$$\widehat{H}(s) = \int_0^\infty dr\, e^{-rs} h(r) \tag{5.26}$$

of the total correlation function $h(r) = g(r) - 1$ is $\widehat{H}(s) = \widehat{G}(s) - s^{-1}$, and thus the Fourier transform can be obtained as

$$\tilde{h}(k) = \left[\widehat{H}(s) + \widehat{H}(-s)\right]_{s=ik} = \left[\widehat{G}(s) + \widehat{G}(-s)\right]_{s=ik}. \tag{5.27}$$

In particular, the zero wavenumber limit is

$$\int dr\, h(r) = 2 \lim_{s\to 0}\left[\widehat{G}(s) - \frac{1}{s}\right] = 2\left[\frac{\widehat{\Omega}'(\beta p)}{\widehat{\Omega}(\beta p)} - \frac{\widehat{\Omega}''(\beta p)}{2\widehat{\Omega}'(\beta p)}\right], \tag{5.28}$$

so that

$$1 + n\int dr\, h(r) = 1 - 2\frac{\widehat{\Omega}(\beta p)}{\widehat{\Omega}'(\beta p)}\left[\frac{\widehat{\Omega}'(\beta p)}{\widehat{\Omega}(\beta p)} - \frac{\widehat{\Omega}''(\beta p)}{2\widehat{\Omega}'(\beta p)}\right] = -1 + \frac{\widehat{\Omega}(\beta p)\widehat{\Omega}''(\beta p)}{\left[\widehat{\Omega}'(\beta p)\right]^2}. \tag{5.29}$$

Comparison between (5.24) and (5.29) shows that (4.29) is indeed satisfied.

5.4   Extension to Mixtures

In the case of one-dimensional mixtures the arguments outlined above can be extended without special difficulties [3–5]. Now, instead of $p^{(\ell)}(r)dr$ one defines $p_{\alpha\gamma}^{(\ell)}(r)dr$ as the conditional probability that the ℓth neighbor to the right of a reference particle of species α is located at a distance between r and $r + dr$ and belongs to species γ. The counterparts of (5.2), (5.3), and (5.7) are

$$\sum_\gamma \int_0^\infty dr\, p_{\alpha\gamma}^{(\ell)}(r) = 1, \tag{5.30a}$$

$$p_{\alpha\gamma}^{(\ell)}(r) = \sum_\nu \int_0^r dr'\, p_{\alpha\nu}^{(1)}(r') p_{\nu\gamma}^{(\ell-1)}(r - r'), \tag{5.30b}$$

$$n x_\gamma g_{\alpha\gamma}(r) = \sum_{\ell=1}^\infty p_{\alpha\gamma}^{(\ell)}(r). \tag{5.30c}$$

Next, by defining the Laplace transforms $\widehat{P}_{\alpha\gamma}^{(\ell)}(s)$ and $\widehat{G}_{\alpha\gamma}(s)$ of $p_{\alpha\gamma}^{(\ell)}(r)$ and $g_{\alpha\gamma}(r)$, respectively, one easily arrives at

$$\widehat{G}_{\alpha\gamma}(s) = \frac{1}{nx_\gamma}\left(\widehat{\mathsf{P}}^{(1)}(s)\cdot\left[\mathsf{I}-\widehat{\mathsf{P}}^{(1)}(s)\right]^{-1}\right)_{\alpha\gamma}, \tag{5.31}$$

where $\widehat{\mathsf{P}}^{(1)}(s)$ is the matrix of elements $\widehat{P}_{\alpha\gamma}^{(1)}(s)$.

The nearest-neighbor probability distribution is again derived in the isothermal–isobaric ensemble with the result

$$p_{\alpha\gamma}^{(1)}(r) = x_\gamma K_{\alpha\gamma}e^{-\beta\phi_{\alpha\gamma}(r)}e^{-\beta pr}, \tag{5.32}$$

so that

$$\widehat{P}_{\alpha\gamma}^{(1)}(s) = x_\gamma K_{\alpha\gamma}\widehat{\Omega}_{\alpha\gamma}(s+\beta p), \tag{5.33}$$

where $\widehat{\Omega}_{\alpha\gamma}(s)$ is the Laplace transform of $e^{-\beta\phi_{\alpha\gamma}(r)}$. The normalization condition (5.30a) imposes the following relationship for the constants $K_{\alpha\gamma} = K_{\gamma\alpha}$:

$$\sum_\gamma x_\gamma K_{\alpha\gamma}\widehat{\Omega}_{\alpha\gamma}(\beta p) = 1. \tag{5.34}$$

To complete the determination of $K_{\alpha\gamma}$, we can make use of the physical condition stating that $\lim_{r\to\infty}p_{\alpha\gamma}^{(1)}(r)/p_{\alpha\nu}^{(1)}(r)$ must be independent of the identity α of the species the reference particle belongs to, so that $K_{\alpha\gamma}/K_{\alpha\nu}$ is independent of α. It is easy to see that such a condition implies

$$K_{\alpha\gamma}^2 = K_{\alpha\alpha}K_{\gamma\gamma}, \quad \forall\alpha,\gamma. \tag{5.35}$$

The EoS $n(T,p,\{x_\nu\})$ is determined, as in the one-component case, from the condition

$$\lim_{r\to\infty}g_{\alpha\gamma}(r) = 1 \Rightarrow \lim_{s\to0}s\widehat{G}_{\alpha\gamma}(s) = 1. \tag{5.36}$$

5.4.1 Binary Case

As a more explicit situation, here we particularize to a binary mixture. In that case, (5.31) yields

$$\widehat{G}_{11}(s) = \frac{\widehat{Q}_{11}(s)\left[1 - \widehat{Q}_{22}(s)\right] + \widehat{Q}_{12}^2(s)}{nx_1\widehat{D}(s)}, \tag{5.37a}$$

$$\widehat{G}_{22}(s) = \frac{\widehat{Q}_{22}(s)\left[1 - \widehat{Q}_{11}(s)\right] + \widehat{Q}_{12}^2(s)}{nx_2\widehat{D}(s)}, \tag{5.37b}$$

$$\widehat{G}_{12}(s) = \frac{\widehat{Q}_{12}(s)}{n\sqrt{x_1x_2}\widehat{D}(s)}, \tag{5.37c}$$

where

$$\widehat{Q}_{\alpha\gamma}(s) \equiv \sqrt{\frac{x_\alpha}{x_\gamma}}\widehat{P}_{\alpha\gamma}^{(1)}(s) = \sqrt{x_\alpha x_\gamma}K_{\alpha\gamma}\widehat{\Omega}_{\alpha\gamma}(s + \beta p), \tag{5.38a}$$

$$\widehat{D}(s) \equiv \left[1 - \widehat{Q}_{11}(s)\right]\left[1 - \widehat{Q}_{22}(s)\right] - \widehat{Q}_{12}^2(s). \tag{5.38b}$$

The parameters $K_{\alpha\gamma}$ are obtained from (5.34) and (5.35). First, K_{11} and K_{22} can be expressed in terms of K_{12} as

$$K_{11} = \frac{1 - x_2 K_{12}\widehat{\Omega}_{12}(\beta p)}{x_1\widehat{\Omega}_{11}(\beta p)}, \quad K_{22} = \frac{1 - x_1 K_{12}\widehat{\Omega}_{12}(\beta p)}{x_2\widehat{\Omega}_{22}(\beta p)}. \tag{5.39}$$

The remaining parameter K_{12} satisfies a quadratic equation whose solution is

$$K_{12} = \frac{1 - \sqrt{1 - 4x_1x_2R}}{2x_1x_2R\widehat{\Omega}_{12}(\beta p)}, \quad R \equiv 1 - \frac{\widehat{\Omega}_{11}(\beta p)\widehat{\Omega}_{22}(\beta p)}{\widehat{\Omega}_{12}^2(\beta p)}. \tag{5.40}$$

Finally, the EoS becomes

$$\frac{1}{n(T, p, x_1)} = -\left[x_1^2 K_{11}\widehat{\Omega}_{11}'(\beta p) + x_2^2 K_{22}\widehat{\Omega}_{22}'(\beta p) + 2x_1x_2 K_{12}\widehat{\Omega}_{12}'(\beta p)\right]. \tag{5.41}$$

As in the one-component case, we can use the thermodynamic relation (1.16b) to obtain the exact Gibbs free energy of the mixture as

$$
G(T,p,N_1,N_2) = N k_B T \left[x_1 \ln \frac{x_1 \Lambda_1}{\widehat{\Omega}_{11}} + x_2 \ln \frac{x_2 \Lambda_2}{\widehat{\Omega}_{22}} - \ln \frac{1 + \sqrt{1 - 4x_1 x_2 R}}{2\sqrt{1-R}} \right.
$$

$$
\left. + |x_1 - x_2| \ln \frac{|x_1 - x_2| + \sqrt{1 - 4x_1 x_2 R}}{(|x_1 - x_2| + 1)\sqrt{1-R}} \right], \tag{5.42}
$$

where henceforth the absence of an argument in $\widehat{\Omega}_{\alpha\gamma}$, $\widehat{\Omega}'_{\alpha\gamma}$, or $\widehat{\Omega}''_{\alpha\gamma}$ means that those functions are evaluated at $s = \beta p$. From (1.16c) we can obtain the chemical potential of species α as

$$
\beta\mu_\alpha = \ln \frac{x_\alpha \Lambda_\alpha}{\widehat{\Omega}_{\alpha\alpha}} - \ln \frac{1 + \sqrt{1 - 4x_1 x_2 R}}{2\sqrt{1-R}} + \text{sgn}\,(2x_\alpha - 1) \ln \frac{|x_1 - x_2| + \sqrt{1 - 4x_1 x_2 R}}{(|x_1 - x_2| + 1)\sqrt{1-R}}, \tag{5.43}
$$

where the sign function is $\text{sgn}(x) = +1$ if $x > 0$ and -1 otherwise.

In analogy with (5.28), the Kirkwood–Buff integrals [5, 7, 8] are

$$
\tilde{h}_{\alpha\gamma}(0) = 2 \lim_{s \to 0} \left[\widehat{G}_{\alpha\gamma}(s) - \frac{1}{s} \right], \tag{5.44}
$$

with the results

$$
\tilde{h}_{11}(0) = n\widehat{J} - 2\frac{x_2 K_{22} \widehat{\Omega}'_{22}}{x_1 K_{12} \widehat{\Omega}_{12}} - \frac{2}{nx_1}, \tag{5.45a}
$$

$$
\tilde{h}_{22}(0) = n\widehat{J} - 2\frac{x_1 K_{11} \widehat{\Omega}'_{11}}{x_2 K_{12} \widehat{\Omega}_{12}} - \frac{2}{nx_2}, \tag{5.45b}
$$

$$
\tilde{h}_{12}(0) = n\widehat{J} + 2\frac{\widehat{\Omega}'_{12}}{\widehat{\Omega}_{12}}, \tag{5.45c}
$$

where

$$
\widehat{J} \equiv x_1^2 K_{11} \widehat{\Omega}''_{11} + x_2^2 K_{22} \widehat{\Omega}''_{22} + 2x_1 x_2 K_{12} \left[\widehat{\Omega}''_{12} - \frac{\widehat{\Omega}'_{11} \widehat{\Omega}'_{22} - (\widehat{\Omega}'_{12})^2}{\widehat{\Omega}_{12}} \right]. \tag{5.46}
$$

The knowledge of the Kirkwood–Buff integrals allows us to obtain the isothermal susceptibility via (4.80). The numerator in χ_T^{-1} is a positive definite quantity since

$$
1 + nx_1 x_2 \left[\tilde{h}_{11}(0) + \tilde{h}_{22}(0) - 2\tilde{h}_{12}(0) \right] = \sqrt{1 - 4x_1 x_2 R}. \tag{5.47}
$$

Fig. 5.5 Léon Charles
Prudent van Hove
(1924–1990)
(Photograph reproduced with
permission from CERN,
https://cds.cern.ch/record/
917605)

Therefore, χ_T^{-1} never vanishes, what confirms the classical proof [9] by van Hove
(see Fig. 5.5) about the absence of phase transitions in one-dimensional nearest-
neighbor models.

5.5 Examples

5.5.1 Square Well

As a first application for one-component systems, let us consider the SW potential of
core diameter σ, range σ', and well depth ε (see Table 3.1). The Laplace transforms
of the pair Boltzmann factor $e^{-\beta\phi_{\mathrm{SW}}(r)}$ and of the product $\phi_{\mathrm{SW}}(r)e^{-\beta\phi_{\mathrm{SW}}(r)}$ are

$$\widehat{\Omega}(s) = \frac{e^{\beta\varepsilon}e^{-\sigma s}}{s}\left[1 - \left(1 - e^{-\beta\varepsilon}\right)e^{-(\sigma'-\sigma)s}\right], \qquad (5.48a)$$

$$\widehat{\Upsilon}(s) = -\varepsilon\frac{e^{\beta\varepsilon}e^{-\sigma s}}{s}\left[1 - e^{-(\sigma'-\sigma)s}\right], \qquad (5.48b)$$

respectively. In order to apply the exact results for one-dimensional systems of
Sect. 5.3, we must prevent the SW interaction from extending beyond nearest
neighbors. This implies the constraint $\sigma' \leq 2\sigma$.

According to (5.18) and (5.22a), the number density and the internal energy per
particle are

$$\frac{1}{n} = \frac{1}{\beta p} + \frac{\sigma - \sigma'\left(1 - e^{-\beta\varepsilon}\right)e^{-(\sigma'-\sigma)\beta p}}{1 - \left(1 - e^{-\beta\varepsilon}\right)e^{-(\sigma'-\sigma)\beta p}}, \qquad (5.49a)$$

$$\beta u \equiv \frac{\langle E\rangle}{Nk_B T} = \frac{1}{2} - \beta\varepsilon\frac{1 - e^{-(\sigma'-\sigma)\beta p}}{1 - \left(1 - e^{-\beta\varepsilon}\right)e^{-(\sigma'-\sigma)\beta p}}. \qquad (5.49b)$$

It can be easily checked that the Maxwell relation [see (2.52)]

$$\left(\frac{\partial(\beta u)}{\partial(\beta p)}\right)_\beta = \beta\left(\frac{\partial n^{-1}}{\partial\beta}\right)_{\beta p} \qquad (5.50)$$

Fig. 5.6 Density dependence (where $n^* \equiv n\sigma$) of the compressibility factor $Z = \beta p/n$ of the one-dimensional SW fluid for several reduced temperatures $T^* \equiv k_B T/\varepsilon$ at $\sigma'/\sigma = 1.4$

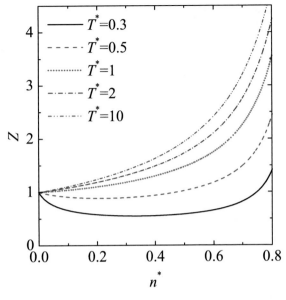

Fig. 5.7 Density dependence of the excess internal energy $\langle E \rangle^{\mathrm{ex}}/N = \langle E \rangle/N - \frac{1}{2}k_B T$ per particle of the one-dimensional SW fluid for several temperatures at $\sigma'/\sigma = 1.4$

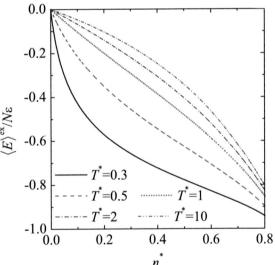

is fulfilled. As illustrations, Figs. 5.6 and 5.7 show the density dependence of the compressibility factor $Z = \beta p/n$ and the excess internal energy $\langle E \rangle^{\mathrm{ex}} = \langle E \rangle - \frac{1}{2}k_B T$ per particle, respectively, for several temperatures at $\sigma'/\sigma = 1.4$. As expected, the pressure and the internal energy decrease as temperature decreases.

As for the RDF, it is convenient to rewrite (5.16) in the series form (5.9), i.e.,

$$\widehat{G}(s) = \frac{1}{n} \sum_{\ell=1}^{\infty} \left[\frac{\widehat{\Omega}(s+\beta p)}{\widehat{\Omega}(\beta p)} \right]^{\ell} = \frac{1}{n} \sum_{\ell=1}^{\infty} \frac{e^{\ell \beta \varepsilon}}{\left[\widehat{\Omega}(\beta p) \right]^{\ell}} (s+\beta p)^{-\ell}$$

$$\times \sum_{j=0}^{\ell} (-1)^j \binom{\ell}{j} \left(1 - e^{-\beta \varepsilon} \right)^j e^{-[j\sigma' + (\ell-j)\sigma](s+\beta p)} . \tag{5.51}$$

Now, taking into account the mathematical property [10]

$$\mathscr{L}^{-1} \left[\frac{e^{-as}}{(s+b)^{\ell}} \right] = \frac{(r-a)^{\ell-1}}{(\ell-1)!} e^{-b(r-a)} \Theta(r-a) , \tag{5.52}$$

where $\mathscr{L}^{-1}[\cdots]$ denotes the inverse Laplace transform, one gets

$$g(r) = \frac{e^{-\beta pr}}{n} \sum_{\ell=1}^{\infty} \sum_{j=0}^{\ell} \Psi_{\ell}^{(j)}(r) , \tag{5.53}$$

with

$$\Psi_{\ell}^{(j)}(r) \equiv \frac{(-1)^j e^{\ell \beta \varepsilon} \ell \left(1 - e^{-\beta \varepsilon} \right)^j}{j!(\ell-j)! \left[\widehat{\Omega}(\beta p) \right]^{\ell}} \left[r - j\sigma' - (\ell-j)\sigma \right]^{\ell-1} \Theta \left(r - j\sigma' - (\ell-j)\sigma \right) . \tag{5.54}$$

This shows that $g(r)$ is discontinuous at $r = \sigma$ and σ', while it presents a kink at $r = 2\sigma, \sigma + \sigma'$, and $2\sigma'$.

Note that, although an infinite number of terms formally appear in (5.53), only the terms up to $\ell = \ell_{\max}$ are actually needed if one is interested in $g(r)$ in the range $r < (\ell_{\max} + 1)\sigma$. In particular, in the range $\sigma < r < 2\sigma$,

$$g(r) = \frac{e^{-\beta pr}}{n\widehat{\Omega}(\beta p)} \begin{cases} e^{\beta \varepsilon} , & \sigma < r < \sigma' , \\ 1 , & \sigma' < r < 2\sigma . \end{cases} \tag{5.55}$$

Figures 5.8 and 5.9 show the RDF for several temperatures (at $n^* = 0.6$) and several densities (at $T^* = 1$), respectively, in the case $\sigma'/\sigma = 1.4$. Needless to say, the RDF becomes more structured as density increases and/or temperature decreases.

The Fourier transform $\tilde{h}(k)$ of the total correlation function is directly obtained from the Laplace transform $\widehat{G}(s)$ via (5.27). This in turn allows one to get the Fourier

Fig. 5.8 Plot of the RDF of
the one-dimensional SW fluid
for several temperatures at
$n^* = 0.6$ and $\sigma'/\sigma = 1.4$

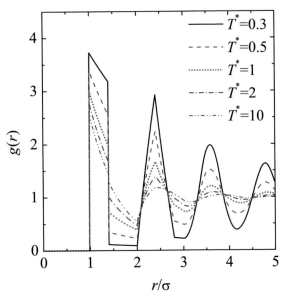

Fig. 5.9 Plot of the RDF of
the one-dimensional SW fluid
for several densities at
$T^* = 1$ and $\sigma'/\sigma = 1.4$

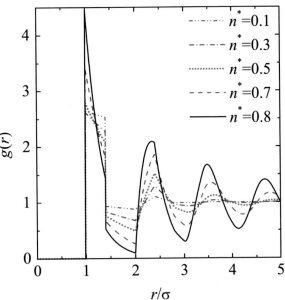

transform $\tilde{c}(k)$ of the DCF through the OZ relation (4.27). The result is

$$n\tilde{c}(k) = 2 - \frac{\bar{A}\left[\frac{k^2}{(\beta p)^2} + 1\right] + A\left(\cos\sigma'k - \frac{k}{\beta p}\sin\sigma'k\right) - \cos\sigma k + \frac{k}{\beta p}\sin\sigma k}{\bar{A}\frac{k^2}{2(\beta p)^2} - (A/\bar{A})\left[1 - \cos(\sigma' - \sigma)k\right]},$$

(5.56)

Fig. 5.10 Plot of the DCF of the one-dimensional SW fluid for several temperatures at $n^* = 0.6$ and $\sigma'/\sigma = 1.4$

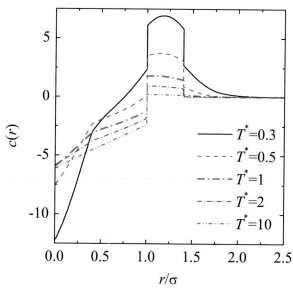

Fig. 5.11 Plot of the DCF of the one-dimensional SW fluid for several densities at $T^* = 1$ and $\sigma'/\sigma = 1.4$

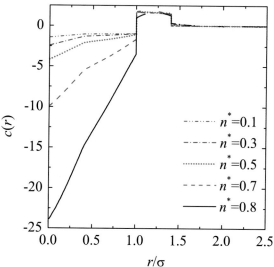

where $A \equiv e^{-(\sigma'-\sigma)\beta p}\left(1 - e^{-\beta\varepsilon}\right)$ and $\bar{A} \equiv 1 - A$. A numerical inverse Fourier transform then yields $c(r)$, while the structure factor $\widetilde{S}(k)$ can be obtained from (4.28). Figures 5.10, 5.11, 5.12, and 5.13 display $c(r)$ and $\widetilde{S}(k)$ for the same states as in Figs. 5.8 and 5.9. We can observe that $c(r)$ has a kink at $r = \sigma' - \sigma$ and is discontinuous at $r = \sigma$ and σ'. Also, the DCF is generally small, but nonzero, in the region $r > \sigma'$, i.e., beyond the range of the interaction potential.

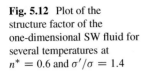

Fig. 5.12 Plot of the structure factor of the one-dimensional SW fluid for several temperatures at $n^* = 0.6$ and $\sigma'/\sigma = 1.4$

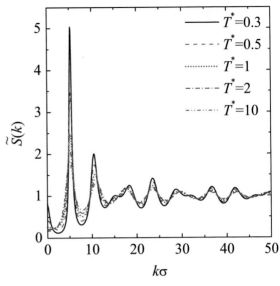

Fig. 5.13 Plot of the structure factor of the one-dimensional SW fluid for several densities at $T^* = 1$ and $\sigma'/\sigma = 1.4$

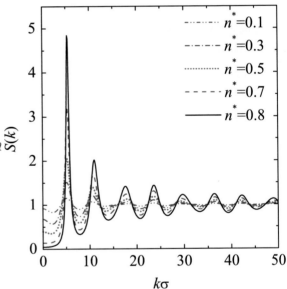

5.5.2   Square Shoulder

As seen from Table 3.1, the SS potential is *formally* equivalent to the SW potential with the change $\varepsilon \to -\varepsilon$. Therefore, the exact one-dimensional solution is still given by (5.48)–(5.56), except that $\varepsilon \to -\varepsilon$. In spite of this simple change, the physics behind the purely repulsive SS potential is quite different from that of the SW potential. This is clearly illustrated by Figs. 5.14–5.21 which are the counterparts of Figs. 5.6–5.13, respectively.

Fig. 5.14 Density
dependence of the
compressibility factor
$Z = \beta p/n$ of the
one-dimensional SS fluid for
several temperatures at
$\sigma'/\sigma = 1.4$

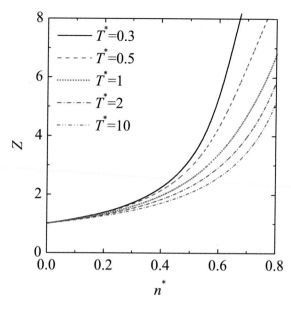

Fig. 5.15 Density
dependence of the excess
internal energy
$\langle E \rangle^{\mathrm{ex}}/N = \langle E \rangle/N - \frac{1}{2}k_B T$
per particle of the
one-dimensional SS fluid for
several temperatures at
$\sigma'/\sigma = 1.4$

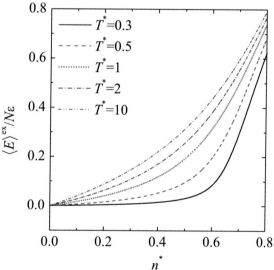

5.5.3 Hard Rods and Sticky Hard Rods

We consider now the sticky-hard-rod (SHR) fluid, which is the one-dimensional
version of the SHS fluid [see (3.4) and Table 3.1].

From (3.5) we have

$$e^{-\beta\phi(r)} = \Theta(r - \sigma) + \tau^{-1}\sigma\delta(r - \sigma) . \qquad (5.57)$$

Fig. 5.16 Plot of the RDF of
the one-dimensional SS fluid
for several temperatures at
$n^* = 0.6$ and $\sigma'/\sigma = 1.4$

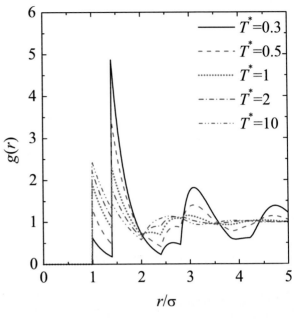

Fig. 5.17 Plot of the RDF of
the one-dimensional SS fluid
for several densities at
$T^* = 1$ and $\sigma'/\sigma = 1.4$

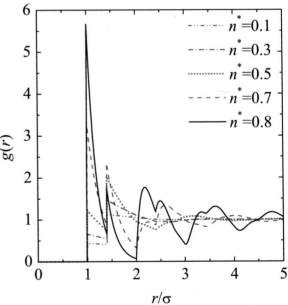

Therefore,

$$\widehat{\Omega}(s) = \left(\tau^{-1}\sigma + \frac{1}{s}\right)e^{-\sigma s} . \tag{5.58}$$

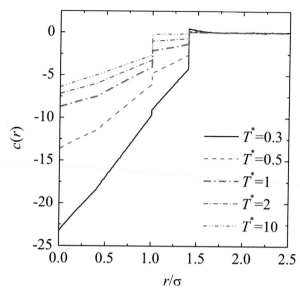

Fig. 5.18 Plot of the DCF of the one-dimensional SS fluid for several temperatures at $n^* = 0.6$ and $\sigma'/\sigma = 1.4$

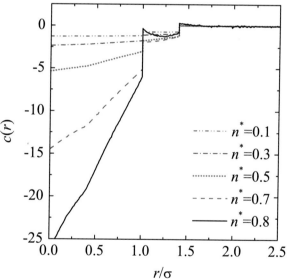

Fig. 5.19 Plot of the DCF of the one-dimensional SS fluid for several densities at $T^* = 1$ and $\sigma'/\sigma = 1.4$

The EoS (5.18) becomes a quadratic equation for the pressure, whose physical solution is

$$Z \equiv \frac{\beta p}{n} = \frac{\sqrt{1 + 4\tau^{-1}n^*/(1 - n^*)} - 1}{2\tau^{-1}n^*}. \tag{5.59}$$

Fig. 5.20 Plot of the structure factor of the one-dimensional SS fluid for several temperatures at $n^* = 0.6$ and $\sigma'/\sigma = 1.4$

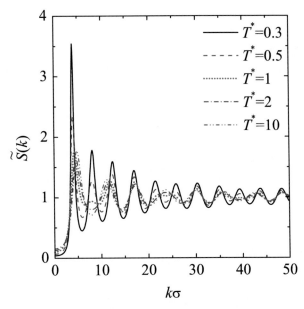

Fig. 5.21 Plot of the structure factor of the one-dimensional SS fluid for several densities at $T^* = 1$ and $\sigma'/\sigma = 1.4$

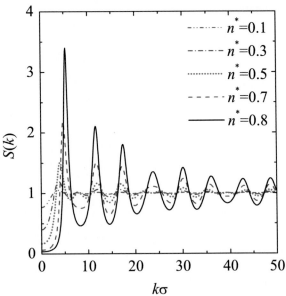

In what concerns the SHR RDF, application of (5.16) gives

$$\widehat{G}(s) = \frac{1}{n} \frac{\left(\tau^{-1} + \frac{\sigma^{-1}}{s+\beta p}\right) e^{-\sigma s}}{\tau^{-1} + \frac{\sigma^{-1}}{\beta p} - \left(\tau^{-1} + \frac{\sigma^{-1}}{s+\beta p}\right) e^{-\sigma s}} = \frac{1}{n} \sum_{\ell=1}^{\infty} \frac{\left(\tau^{-1} + \frac{\sigma^{-1}}{s+\beta p}\right)^{\ell}}{\left(\tau^{-1} + \frac{\sigma^{-1}}{\beta p}\right)^{\ell}} e^{-\ell\sigma s} .$$

$$(5.60)$$

Analogously to the case of (5.51), the last equality in (5.60) allows one to perform the inverse Laplace transform term by term with the result

$$g(r) = \frac{e^{-\beta pr}}{n} \sum_{\ell=1}^{\infty} \Psi_\ell(r) , \qquad (5.61)$$

where

$$\Psi_\ell(r) = \frac{e^{\beta p \ell \sigma}}{\left(\tau^{-1} + \frac{\sigma^{-1}}{\beta p}\right)^\ell} \left[\tau^{-\ell} \delta(r - \ell\sigma) \right.$$

$$\left. + \sigma^{-1} \sum_{j=1}^{\ell} \binom{\ell}{j} \frac{\tau^{-(\ell-j)}}{(j-1)!} \left(\frac{r}{\sigma} - \ell\right)^{j-1} \Theta(r - \ell\sigma) \right]. \qquad (5.62)$$

Thus, the RDF includes a "comb" of Dirac deltas at $r = \sigma, 2\sigma, 3\sigma, \ldots$, plus a "regular" part that otherwise is discontinuous at $r = \sigma, 2\sigma, 3\sigma, \ldots$. Again, only the first $\ell_{\max}$ terms are needed in (5.61) if one is interested in the range $1 \leq r/\sigma < \ell_{\max} + 1$.

Using (5.57), it is straightforward to see that the RDF and the cavity function in the SHR model are related by

$$g(r) = \tau^{-1} \sigma y(\sigma) \delta(r - \sigma) + y(r) \Theta(r - \sigma) . \qquad (5.63)$$

This, together with (5.61) and (5.62), implies the contact value

$$y(\sigma) = \frac{\beta p}{n(1 + \tau^{-1}\beta p\sigma)} . \qquad (5.64)$$

This is useful to obtain the mean potential energy per particle,

$$\frac{u^{\text{ex}}}{\varepsilon} \equiv \frac{\langle E \rangle^{\text{ex}}}{N\varepsilon} = -n\sigma\tau^{-1} y(\sigma) = -\frac{1}{1 + \tau/\beta p\sigma} , \qquad (5.65)$$

where the energy route (4.33) has been particularized to our system.

From (4.27) and (5.27) it is easy to obtain the Fourier transform of the DCF and its inverse transform as [11]

$$\tilde{c}(k) = -\frac{2\beta p}{n} \frac{1 + \tau^{-1}\beta p\sigma}{1 + 2\tau^{-1}\beta p\sigma} \left[\frac{\tau^{-2}\beta p\sigma^2}{1 + \tau^{-1}\beta p\sigma} - \tau^{-1}\sigma \cos\sigma k + \frac{\sin\sigma k}{k} \right.$$

$$\left. + \left(1 + \tau^{-1}\beta p\sigma\right) \beta p \frac{1 - \cos\sigma k}{k^2} \right], \qquad (5.66a)$$

$$c(r) = -\frac{\beta p}{n}\frac{1 + \tau^{-1}\beta p\sigma}{1 + 2\tau^{-1}\beta p\sigma}\left\{\frac{\tau^{-2}\beta p\sigma^2}{1 + \tau^{-1}\beta p\sigma}\delta(r) - \tau^{-1}\sigma\delta(r - \sigma)\right.$$

$$\left.\left[1 + \left(1 + \tau^{-1}\beta p\sigma\right)\beta p\left(\sigma - r\right)\right]\Theta(\sigma - r)\right\}. \tag{5.66b}$$

Note that, in contrast to the SW case [see (5.56)], the inverse Fourier transform $\tilde{c}(k) \to c(r)$ can be performed analytically.

The hard-rod (HR) model is the one-dimensional version of the HS system. It can be seen as a special case of the SHR model with zero stickiness ($\tau^{-1} \to 0$). In that limit, (5.59), (5.62), and (5.66) simply become

$$Z = \frac{1}{1 - n\sigma}, \tag{5.67}$$

$$\Psi_\ell(r) = e^{\beta p\ell\sigma}(\beta p)^\ell\frac{(r - \ell\sigma)^{\ell - 1}}{(\ell - 1)!}\Theta(r - \ell\sigma), \tag{5.68}$$

$$\tilde{c}(k) = -\frac{2\beta p}{n}\left(\frac{\sin\sigma k}{k} + \beta p\frac{1 - \cos\sigma k}{k^2}\right), \tag{5.69a}$$

$$c(r) = -\frac{\beta p}{n}\left[1 + \beta p\left(\sigma - r\right)\right]\Theta(\sigma - r), \tag{5.69b}$$

respectively. From (5.68) we see that $g(r)$ is discontinuous at $r = \sigma$ and has a kink at $r = 2\sigma$. The cavity function $y(r)$ in the hard-core region $0 \le r \le \sigma$ turns out to be given by the same expression as that of $g(r)$ in the first coordination shell $\sigma < r < 2\sigma$ [12], i.e., $y(r) = e^{-\beta p(r-\sigma)}/(1 - n^*)$, $0 \le r \le \sigma$. This implies that $y(r)$ is analytic at $r = \sigma$.

The exact EoS (5.67) for the HR system was obtained independently by Lord Rayleigh (see Fig. 5.22) [13] and Korteweg (see Fig. 5.23) [14] in 1891, and rederived much later by Herzfeld and Goeppert-Mayer (see Fig. 3.2) [15] and Tonks [16].

Figures 5.24, 5.25, 5.26, and 5.27 show $g(r)$ and $\widetilde{S}(k)$ for a HR fluid ($\tau^{-1} = 0$) and for a representative case of a SHR fluid ($\tau = 5$) at several densities.

5.5.4 Mixtures of Nonadditive Hard Rods

As an illustrative example of a one-dimensional mixture, we consider here a nonadditive hard-rod (NAHR) binary mixture [see (3.7)]. The nearest-neighbor interaction condition requires $\sigma_{\alpha\omega} \le \sigma_{\alpha\gamma} + \sigma_{\gamma\omega}$, $\forall(\alpha, \gamma, \omega)$, as illustrated by Fig. 5.28. In the binary case, this condition implies $2\sigma_{12} \ge \max(\sigma_1, \sigma_2)$.

Fig. 5.22 John William
Strutt, 3rd Baron Rayleigh
(1842–1919)
(Photograph from Wikimedia
Commons, https://commons.
wikimedia.org/wiki/File:
Robert_John_Strutt,
_Lord_Rayleigh.
_Photograph_by_Elliott_
%26_F_Wellcome_V0027060.
jpg)

Fig. 5.23 Diederik Johannes
Korteweg (1848–1941)
(Photograph from Wikimedia
Commons, https://en.
wikipedia.org/wiki/File:D.J.
Korteweg.JPG)

The Laplace transform of $e^{-\beta\phi_{\alpha\gamma}(r)}$ is

$$\widehat{\Omega}_{\alpha\gamma}(s) = \frac{e^{-\sigma_{\alpha\gamma}s}}{s} . \tag{5.70}$$

The EoS is obtained by application of (5.39)–(5.41). Figures 5.29 and 5.30 present
the compressibility factor for representative cases of a symmetric and an asymmetric
mixture, respectively.

In the special case of an *additive* mixture, i.e., $\sigma_{12} = \frac{1}{2}(\sigma_1 + \sigma_2)$, one has
$\widehat{\Omega}_{11}(s)\widehat{\Omega}_{22}(s) = \widehat{\Omega}_{12}^2(s)$. Thus, by taking the limit $R \to 0$ in (5.40), one gets

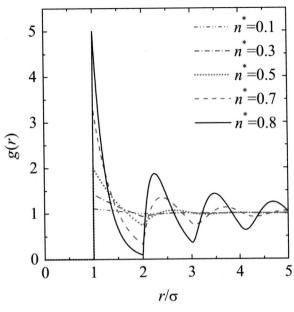

Fig. 5.24 Plot of the RDF of the one-dimensional HR fluid for several densities

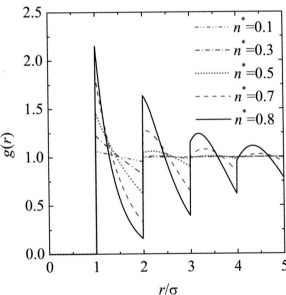

Fig. 5.25 Plot of the "regular" part of the RDF of the one-dimensional SHR fluid for several densities at $\tau = 5$. The complete RDF requires adding $\sum_{\ell=1}^{\infty} \delta(r - \ell\sigma)/n(1 + \tau/\beta p\sigma)^\ell$ to the regular part

$K_{\alpha\gamma} = 1/\widehat{\Omega}_{\alpha\gamma}(\beta p)$. In that case, (5.41) becomes

$$Z = \frac{1}{1 - \eta}, \qquad (5.71)$$

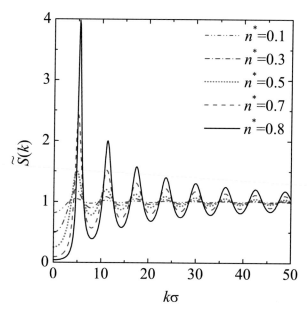

Fig. 5.26 Plot of the structure factor of the one-dimensional HR fluid for several densities

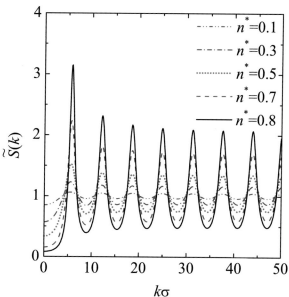

Fig. 5.27 Plot of the structure factor of the one-dimensional SHR fluid for several densities at $\tau = 5$

where $\eta = n(x_1\sigma_1 + x_2\sigma_2)$ is the packing fraction of the one-dimensional mixture. The EoS (5.71) is plotted in Figs. 5.29 and 5.30 as the curves corresponding to $\sigma_{12}/\sigma_1 = 1$ and $\sigma_{12}/\sigma_1 = 0.75$, respectively. It is worth noting that the compressibility factor (5.71) in the additive case is independent of the composition

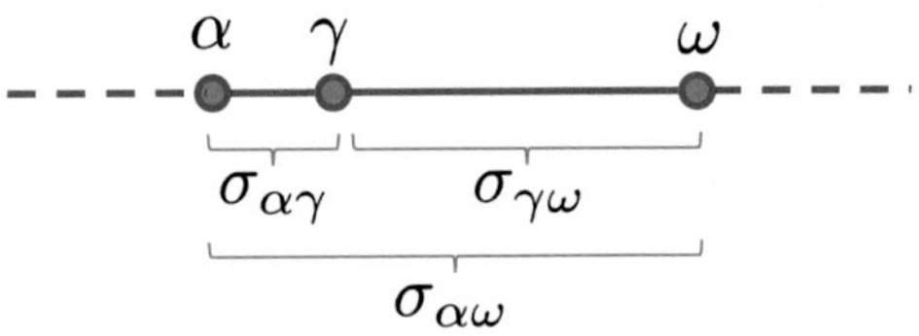

Fig. 5.28 Threshold situation ($\sigma_{\alpha\omega} = \sigma_{\alpha\gamma} + \sigma_{\gamma\omega}$) for nearest-neighbor interaction

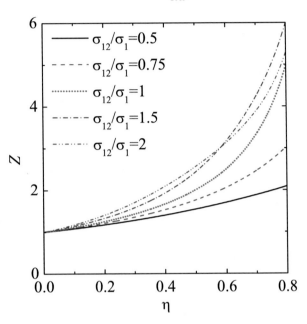

Fig. 5.29 Dependence on the packing fraction $\eta = n(x_1\sigma_1 + x_2\sigma_2)$ of the compressibility factor $Z = \beta p/n$ of an equimolar ($x_1 = x_2 = \frac{1}{2}$) and symmetric ($\sigma_2/\sigma_1 = 1$) one-dimensional NAHR fluid for several values of σ_{12}/σ_1. Note that the cases with $\sigma_{12}/\sigma_1 < 1$ correspond to negative nonadditivity, while those with $\sigma_{12}/\sigma_1 > 1$ correspond to positive nonadditivity (see p. 39)

of the mixture at a fixed packing fraction and, therefore, it is the same as for the one-component HR system [see (5.67)].

As for the structural properties, the recipe described by (5.37) and (5.38) can be easily implemented [4]. In order to obtain the RDFs $g_{\alpha\gamma}(r)$ in real space, we first note that, according to (5.38b),

$$\frac{1}{\widehat{D}(s)} = \sum_{\ell=0}^{\infty} \left[\widehat{Q}_{11}(s) + \widehat{Q}_{22}(s) + \widehat{Q}_{12}^2(s) - \widehat{Q}_{11}(s)\widehat{Q}_{22}(s) \right]^\ell . \tag{5.72}$$

When this is inserted into (5.37), one can express $\widehat{G}_{\alpha\gamma}(s)$ as linear combinations of terms of the form $\widehat{Q}_{11}^{j_1}(s)\widehat{Q}_{22}^{j_2}(s)\widehat{Q}_{12}^{j_3}(s)$. The inverse Laplace transforms $g_{\alpha\gamma}(r) = \mathscr{L}^{-1}\left[\widehat{G}_{\alpha\gamma}(s)\right]$ are readily evaluated by using the property (5.52). The final result can be written as

$$g_{11}(r) = \frac{e^{-\beta pr}}{nx_1} \sum_{\ell=0}^{\infty}\sum_{j_1=0}^{\ell}\sum_{j_2=0}^{\ell-j_1}\sum_{j_3=0}^{\ell-j_1-j_2} \frac{(-1)^{\ell-j_1-j_2-j_3}\ell!}{j_1!j_2!j_3!(\ell-j_1-j_2-j_3)!}$$

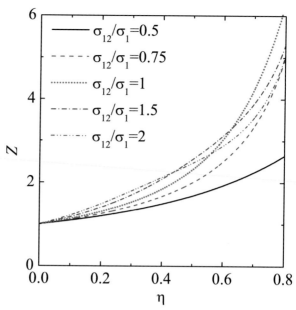

Fig. 5.30 Dependence on the packing fraction $\eta = n(x_1\sigma_1 + x_2\sigma_2)$ of the compressibility factor $Z = \beta p/n$ of an equimolar $(x_1 = x_2 = \frac{1}{2})$ and asymmetric $(\sigma_2/\sigma_1 = 0.5)$ one-dimensional NAHR fluid for several values of σ_{12}/σ_1. Note that the case with $\sigma_{12}/\sigma_1 < 0.75$ corresponds to negative nonadditivity, while those with $\sigma_{12}/\sigma_1 > 0.75$ correspond to positive nonadditivity (see p. 39)

$$\times \left[(1 - \delta_{\ell,0})\Psi^{(2j_3)}_{\ell-j_1-j_3,\ell-j_2-j_3}(r) - \Psi^{(2j_3)}_{\ell-j_1-j_3,\ell+1-j_2-j_3}(r)\right] , \quad (5.73\text{a})$$

$$g_{22}(r) = \frac{e^{-\beta pr}}{nx_2} \sum_{\ell=0}^{\infty}\sum_{j_1=0}^{\ell}\sum_{j_2=0}^{\ell-j_1}\sum_{j_3=0}^{\ell-j_1-j_2} \frac{(-1)^{\ell-j_1-j_2-j_3}\ell!}{j_1!j_2!j_3!(\ell-j_1-j_2-j_3)!}$$

$$\times \left[(1 - \delta_{\ell,0})\Psi^{(2j_3)}_{\ell-j_1-j_3,\ell-j_2-j_3}(r) - \Psi^{(2j_3)}_{\ell+1-j_1-j_3,\ell-j_2-j_3}(r)\right] , \quad (5.73\text{b})$$

$$g_{12}(r) = \frac{e^{-\beta pr}}{n\sqrt{x_1x_2}} \sum_{\ell=0}^{\infty}\sum_{j_1=0}^{\ell}\sum_{j_2=0}^{\ell-j_1}\sum_{j_3=0}^{\ell-j_1-j_2} \frac{(-1)^{\ell-j_1-j_2-j_3}\ell!}{j_1!j_2!j_3!(\ell-j_1-j_2-j_3)!}$$

$$\times \Psi^{(2j_3+1)}_{\ell-j_1-j_3,\ell-j_2-j_3}(r) , \quad (5.73\text{c})$$

where

$$\Psi^{(j_3)}_{j_1,j_2}(r) = (x_1K_{11})^{j_1+j_3/2}(x_2K_{22})^{j_2+j_3/2} \frac{(r - j_1\sigma_1 - j_2\sigma_2 - j_3\sigma_{12})^{j_1+j_2+j_3-1}}{(j_1+j_2+j_3-1)!}$$

$$\times \Theta(r - j_1\sigma_1 - j_2\sigma_2 - j_3\sigma_{12}) . \quad (5.74)$$

Analogously to the cases of (5.53) and (5.61), only the terms $\Psi^{(j_3)}_{j_1,j_2}(r)$ such that $j_1\sigma_1 + j_2\sigma_2 + j_3\sigma_{12} < r_{\max}$ are needed if one is interested in distances $r < r_{\max}$.

Figures 5.31 and 5.32 show $g_{\alpha\gamma}(r)$ for particular binary mixtures with negative and positive nonadditivities, respectively. Although the composition, density, and

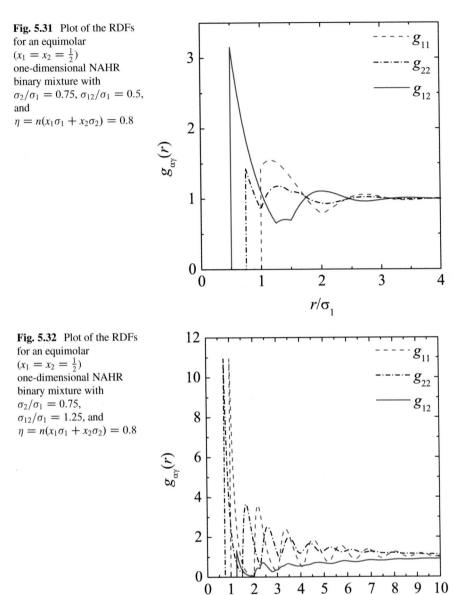

Fig. 5.31 Plot of the RDFs for an equimolar $(x_1 = x_2 = \frac{1}{2})$ one-dimensional NAHR binary mixture with $\sigma_2/\sigma_1 = 0.75$, $\sigma_{12}/\sigma_1 = 0.5$, and $\eta = n(x_1\sigma_1 + x_2\sigma_2) = 0.8$

Fig. 5.32 Plot of the RDFs for an equimolar $(x_1 = x_2 = \frac{1}{2})$ one-dimensional NAHR binary mixture with $\sigma_2/\sigma_1 = 0.75$, $\sigma_{12}/\sigma_1 = 1.25$, and $\eta = n(x_1\sigma_1 + x_2\sigma_2) = 0.8$

sizes σ_1 and σ_2 are common in both cases, a dramatic effect can be observed when changing the cross distance σ_{12}.

Exercises

5.1 Justify (5.10) in view of Fig. 5.4.

5.2 Check (5.11).

5.3 Derive (5.18) and (5.20).

5.4 Derive (5.22).

5.5 Use (5.21), together with $\lim_{s\to 0} s^2 \widehat{\Omega}'(s) = -1$ and $\lim_{s\to 0} \widehat{\Upsilon}(s) = 0$, to check that the set of Eq. (5.22) yields ideal-gas expressions (see Table 2.2) in the limit $p \to 0$.

5.6 Derive (5.27) and (5.28).

5.7 Justify (5.30) physically.

5.8 Derive (5.31).

5.9 Making use of (5.32), prove that (5.35) is implied by the physical condition $\lim_{r\to\infty} p_{\alpha\gamma}^{(1)}(r)/p_{\alpha\nu}^{(1)}(r) =$ independent of α.

5.10 Check that (5.31) reduces to (5.37) and (5.38) in the binary case.

5.11 Derive (5.39) and (5.40) from (5.34) and (5.35).

5.12 Making use of (5.38)–(5.40), check that $\lim_{s\to 0} \widehat{D}(s) = 0$ and that (5.41) can be derived by applying the condition (5.36) to *any* of the three functions (5.37).

5.13 Consider a binary mixture where all the interactions are identical, i.e., $\widehat{\Omega}_{11}(s) = \widehat{\Omega}_{22}(s) = \widehat{\Omega}_{12}(s) = \widehat{\Omega}(s)$. Taking the limit $R \to 0$ in (5.40), check that $K_{11} = K_{22} = K_{12} = K = 1/\widehat{\Omega}(\beta p)$, $\widehat{Q}_{\alpha\gamma}(s) = \sqrt{x_\alpha x_\gamma} K\widehat{\Omega}(s + \beta p)$, $\widehat{D}(s) = 1 - K\widehat{\Omega}(s + \beta p)$. Then, check that (5.37) and (5.41) reduce to (5.16) and (5.18), respectively.

5.14 Using (1.16b), (5.39), and (5.40), check that (5.41) can be obtained from (5.42).

5.15 Taking into account that $\lim_{s\to 0} s\widehat{\Omega}_{\alpha\gamma}(s) = 1$ [sce (5.21)], prove that (5.41) and (5.42) become $n^{\text{id}} = \beta p$ and $G^{\text{id}} = Nk_BT\,[x_1 \ln(\beta p x_1 \Lambda_1) + x_2 \ln(\beta p x_2 \Lambda_2)]$, respectively, in the ideal-gas limit $p \to 0$.

5.16 Derive (5.43).

5.17 Check that (5.42) and (5.43) satisfy the thermodynamic relation (1.14).

5.18 Check the correctness of (5.45).

5.19 Consider again a binary mixture where all the interactions are identical, i.e., $\widehat{\Omega}_{11}(s) = \widehat{\Omega}_{22}(s) = \widehat{\Omega}_{12}(s) = \widehat{\Omega}(s)$. Check that in that case the three equations (5.45) reduce to (5.28).

5.20 Using (5.41) and inserting (5.45) into (4.80), check that the thermodynamic relation $\chi_T = k_B T (\partial n / \partial p)_{T, x_1}$ is indeed satisfied.

5.21 Check (5.47).

5.22 Show that (5.49a) becomes $n^{-1} = (\beta p)^{-1} + B_2^{\mathrm{SW}} + \mathcal{O}(\beta p)$ in the limit of low pressure, where the second virial coefficient B_2^{SW} can be found in Table 3.6.

5.23 Check that (5.49) satisfy the Maxwell relation (5.50).

5.24 Starting from the EoS of the one-dimensional SW fluid, check that the EoS (5.67) of the HR fluid is obtained either in the limit of vanishing well width ($\sigma' \rightarrow \sigma$) or in the limit of infinite temperature ($\beta \varepsilon \rightarrow 0$).

5.25 By using (5.55), check that (5.49b) can be reobtained from the energy route (4.32).

5.26 From (5.53)–(5.55), check that the cavity function $y(r) = e^{\beta \phi(r)} g(r)$ of the one-dimensional SW potential is everywhere continuous.

5.27 Derive (5.56).

5.28 Take the zero wavenumber limit $k \rightarrow 0$ in (5.56) and apply the compressibility route (4.30) to obtain the isothermal susceptibility of the one-dimensional SW fluid. Check that the same result is obtained from (5.24).

5.29 Download and install the Wolfram CDF Player (http://www.wolfram.com/cdf-player) in your computer. Play with the Demonstration of reference [17] to explore how the RDF and the structure factor of the one-dimensional SW and SS fluids change with the well or shoulder width, the reduced temperature $T^* = k_B T / \varepsilon$, and the packing fraction $\eta = n^* = n\sigma$.

5.30 Make the change $\varepsilon \rightarrow -\varepsilon$ in (5.49a) to obtain the EoS of the one-dimensional SS fluid. Then, check that the EoS (5.67) of the HR fluid of length σ' is obtained in the limit of vanishing temperature ($\beta \varepsilon \rightarrow \infty$).

5.31 Make the change $e^{\beta \varepsilon} \rightarrow \tau^{-1} / (\sigma'/\sigma - 1)$ in (5.48a). Then, take the limit $\sigma' \rightarrow \sigma$ to reobtain (5.58).

5.32 Derive (5.59).

5.33 Derive (5.61) and (5.62).

5.34 Make the change $e^{\beta \varepsilon} \rightarrow \tau^{-1} / (\sigma'/\sigma - 1)$ in (5.49b). Then, take the limit $\sigma' \rightarrow \sigma$ to reobtain (5.65).

5.35 Insert (5.58) into (5.20) to obtain

$$\frac{\beta G(T, p, N)}{N} = \beta p \sigma - \ln \frac{\tau^{-1} \sigma + 1/\beta p}{\Lambda(\beta)}$$

for the Gibbs free energy of the SHR model. Next, from the thermodynamic relations of Table 1.1, prove that the internal energy in the isothermal–isobaric ensemble is given by

$$E = \frac{\partial(\beta G)}{\partial \beta} - p\frac{\partial G}{\partial p}.$$

Finally, using the property $\partial \tau^{-1}/\partial\beta = \varepsilon\tau^{-1}$, check that the excess internal energy per particle coincides with (5.65).

5.36 Derive (5.66).

5.37 Make the change $e^{\beta\varepsilon} \rightarrow \tau^{-1}/(\sigma'/\sigma-1)$ in (5.56). Then, take the limit $\sigma' \rightarrow \sigma$ to reobtain (5.66a).

5.38 One can always use (3.4) to assign an *effective* stickiness τ^{-1} to the SW potential. Using (5.59) and (5.65), compare the compressibility factor and the excess internal energy per particle of the one-dimensional SW fluid for the cases of Figs. 5.6 and 5.7 with those of the one-dimensional SHR fluid with the corresponding values of the effective stickiness τ^{-1}.

5.39 Download and install the Wolfram CDF Player (http://www.wolfram.com/cdf-player) in your computer. Play with the Demonstration of reference [18] to explore how the RDF of the one-dimensional SHR fluid changes with the stickiness τ^{-1} and the packing fraction $\eta = n^* = n\sigma$.

5.40 Check (5.71).

5.41 Derive (5.73).

5.42 Check from (5.73) and (5.74) that the contact values of the cavity functions of an NAHR mixture are $y_{\alpha\gamma}(\sigma_{\alpha\gamma}) = g_{\alpha\gamma}(\sigma^+_{\alpha\gamma}) = n^{-1}K_{\alpha\gamma}e^{-\beta p\sigma_{\alpha\gamma}}$. Then, using (5.39), check that in an equimolar binary mixture ($x_1 = x_2 = \frac{1}{2}$) one has $y_{11}(\sigma_1) = y_{22}(\sigma_2)$.

5.43 Consider again a NAHR mixture. Insert the contact values $y_{\alpha\gamma}(\sigma_{\alpha\gamma}) = n^{-1}K_{\alpha\gamma}e^{-\beta p\sigma_{\alpha\gamma}}$ into the virial route (4.88) to check that the resulting expression is equivalent to (5.41).

5.44 Download and install the Wolfram CDF Player (http://www.wolfram.com/cdf-player) in your computer. Play with the Demonstration of reference [19] to explore how the RDFs of the one-dimensional NAHR mixture change with the packing fraction, the composition, and the size ratios.

References

1. Z.W. Salsburg, R.W. Zwanzig, J.G. Kirkwood, J. Chem. Phys. **21**, 1098 (1953)
2. J.L. Lebowitz, D. Zomick, J. Chem. Phys. **54**, 3335 (1971)
3. M. Heying, D.S. Corti, Fluid Phase Equilib. **220**, 85 (2004)

4. A. Santos, Phys. Rev. E **76**, 062201 (2007)
5. A. Ben-Naim, A. Santos, J. Chem. Phys. **131**, 164 (2009)
6. G.B. Rybicki, Astrophys. Space Sci. **14**, 56 (1971)
7. J.G. Kirkwood, F.P. Buff, J. Chem. Phys. **19**, 774 (1951)
8. A. Ben-Naim, *Molecular Theory of Solutions* (Oxford University Press, Oxford, 2006)
9. L. van Hove, Physica **16**, 137 (1950)
10. M. Abramowitz, I.A. Stegun (eds.), *Handbook of Mathematical Functions* (Dover, New York, 1972)
11. S.B. Yuste, A. Santos, J. Stat. Phys. **72**, 703 (1993)
12. A. Malijevský, A. Santos, J. Chem. Phys. **124**, 074508 (2006)
13. L. Rayleigh, Nature **45**, 80 (1891)
14. D.T. Korteweg, Nature **45**, 152 (1891)
15. K.F. Herzfeld, M. Goeppert-Mayer, J. Chem. Phys. **2**, 38 (1934)
16. L. Tonks, Phys. Rev. **50**, 955 (1936)
17. A. Santos, Radial distribution function for one-dimensional square-well and square-shoulder fluids. Wolfram Demonstrations Project (2015), http://demonstrations.wolfram.com/RadialDistributionFunctionForOneDimensionalSquareWellAndSqua/
18. A. Santos, Radial distribution function for sticky hard rods. Wolfram Demonstrations Project (2012), http://demonstrations.wolfram.com/RadialDistributionFunctionForStickyHardRods/
19. A. Santos, Radial distribution functions for nonadditive hard-rod mixtures. Wolfram Demonstrations Project (2015), http://demonstrations.wolfram.com/RadialDistributionFunctionsForNonadditiveHardRodMixtures/

Chapter 6
Density Expansion of the Radial Distribution Function and Approximate Integral Equations

This chapter deals with the derivation of the coefficients of the radial distribution function in its expansion in powers of density. As in Chap. 3, the main steps involving diagrammatic manipulations are justified with simple examples. The classification of diagrams depending on their topology leads to the introduction of the hypernetted-chain and Percus–Yevick approximations, plus other approximate integral equations. The chapter ends with some relations in connection with the internal consistency among different thermodynamic routes in approximate theories.

6.1 Introduction

In analogy with what was said for the EoS $p(n, T)$ in Chap. 3, it is not possible to derive the exact RDF $g(r)$ for a general interaction potential $\phi(r)$ and arbitrary density n and temperature T. As discussed in Chap. 5, an exceptional case is that of one-dimensional systems with interactions restricted to nearest neighbors.

On the other hand, again in analogy with Chap. 3 [see (3.8)], the problem simplifies significantly if the RDF is represented by an expansion in powers of the number density:

$$g(r) = g_0(r) + g_1(r)n + g_2(r)n^2 + \cdots . \tag{6.1}$$

The aim of this chapter is to derive expressions for the *virial* coefficients $g_k(r)$ as functions of T for any (short-range) interaction potential $\phi(r)$. First, a note of caution: although for an ideal gas one has $g^{\mathrm{id}}(r) = 1$ (and $\lim_{n \to 0} Z = Z^{\mathrm{id}} = B_1 = 1$), in a real gas $\lim_{n \to 0} g(r) = g_0(r) \neq 1$. This is so because, even if the density is extremely small, interactions create correlations among particles. For instance, in a HS fluid, $g(r) = 0$ for $r < \sigma$, no matter how large or small the density is (see Fig. 4.4).

© Springer International Publishing Switzerland 2016
A. Santos, *A Concise Course on the Theory of Classical Liquids*,
Lecture Notes in Physics 923, DOI 10.1007/978-3-319-29668-5_6

6.2   External Force: Functional Analysis

Because of the same reasons as in Chap. 3, the most convenient framework to derive the coefficients $g_k(r)$ in (6.1) is provided by the grand canonical ensemble.

As seen from (2.36)–(2.39), the thermodynamic quantities can be obtained in the grand canonical ensemble from derivatives of $\ln \Xi$ with respect to the temperature parameter $\beta = 1/k_B T$ and the chemical potential parameter $\alpha = -\beta\mu$. On the other hand, the correlation functions $n_s(\mathbf{r}^s)$ are given by (4.7b) and it is not obvious at all how they can be related to a derivative of $\ln \Xi$. This is possible, however, by means of a trick consisting in assuming that an *external* potential $u_{\text{ext}}(\mathbf{r})$ is introduced in the system. In that case, the potential energy function becomes

$$\Phi_N(\mathbf{r}^N) \rightarrow \Phi_N(\mathbf{r}^N|u_{\text{ext}}) = \Phi_N(\mathbf{r}^N) + \sum_{i=1}^{N} u_{\text{ext}}(\mathbf{r}_i) . \tag{6.2}$$

The notation $\Phi_N(\mathbf{r}^N|u_{\text{ext}})$ means that Φ_N is a *functional* of the external potential $u_{\text{ext}}(\mathbf{r})$. The associated grand canonical partition function becomes, in analogy with (3.13) and (3.17),

$$\Xi(\alpha, \beta, V|\theta) = 1 + \sum_{N=1}^{\infty} \frac{\hat{z}^N}{N!} \int d\mathbf{r}^N W_N(1, 2, \ldots, N|\theta) , \tag{6.3a}$$

$$\ln \Xi(\alpha, \beta, V|\theta) = \sum_{\ell=1}^{\infty} \frac{\hat{z}^\ell}{\ell!} \int d\mathbf{r}^\ell U_\ell(1, 2, \ldots, \ell|\theta) , \tag{6.3b}$$

where

$$\theta(\mathbf{r}) \equiv e^{-\beta u_{\text{ext}}(\mathbf{r})} , \tag{6.4a}$$

$$W_N(\mathbf{r}^N|\theta) \equiv W_N(\mathbf{r}^N) \prod_{i=1}^{\ell} \theta(\mathbf{r}_i) , \tag{6.4b}$$

$$U_\ell(\mathbf{r}^\ell|\theta) \equiv U_\ell(\mathbf{r}^\ell) \prod_{i=1}^{\ell} \theta(\mathbf{r}_i) . \tag{6.4c}$$

To proceed, we will need the simple functional derivative rule

$$\frac{\delta}{\delta\theta(\mathbf{r})}\theta(\mathbf{r}_1) = \delta(\mathbf{r}_1 - \mathbf{r}) . \tag{6.5}$$

This implies

$$\frac{\delta}{\delta\theta(\mathbf{r})}\prod_{k=1}^{N}\theta(\mathbf{r}_k) = \left[\prod_{k=1}^{N}\theta(\mathbf{r}_k)\right]\sum_{i=1}^{N}\frac{\delta(\mathbf{r}_i - \mathbf{r})}{\theta(\mathbf{r}_i)}, \tag{6.6a}$$

$$\frac{\delta^2}{\delta\theta(\mathbf{r})\delta\theta(\mathbf{r}')}\prod_{k=1}^{N}\theta(\mathbf{r}_k) = \left[\prod_{k=1}^{N}\theta(\mathbf{r}_k)\right]\sum_{i\neq j}\frac{\delta(\mathbf{r}_i - \mathbf{r})\delta(\mathbf{r}_j - \mathbf{r}')}{\theta(\mathbf{r}_i)\theta(\mathbf{r}_j)}, \tag{6.6b}$$

plus similar relations for higher order functional derivatives of $\prod_{k=1}^{N}\theta(\mathbf{r}_k)$. It is then straightforward to obtain the s-body configurational distribution function n_s in the absence of external force ($\theta = 1$) [see (4.7b)] as the sth-order functional derivative of $\varXi(\theta)$ at $\theta = 1$, divided by $\varXi$, i.e.,

$$n_s(\mathbf{r}^s) = \frac{1}{\varXi}\frac{\delta^s\varXi(\theta)}{\delta\theta(\mathbf{r}_1)\delta\theta(\mathbf{r}_2)\cdots\delta\theta(\mathbf{r}_s)}\bigg|_{\theta=1}. \tag{6.7}$$

In particular,

$$n_1(\mathbf{r}_1) = \frac{\delta\ln\varXi(\theta)}{\delta\theta(\mathbf{r}_1)}\bigg|_{\theta=1}, \tag{6.8a}$$

$$n_2(\mathbf{r}_1, \mathbf{r}_2) = \frac{\delta\ln\varXi(\theta)}{\delta\theta(\mathbf{r}_1)}\frac{\delta\ln\varXi(\theta)}{\delta\theta(\mathbf{r}_2)}\bigg|_{\theta=1} + \frac{\delta^2\ln\varXi(\theta)}{\delta\theta(\mathbf{r}_1)\delta\theta(\mathbf{r}_2)}\bigg|_{\theta=1}$$

$$= n_1(\mathbf{r}_1)n_1(\mathbf{r}_2) + \frac{\delta^2\ln\varXi(\theta)}{\delta\theta(\mathbf{r}_1)\delta\theta(\mathbf{r}_2)}\bigg|_{\theta=1}. \tag{6.8b}$$

In (6.8), $n_1(\mathbf{r}) = n = \langle N\rangle/V$ is actually independent of the position $\mathbf{r}$ of the particle, but it is convenient for the moment to keep the notation $n_1(\mathbf{r})$.

6.3 Root and Field Points

Taking into account (6.4c), application of (6.6) yields

$$\frac{\delta}{\delta\theta(\mathbf{r})}\int d\mathbf{r}^\ell U_\ell(\mathbf{r}^\ell|\theta)\bigg|_{\theta=1} = \ell\int d\mathbf{r}_2\cdots d\mathbf{r}_\ell\, U_\ell(\mathbf{r}; \mathbf{r}_2, \ldots, \mathbf{r}_\ell), \tag{6.9a}$$

$$\frac{\delta^2}{\delta\theta(\mathbf{r})\delta\theta(\mathbf{r}')}\int d\mathbf{r}^\ell U_\ell(\mathbf{r}^\ell|\theta)\bigg|_{\theta=1} = \ell(\ell-1)\int d\mathbf{r}_3\cdots d\mathbf{r}_\ell\, U_\ell(\mathbf{r}, \mathbf{r}'; \mathbf{r}_3, \ldots, \mathbf{r}_\ell). \tag{6.9b}$$

Therefore, using (6.3b) and (6.8), we have

$$n_1(\mathbf{r}_1) = \hat{z} + \sum_{\ell=2}^{\infty} \frac{\hat{z}^{\ell}}{(\ell-1)!} \int d\mathbf{r}_2 \cdots d\mathbf{r}_{\ell}\, U_{\ell}(1; 2, \ldots, \ell)\,, \qquad (6.10a)$$

$$n_2(\mathbf{r}_1, \mathbf{r}_2) = n_1(\mathbf{r}_1) n_1(\mathbf{r}_2) + \hat{z}^2 U_2(1, 2)$$

$$+ \sum_{\ell=3}^{\infty} \frac{\hat{z}^{\ell}}{(\ell-2)!} \int d\mathbf{r}_3 \cdots d\mathbf{r}_{\ell}\, U_{\ell}(1, 2; 3, \ldots, \ell)\,. \qquad (6.10b)$$

In the above equations we have distinguished between position variables that are integrated out and those which are not. We will call *field* points to the former and *root* points to the latter (see p. 45). Thus,

$$U_{\ell}(\mathbf{r}; \mathbf{r}_2, \ldots, \mathbf{r}_{\ell}) : \quad \text{Cluster function with 1 root point and } \ell - 1 \text{ field points},$$

$$U_{\ell}(\mathbf{r}, \mathbf{r}'; \mathbf{r}_3, \ldots, \mathbf{r}_{\ell}) : \quad \text{Cluster function with 2 root points and } \ell - 2 \text{ field points}.$$

Based on (3.19), the first few one-root cluster diagrams are

$$\mathfrak{b}_1 = U_1(1) = \circ \qquad , \qquad (6.11a)$$

$$2\mathfrak{b}_2 = \int d\mathbf{r}_2\, U_2(1; 2) = \circ\!\!-\!\!\bullet \quad , \qquad (6.11b)$$

$$6\mathfrak{b}_3 = \int d\mathbf{r}_2 \int d\mathbf{r}_3\, U_3(1; 2, 3) = \text{⟨diagram⟩} + 2\,\text{⟨diagram⟩} + \text{⟨diagram⟩}\,, \qquad (6.11c)$$

$$24\mathfrak{b}_4 = \int d\mathbf{r}_2 \int d\mathbf{r}_3 \int d\mathbf{r}_4\, U_4(1; 2, 3, 4)$$

$$= 6\,\text{⟨diagram⟩} + 6\,\text{⟨diagram⟩} + \text{⟨diagram⟩} + 3\,\text{⟨diagram⟩} + 3\,\text{⟨diagram⟩} + 3\,\text{⟨diagram⟩}$$

$$+ 6\,\text{⟨diagram⟩} + 3\,\text{⟨diagram⟩} + 3\,\text{⟨diagram⟩} + 3\,\text{⟨diagram⟩} + \text{⟨diagram⟩}\,. \qquad (6.11d)$$

In agreement with (3.22), a filled circle means that the integration over that field point is carried out. As a consequence, some of the diagrams in (3.19c) and (3.19d) that were topologically equivalent need (in principle) to be disentangled in (6.11c) and (6.11d), respectively, since the new diagrams are invariant under the permutation of two field points but not under the permutation root ↔ field.

We observe from (6.10a) that the expansion of density in powers of fugacity has the structure

$$n_1(\mathbf{r}_1) = n = \sum_{\ell=1}^{\infty} \ell \mathsf{b}_\ell \hat{z}^\ell \,, \qquad (6.12)$$

with

$$\mathsf{b}_\ell = \frac{1}{\ell!} \sum \text{all clusters with 1 root and } \ell - 1 \text{ field points.} \qquad (6.13)$$

As expected, (6.11), (6.12), and (6.13) are equivalent to (3.29), (3.10), and (3.30), respectively. Note that, due to the translational invariance property (2.62), all the one-root diagrams sharing the same topology have a common value.

Again from (3.19), one can realize that the first few two-root cluster diagrams are

$$U_2(1,2) = \boxed{\text{o——o}} \,, \qquad (6.14a)$$

$$\int d\mathbf{r}_3\, U_3(1,2;3) = \text{⟨diagram⟩} + 2\,\text{⟨diagram⟩} + \text{⟨diagram⟩} \,, \qquad (6.14b)$$

$$\int d\mathbf{r}_3 \int d\mathbf{r}_4\, U_4(1,2;3,4) = 2\,\text{⟨diagram⟩} + 4\,\text{⟨diagram⟩} + 2\,\text{⟨diagram⟩} + 4\,\text{⟨diagram⟩}$$

$$+2\,\text{⟨diagram⟩} + 2\,\text{⟨diagram⟩} + 2\,\text{⟨diagram⟩} + 4\,\text{⟨diagram⟩}$$

$$+4\,\text{⟨diagram⟩} + 2\,\text{⟨diagram⟩} + 2\,\text{⟨diagram⟩} + \text{⟨diagram⟩}$$

$$+4\,\text{⟨diagram⟩} + \text{⟨diagram⟩} + \text{⟨diagram⟩} + \text{⟨diagram⟩} \,. \qquad (6.14c)$$

The diagrams in (6.14) framed with a box are those in which a direct bond between the root particles 1 and 2 exists. We will call them *closed* clusters. The other clusters in which the two root particles are not directly linked will be called *open* clusters.

An immediate property of closed clusters is that they factorize into ○—○ times an *open* cluster. For instance,

$$\left[\triangle\right] = \text{○—○} \times \quad \overset{\circ}{\underset{\circ}{\bullet}} \quad , \tag{6.15a}$$

$$\left[\triangle\right] = \text{○—○} \times \quad \overset{\bullet}{\triangle} \quad , \tag{6.15b}$$

$$\left[\sqcup\right] = \text{○—○} \times \quad \overset{\circ}{\underset{\bullet}{\sqcup}} \quad , \tag{6.15c}$$

$$\left[\sqcup\right] = \text{○—○} \times \quad \overset{\text{○—●}}{\underset{\text{○—●}}{}} \quad . \tag{6.15d}$$

In some cases, the root particles 1 and 2 become isolated *after* factorization. This happens, for instance, in (6.15a), (6.15c), and (6.15d).

6.4   Expansion of the Pair Correlation Function in Powers of Fugacity

According to (6.10b), the coefficients of the expansion of $n_2(1, 2)$ come from two sources: the product $n_1(1)n_1(2)$ and the two-root clusters. The first class is represented by two-root diagrams where particles 1 and 2 are fully isolated [see (6.12) and (6.13)]. The second class includes open and closed clusters, the latter ones factorizing as in (6.15). Taking into account all of this, it is easy to see that the first few coefficients in the expansion of $n_2(1, 2)$ in powers of $\hat{z}$ can be factorized as

$$\hat{z}^2 : 1 + \text{○—○} = e^{-\beta\phi_{12}} , \tag{6.16a}$$

$$\hat{z}^3 : (1 + \text{○—○})\left(2 \, \overset{\circ}{\underset{\text{○—●}}{}} + \overset{\bullet}{\triangle} \right)$$

$$\equiv e^{-\beta\phi_{12}}\alpha_3(1, 2) , \tag{6.16b}$$

$$\hat{z}^4 : (1 + \text{○—○})\frac{1}{2}\left(2\,\overset{\text{○—●}}{\underset{\text{○—●}}{}} + 2\,\overset{\bullet}{\underset{\bullet}{\sqcup}}\overset{\circ}{} + 4\,\overset{\circ}{\underset{\bullet}{\sqcup}}\overset{\circ}{} + 2\,\overset{\circ}{\triangle}\overset{\circ}{} + 2\,\overset{\circ}{\underset{\bullet}{\sqcup}}\overset{\circ}{} \right.$$

$$\left. + 4\,\overset{\bullet}{\underset{\circ}{\sqcup}}\overset{\bullet}{} + 2\,\triangleright\!\!\triangleleft + 4\,\boxtimes + \boxdot + \boxtimes \right)$$

$$\equiv e^{-\beta\phi_{12}}\alpha_4(1, 2) . \tag{6.16c}$$

It can be proved that this factorization scheme extends to all orders. Thus, in general,

$$n_2(\mathbf{r}_1,\mathbf{r}_2) = e^{-\beta\phi(\mathbf{r}_1,\mathbf{r}_2)} \sum_{\ell=2}^{\infty} \alpha_\ell(\mathbf{r}_1,\mathbf{r}_2)\hat{z}^\ell , \qquad (6.17)$$

where

$$\alpha_\ell(\mathbf{r}_1,\mathbf{r}_2) = \frac{1}{(\ell-2)!} \sum \text{all } open \text{ clusters with 2 root points and } \ell-2 \text{ field points.}$$

A note of caution about the nomenclature employed is in order. We say that the diagrams in α_ℓ are *open* because the two root particles are not directly linked. But they are also *clusters* because either the group of ℓ particles are connected or they would be connected if we imagine a bond between the two roots. Having this in mind, we can apply the same criterion as in Chap. 3 [see (3.20) and (3.21)] and classify the open clusters into open reducible clusters and open irreducible clusters (or open stars). Of course, all open clusters with particles 1 and 2 isolated are reducible. In general, the open reducible clusters factorize into products of open irreducible clusters. For instance,

$$(6.18a)$$
$$(6.18b)$$
$$(6.18c)$$
$$(6.18d)$$

Examples of two-root open *irreducible* clusters (open "stars") are

$$(6.19)$$

6.5 Expansion of the Radial Distribution Function in Powers of Density

Equation (6.17) has the structure of (3.9) with $X \to e^{\beta\phi} n_2$, $\bar{X}_0 = \bar{X}_1 = 0$, and $\bar{X}_\ell \to \alpha_\ell$. Elimination of fugacity in favor of density, as in (3.11), allows us to write

$$n_2(\mathbf{r}_1, \mathbf{r}_2) = e^{-\beta\phi(\mathbf{r}_1,\mathbf{r}_2)} \sum_{k=2}^{\infty} \gamma_k(\mathbf{r}_1, \mathbf{r}_2) n^k , \qquad (6.20)$$

where the role of the coefficients X_k is played by γ_k. Now, using the relationship (3.12), we obtain

$$\gamma_2 = 1 , \qquad (6.21a)$$

$$\gamma_3 = \alpha_3 - 4\mathfrak{b}_2 = \qquad (6.21b)$$

$$= \frac{1}{2}\left(2 \quad + 4 \quad + \quad + \quad \right) . \qquad (6.21c)$$

Here we have taken into account that $\mathfrak{b}_1 = \alpha_2 = 1$. The explicit diagrams displayed in (6.21b) and (6.21c) are the ones surviving after making use of (6.11b), (6.11c), (6.16b), (6.16c), and the factorization properties (6.18). In general,

$$\gamma_k(\mathbf{r}_1, \mathbf{r}_2) = \frac{1}{(k-2)!} \sum \text{ all open } \textit{stars} \text{ with 2 root points and } k-2 \text{ field points.}$$
$$(6.22)$$

In analogy with Table 3.3, a summary of the "distillation" process leading to (6.20) is presented in Table 6.1.

Table 6.1 Summary of diagrams contributing to different quantities

Quantity	Expansion in powers of	Coefficient	Diagrams	Equation
$\Xi(\theta)$	Fugacity $(\hat{z})$	$W_N(\theta)/N!$	All (disconnected+clusters)	(6.3a)
$\ln \Xi(\theta)$	Fugacity $(\hat{z})$	$U_\ell(\theta)/\ell!$	Clusters (reducible+stars)	(6.3b)
$n_1(\mathbf{r}_1)$	Fugacity $(\hat{z})$	$\ell \mathfrak{b}_\ell$	1-root open clusters (reducible+stars)	(6.12)
$n_2(\mathbf{r}_1, \mathbf{r}_2)$	Fugacity $(\hat{z})$	$\alpha_\ell(\mathbf{r}_1, \mathbf{r}_2)$	2-root open clusters (reducible+stars)	(6.17)
$n_2(\mathbf{r}_1, \mathbf{r}_2)$	Density (n)	$\gamma_k(\mathbf{r}_1, \mathbf{r}_2)$	2-root open stars	(6.20)

Taking into account the definitions (4.14) (with $s = 2$) and (4.25) of the RDF and the cavity function, respectively, (6.20) can be rewritten as

$$g(r) = e^{-\beta\phi(r)} \left[1 + \sum_{k=1}^{\infty} \gamma_{k+2}(r)n^k \right] , \quad (6.23a)$$

$$y(r) = 1 + \sum_{k=1}^{\infty} \gamma_{k+2}(r)n^k . \quad (6.23b)$$

Thus, the functions $g_k(r)$ in (6.1) are given by $g_k(r) = e^{-\beta\phi(r)}\gamma_{k+2}(r)$. In particular, in the limit $n \to 0$, $g(r) \to g_0(r) = e^{-\beta\phi(r)}$, which differs from the ideal-gas function $g^{id}(r) = 1$, as anticipated. However, $\lim_{n\to 0} y(r) = 1$.

The formal extension of the result $g_0(r) = e^{-\beta\phi(r)}$ to any order in density *defines* the so-called *potential of mean force* $\psi(r)$ from

$$g(r) = e^{-\beta\psi(r)} \Rightarrow \psi(r) = -k_B T \ln g(r) . \quad (6.24)$$

Obviously, $\psi(r) \neq \phi(r)$, except in the limit $n \to 0$. In general,

$$\beta\psi(r) = \beta\phi(r) - \ln y(r) . \quad (6.25)$$

Table 6.2 shows the diagrams contributing to $\gamma_2(r)-\gamma_5(r)$ [1]. As the order k increases, the number of diagrams and their complexity increase dramatically. The simplest diagram (of course, apart from $\gamma_2 = 1$) is the one corresponding to γ_3. More explicitly [see (6.21b)],

$$\gamma_3(r_{12}) = \int d\mathbf{r}_3 f(r_{13})f(r_{23}) . \quad (6.26)$$

6.5.1 Some Examples

In the special case of HSs, where $f(r) = -\Theta(\sigma - r)$ (see Table 3.1), it can be seen from (6.26) that $\gamma_3(r)$ coincides with the intersection volume of two spheres of radius σ whose centers are separated a distance r [2], i.e.,

$$\gamma_3^{HS}(r) = \mathcal{V}_{\sigma,\sigma}(r)$$

$$= \frac{2^{d-1} (\pi/4)^{(d-1)/2}}{\Gamma\left(\frac{d+1}{2}\right)} \sigma^d \Theta(2\sigma - r) B_{1-r^2/4\sigma^2}\left(\frac{d+1}{2}, \frac{1}{2}\right) , \quad (6.27)$$

Table 6.2 Diagrams contributing to $\gamma_2(r)$, $\gamma_3(r)$, $\gamma_4(r)$, and $\gamma_5(r)$ in the expansions (6.23)

Density term	Coefficient	Diagrams
n^0	$\gamma_2 = 1$	
n	γ_3	
$\dfrac{n^2}{2}$	$2\gamma_4$	
$\dfrac{n^3}{6}$	$6\gamma_5$	

where we recall that $B_x(a, b)$ [see (3.64)] is the incomplete beta function [3, 4]. In particular, (3.81a) and (3.82) imply

$$\gamma_3^{\text{HS}}(r) = \sigma^2 \left[2\cos^{-1}\frac{r}{2\sigma} - \frac{r}{\sigma}\sqrt{1 - \frac{r^2}{4\sigma^2}} \right] \Theta(2\sigma - r) , \quad (d = 2) , \quad (6.28a)$$

$$\gamma_3^{\text{HS}}(r) = \frac{\pi}{12}\sigma^3 \left(2 - \frac{r}{\sigma}\right)^2 \left(4 + \frac{r}{\sigma}\right) \Theta(2\sigma - r) , \quad (d = 3) . \quad (6.28b)$$

Let us consider now the SW potential (see Table 3.1). One can check that its associated Mayer function can be written as

$$f_{\text{SW}}(r) = (1 + x_{\text{SW}})f_{\text{HS}}(r) - x_{\text{SW}}f_{\text{HS}'}(r) , \quad x_{\text{SW}} \equiv e^{\beta\varepsilon} - 1 , \quad (6.29)$$

where $f_{\text{HS}}(r)$ and $f_{\text{HS}'}(r)$ are the Mayer functions of HS fluids with diameters σ and σ', respectively. Therefore, from (6.26) we obtain [5]

$$\gamma_3^{\text{SW}}(r) = (1 + x_{\text{SW}})^2 \mathcal{V}_{\sigma,\sigma}(r) - 2x_{\text{SW}}(1 + x_{\text{SW}})\mathcal{V}_{\sigma,\sigma'}(r) + x_{\text{SW}}^2 \mathcal{V}_{\sigma',\sigma'}(r) . \quad (6.30)$$

We recall here that the intersection volume $\mathcal{V}_{a,b}(r)$ is given by (3.80), complemented by (3.82), (3.81a), and (3.81b) for $d = 2$, 3, and 5, respectively. The result for the

SS potential (see again Table 3.1) is obtained from the SW result by the formal replacement $\varepsilon \to -\varepsilon$, i.e., $x_{SW} \to -x_{SS} \equiv -(1 - e^{-\beta\varepsilon})$.

The SHS function $\gamma_3^{SHS}(r)$ can be derived from (6.30) by expanding in powers of $\sigma' - \sigma$ and taking the limit $x_{SW} \to \infty$ with $\tau =$ finite [see (3.4)]. The result is

$$\gamma_3^{SHS}(r) = \mathcal{V}_{\sigma,\sigma}(r) - \frac{2\sigma}{d2^{d-1}\tau}\left[\frac{\partial \mathcal{V}_{a,b}(r)}{\partial b}\right]_{a=b=\sigma} + \left(\frac{\sigma}{d2^{d-1}\tau}\right)^2\left[\frac{\partial^2 \mathcal{V}_{a,b}(r)}{\partial b \partial a}\right]_{a=b=\sigma}.$$

(6.31)

For three-dimensional SHS fluids,

$$\gamma_3^{SHS}(r) = \gamma_3^{HS}(r) - \frac{\pi}{72}\sigma^3\tau^{-1}\left[12\left(2 - \frac{r}{\sigma}\right) - \tau^{-1}\frac{\sigma}{r}\right]\Theta(2\sigma - r), \quad (d = 3).$$

(6.32)

The exact evaluation of higher-order functions $\gamma_k(r)$ becomes much more complicated. On the other hand, each one of the diagrams contributing to $\gamma_4(r)$ [see (6.21c)] has been evaluated for three-dimensional HSs [6–8]. The results are

$$\text{⟨diagram⟩} = \frac{\pi^2}{36}\frac{3(r-1)^4}{35r}(r^3 + 4r^2 - 53r - 162)\Theta(1 - r)$$
$$- \frac{\pi^2}{36}\frac{(r-3)^4}{35r}(r^3 + 12r^2 + 27r - 6)\Theta(3 - r),$$

(6.33a)

$$\text{⟨diagram⟩} = -\frac{\pi^2}{36}\frac{2(r-1)^4}{35r}(r^3 + 4r^2 - 53r - 162)\Theta(1 - r)$$
$$+ \frac{\pi^2}{36}\frac{(r-2)^2}{35r}(r^5 + 4r^4 - 51r^3 - 10r^2 + 479r - 81)\Theta(2 - r),$$

(6.33b)

$$\text{⟨diagram⟩} = \left[\gamma_3^{HS}(r)\right]^2,$$

(6.33c)

$$\text{⟨diagram⟩} = \chi_A(r)\Theta(1 - r) + \chi_B(r)\Theta(\sqrt{3} - r) - \left[\gamma_3^{HS}(r)\right]^2,$$

(6.33d)

where

$$\chi_A(r) \equiv \frac{\pi^2}{630}(r - 1)^4\left(r^2 + 4r - 53 - 162r^{-1}\right)$$
$$- 2\pi\left(\frac{3r^6}{560} - \frac{r^4}{15} + \frac{r^2}{2} - \frac{2r}{15} + \frac{9}{35r}\right)\cos^{-1}\frac{-r^2 + r + 3}{\sqrt{3(4 - r^2)}},$$

(6.34a)

$$\chi_B(r) \equiv \pi\left[-r^2\left(\frac{3r^2}{280} - \frac{41}{420}\right)\sqrt{3 - r^2} - \left(\frac{23}{15}r - \frac{36}{35r}\right)\cos^{-1}\frac{r}{\sqrt{3(4 - r^2)}}\right.$$

$$+\left(\frac{3r^6}{560}-\frac{r^4}{15}+\frac{r^2}{2}+\frac{2r}{15}-\frac{9}{35r}\right)\cos^{-1}\frac{r^2+r-3}{\sqrt{3(4-r^2)}}$$

$$+\left(\frac{3r^6}{560}-\frac{r^4}{15}+\frac{r^2}{2}-\frac{2r}{15}+\frac{9}{35r}\right)\cos^{-1}\frac{-r^2+r+3}{\sqrt{3(4-r^2)}}\right]. \qquad (6.34b)$$

The separate knowledge of each one of the diagrams contributing to $\gamma_k(r)$ for HS fluids allows one to immediately know those diagrams in the case of PS fluids. From Table 3.1 we see that

$$f_{PS}(r) = x_{PS}f_{HS}(r), \quad x_{PS} \equiv 1 - e^{-\beta\varepsilon}. \qquad (6.35)$$

Thus, one simply needs to multiply a given HS diagram by a factor equal to x_{PS} raised to a power equal to the number of bonds to get the corresponding PS diagram. For example, (6.28b), (6.33a), (6.33b), (6.33c), and (6.33d) need to be multiplied by x_{PS}^2, x_{PS}^3, x_{PS}^4, x_{PS}^4, and x_{PS}^5, respectively.

6.6   Equation of State: Virial Coefficients

The knowledge of the coefficients $\gamma_k(r)$ allows us to obtain the virial coefficients $B_k(T)$ defined in (3.8) in a way alternative to the one worked out in Chap. 3. Any of the thermodynamic routes summarized in Table 4.1 can in principle be used. This is schematically summarized by Fig. 6.1. As long as all the exact diagrams in $\gamma_k(r)$ are incorporated, it does not matter which route is employed to get the virial coefficients. The most straightforward route is the virial one [see (4.41)], according

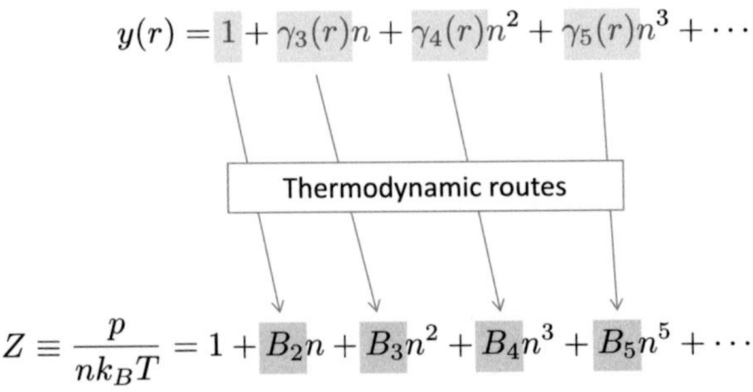

Fig. 6.1 Scheme of the relationship between the functions $\gamma_k(r)$ and the virial coefficients B_k

$$y(r) = 1 + \gamma_3(r)n + \gamma_4(r)n^2 + \gamma_5(r)n^3 + \cdots$$

Thermodynamic routes

$$Z \equiv \frac{p}{nk_BT} = 1 + B_2n + B_3n^2 + B_4n^3 + B_5n^5 + \cdots$$

to which

$$B_k(T) = \frac{1}{2d} \int d\mathbf{r}\, \gamma_k(r) r \frac{\partial f(r)}{\partial r} \,. \tag{6.36}$$

Passing to spherical coordinates and integrating by parts, (6.36) can be rewritten as

$$B_k = 2^{d-1} v_d \int_0^\infty dr\, r^d \gamma_k(r) \frac{\partial f(r)}{\partial r} = -2^{d-1} v_d \int_0^\infty dr\, r^{d-1} f(r) \left[d + r \frac{\partial \gamma_k(r)}{\partial r} \right]$$

$$= -\frac{1}{2} \int d\mathbf{r} f(r) \gamma_k(r) - \frac{1}{2d} \int d\mathbf{r} f(r) \mathbf{r} \cdot \nabla \gamma_k(r) \,, \tag{6.37}$$

where in the last step use has been made of the mathematical property $r\partial/\partial r = \mathbf{r}\cdot\nabla$. In particular, since $\gamma_2(r) = 1$, expression (3.48) for the second virial coefficient is recovered from the second line of (6.37).

As a slightly less simple example, let us consider the third virial coefficient. Taking into account (6.26), we can write

$$\int d\mathbf{r} f(r) \mathbf{r} \cdot \nabla \gamma_3(r) = \int d\mathbf{r}_2 \int d\mathbf{r}_3 f(r_2) f(r_3) \mathbf{r}_2 \cdot \nabla_2 f(r_{23})$$

$$= -\int d\mathbf{r}_2 \int d\mathbf{r}_3 f(r_2) f(r_3) \mathbf{r}_3 \cdot \nabla_2 f(r_{23}) \,. \tag{6.38}$$

In the first step we have chosen $\mathbf{r}_1$ as the origin of coordinates, while in the second step we have exchanged the variables $\mathbf{r}_2 \leftrightarrow \mathbf{r}_3$ and have taken into account that $\nabla_3 f(r_{23}) = -\nabla_2 f(r_{23})$. Next, we express $\int d\mathbf{r} f(r) \mathbf{r} \cdot \nabla \gamma_3(r)$ as the arithmetic mean of the first and second lines, namely

$$\int d\mathbf{r} f(r) \mathbf{r} \cdot \nabla \gamma_3(r) = \frac{1}{2} \int d\mathbf{r}_2 \int d\mathbf{r}_3 f(r_2) f(r_3) \mathbf{r}_{23} \cdot \nabla_{23} f(r_{23})$$

$$= \frac{1}{2} \int d\mathbf{r}_1 \left[\mathbf{r}_1 \cdot \nabla_1 f(r_1) \right] \int d\mathbf{r}_2 f(r_2) f(r_{12}) \,. \tag{6.39}$$

In the second step we have made first the change of variable $\mathbf{r}_3 \to \mathbf{r}_{23}$ and then the change of notation $\mathbf{r}_{23} \to \mathbf{r}_1$. The final result in (6.39) can be recognized from (6.26) and (6.36) as dB_3. Therefore, inserting $\int d\mathbf{r} f(r) \mathbf{r} \cdot \nabla \gamma_3(r) = dB_3$ into (6.37), we get

$$B_3 = -\frac{1}{3} \int d\mathbf{r} f(r) \gamma_3(r) \,, \tag{6.40}$$

in agreement with (3.33b).

Obviously, the equivalence between (6.36), complemented by (6.22), and (3.34) extends to all orders.

As a simple test, let us rederive the first four virial coefficients of three-dimensional HSs. From (6.28b), (6.33), and (6.34), we have

$$
y(\sigma) = 1 + \frac{5}{2}\eta + \frac{1}{4}\left[\frac{2\,707}{70} + \frac{219\sqrt{2}}{35\pi} - \frac{4\,131\cos^{-1}(1/3)}{70\pi}\right]\eta^2 + \cdots, \quad (6.41\text{a})
$$

$$
\widetilde{S}(0) = 1 + 4\pi n \int_0^\infty dr\, r^2 h(r)
$$

$$
= 1 - 8\eta + 34\eta^2 - \left[\frac{6\,534}{35} + \frac{876\sqrt{2}}{35\pi} - \frac{8\,262\cos^{-1}(1/3)}{35\pi}\right]\eta^3 + \cdots.
$$

$$(6.41\text{b})$$

Using the compressibility and virial routes [see (4.29) and (4.88), respectively], it is straightforward to obtain the (rescaled) virial coefficients

$$
b_2 = 4, \quad b_3 = 10, \quad b_4 = \frac{2\,707}{70} + \frac{219\sqrt{2}}{35\pi} - \frac{4\,131\cos^{-1}(1/3)}{70\pi}, \quad (d=3),
$$

$$(6.42)$$

in full agreement with the values presented in Chap. 3 (see Tables 3.8 and 3.11).

6.7 Classification of Open Star Diagrams

We have already seen in (6.22) and (6.23) that all the open star diagrams contribute to the RDF $g(r)$. Now we want to find out which subset of those diagrams contributes to the DCF $c(r)$ defined by the OZ relation (4.26). In the process, we will derive *formally exact* relations between $c(r)$, $h(r)$, and some other functions.

First, we recall from Table 6.2 that

$$
y(r_{12}) = 1 + n\;\triangle\; + \frac{n^2}{2}\left(2\;\sqcup\!\sqcup\; + 4\;\boxtimes\; + \;\square\; + \;\boxtimes\;\right) + \cdots.
$$

$$(6.43)$$

We now introduce the following classification of *open* stars:

- **"Chains" (or nodal diagrams)**, $\mathscr{C}(r)$: Subset of *open* diagrams having at least one *node*. A node is a field particle which must be *necessarily* traversed when going from one root to the other root.

The first few terms in the expansion of $\mathscr{C}(r_{12})$ are

$$\mathscr{C}(r_{12}) = n \; \triangle \; + \frac{n^2}{2}\left(2 \; \sqcup \; + 4 \; \boxtimes \; \right) + \cdots . \tag{6.44}$$

- **Open "parallel" diagrams (or open "bundles"), $\mathscr{P}(r)$**: Subset of *open* diagrams with no nodes, such that there are at least *two* totally independent ("parallel") paths to go from one root to the other root. The existence of parallel paths means that if the roots (together with their bonds) were removed, the resulting diagram would break into two or more pieces.

The function $\mathscr{P}(r)$ is of second order in density:

$$\mathscr{P}(r_{12}) = \frac{n^2}{2} \; \boxdot \; + \cdots . \tag{6.45}$$

- **"Bridge" (or "elementary") diagrams, $\mathscr{B}(r)$**: Subset of *open* diagrams with no nodes, such that there do *not* exist two totally independent ways to go from one root to the other root.

Analogously to $\mathscr{P}(r)$, the bridge function $\mathscr{B}(r)$ is of order n^2:

$$\mathscr{B}(r_{12}) = \frac{n^2}{2} \; \boxtimes \; + \cdots . \tag{6.46}$$

Table 6.3 shows the three classes of star diagrams up to order n^3 [1]. Since the three classes exhaust all the open stars, we can write

$$\boxed{y(r) = 1 + \mathscr{C}(r) + \mathscr{P}(r) + \mathscr{B}(r) .} \tag{6.47}$$

As for the total correlation function, the diagrams contributing to it are

$$h(r_{12}) = (1 + \; \circ\!\!-\!\!\circ \;)y(r_{12}) - 1$$

$$= \; \circ\!\!-\!\!\circ \; + n\left(\triangle \; + \; \triangle \; \right) + \frac{n^2}{2}\left(2 \; \sqcup \; + 4 \; \boxtimes \; + \; \boxdot \right.$$

$$\left. + \; \boxtimes \; + 2 \; \boxdot \; + 4 \; \boxtimes \; + \; \boxtimes \; + \; \boxtimes \; \right) + \cdots . \tag{6.48}$$

Table 6.3 Diagrams contributing to $\gamma_2(r)$, $\gamma_3(r)$, $\gamma_4(r)$, and $\gamma_5(r)$ in the expansions (6.23)

Density term	Coefficient	Diagrams
n^0	$\gamma_2 = 1$	
n	γ_3	
$\dfrac{n^2}{2}$	$2\gamma_4$	
$\dfrac{n^3}{6}$	$6\gamma_5$	

The unframed diagrams, the diagrams framed with an oval, and the diagrams framed with a box contribute to $\mathscr{C}(r)$, $\mathscr{P}(r)$, and $\mathscr{B}(r)$, respectively

In general,

$$h(r) = \sum_{k=0}^{\infty} \frac{n^k}{k!} \sum open \text{ and } closed \text{ stars with 2 roots and } k \text{ field points.}$$

It is not worth classifying the closed diagrams any further. Instead, they join the open bundles to create an augmented class:

- **"Parallel" diagrams (or "bundles")**, $\mathscr{P}^+(r)$: All *closed* diagrams plus the *open* bundles.

The first few ones are

$$\mathscr{P}^{+}(r_{12}) = \text{⚬—⚬} + n \; \triangle + \frac{n^2}{2} \left(\; \square + 2 \; \square + 4 \; \boxed{\diagup} \right.$$

$$\left. + \; \boxed{\diagup} + \; \boxtimes \right) + \cdots \tag{6.49}$$

Obviously,

$$\boxed{h(r) = \mathscr{C}(r) + \mathscr{P}^{+}(r) + \mathscr{B}(r) \,.} \tag{6.50}$$

Why this classification? There are two main reasons. First, open parallel diagrams ($\mathscr{P}$) factorize into products of chains ($\mathscr{C}$) and bridge diagrams ($\mathscr{B}$). For instance,

$$\square = \left(\wedge \right)^2 \,, \tag{6.51a}$$

$$\text{⋀} = \square \times \wedge \,, \tag{6.51b}$$

$$\text{⋈} = \boxed{\diagup} \times \wedge \,, \tag{6.51c}$$

$$\text{⋈} = \square \times \wedge = \left(\wedge \right)^3 \,, \tag{6.51d}$$

$$\text{⋈} = \boxed{\square} \times \wedge \,. \tag{6.51e}$$

As a consequence, it can be proved that

$$\mathscr{P} = \frac{1}{2!}(\mathscr{C} + \mathscr{B})^2 + \frac{1}{3!}(\mathscr{C} + \mathscr{B})^3 + \cdots = e^{\mathscr{C} + \mathscr{B}} - (1 + \mathscr{C} + \mathscr{B}) \,.$$

This is equivalent to

$$\boxed{\mathscr{C} + \mathscr{B} = \ln(1 + \mathscr{C} + \mathscr{P} + \mathscr{B}) \,.} \tag{6.52}$$

Making use of (6.52) in (6.47), we obtain $\ln y = \mathscr{C} + \mathscr{B}$ or, equivalently,

$$\ln g(r) = -\beta\phi(r) + \mathscr{C}(r) + \mathscr{B}(r) .$$

(6.53)

The second important reason for the classification of open stars is that, as we are about to see, the chains ($\mathscr{C}$) *do not* contribute to the DCF $c(r)$. Let us first rewrite (6.48) by overlining the chains:

(6.54)

Next, the OZ relation (4.26) can be iterated to yield

$$c = h - nh * h + n^2 h * h * h - n^3 h * h * h * h + \cdots ,$$

(6.55)

where the asterisk denotes a convolution integral. It turns out that the diagrams representing those convolutions are always chains. For instance,

(6.56a)

(6.56b)

Inserting (6.54) and (6.56) into (6.55), one obtains

(6.57)

Thus, as anticipated, all chain diagrams cancel out! This is not surprising after all since the chains are the open diagrams that more easily can be "stretched out", thus allowing particles 1 and 2 to be correlated via intermediate particles, even if the distance r_{12} is much larger than the interaction range. Note, however, that the DCF is not limited to closed diagrams but also includes the open diagrams with no nodes. Therefore,

$$c(r) = \mathscr{P}^+(r) + \mathscr{B}(r) \, .$$

(6.58)

From (6.47), (6.50), (6.53), and (6.58) we can extract the chain function in three alternative ways:

$$\mathscr{C}(r) = e^{\beta\phi(r)}g(r) - 1 - \mathscr{P}(r) - \mathscr{B}(r) \, ,$$ (6.59a)

$$\mathscr{C}(r) = \ln g(r) + \beta\phi(r) - \mathscr{B}(r) \, ,$$ (6.59b)

$$\mathscr{C}(r) = h(r) - c(r) \, .$$ (6.59c)

Combination of (6.59a) and (6.59c) yields

$$c(r) = g(r)\left[1 - e^{\beta\phi(r)}\right] + \mathscr{P}(r) + \mathscr{B}(r) \, .$$

(6.60)

Similarly, combining (6.59b) and (6.59c) one gets

$$c(r) = g(r) - 1 - \ln g(r) - \beta\phi(r) + \mathscr{B}(r) \, .$$

(6.61)

6.8 Approximate Closures

As already remarked, the OZ (4.26) *defines* $c(r)$. Therefore, it is *not* a closed equation. Likewise, (6.60) and (6.61) are formally exact, but they are not closed either since they have the structure $c(r) = \mathfrak{c}_1[h(r), \mathscr{P}(r) + \mathscr{B}(r)]$ and $c(r) = \mathfrak{c}_2[h(r), \mathscr{B}(r)]$, respectively, where

$$\mathfrak{c}_1[X, Y] \equiv (1 + X)\left(1 - e^{\beta\phi}\right) + Y \, ,$$ (6.62a)

$$\mathfrak{c}_2[X, Y] \equiv X - \ln(1 + X) - \beta\phi + Y \, .$$ (6.62b)

However, if an *approximate closure* of the form $c(r) = \mathfrak{c}_{\text{approx}}[h(r)]$ is assumed, the OZ relation becomes an *approximate closed integral equation*:

$$h(r) = \mathfrak{c}_{\text{approx}}[h(r)] + n \int d\mathbf{r}' \, \mathfrak{c}_{\text{approx}}[h(r')]h(|\mathbf{r} - \mathbf{r}'|) \, .$$

(6.63)

In contrast to a truncated density expansion, a closure $c(r) = \mathfrak{c}_{\text{approx}}[h(r)]$ is applied to *all orders* in density. In most of the cases, it is an ad hoc approximation whose usefulness must be judged a posteriori. The two prototype closures are the

Fig. 6.2 Jerome K. Percus
(b. 1926)
(Photograph courtesy of J.K.
Percus, http://physics.as.nyu.
edu/object/JeromePercus.
html)

Fig. 6.3 George J. Yevick
(1922–2011)
[Photograph courtesy of
David Yevick, from D. Yevick
and H. Yevick, *Fundamental
Math and Physics for
Scientists and Engineers*
(Wiley, Hoboken, NJ, 2014)]

hypernetted-chain (HNC) [9, 10] and the Percus–Yevick (PY, see Figs. 6.2 and 6.3)
[11] ones.

6.8.1   Hypernetted-Chain and Percus–Yevick Approximate Integral Equations

The HNC closure consists in setting $\mathscr{B}(r) = 0$ in (6.61), i.e., $\mathfrak{c}_{\mathrm{HNC}}[h] = \mathfrak{c}_2[h, 0]$, so
that

$$\boxed{c(r) = g(r) - 1 - \ln g(r) - \beta\phi(r) , \quad (\mathrm{HNC}) .} \tag{6.64}$$

Similarly, the PY closure is obtained by setting $\mathscr{P}(r) + \mathscr{B}(r) = 0$ in (6.60), i.e., $\mathfrak{c}_{PY}[h] = \mathfrak{c}_1[h, 0]$, what results in

$$\boxed{c(r) = g(r)\left[1 - e^{\beta\phi(r)}\right], \quad (PY).}$$
(6.65)

By inserting the above closures into the OZ relation (4.26) we obtain the HNC and PY integral equations, respectively:

$$\text{HNC} \Rightarrow \ln\left[g(r)e^{\beta\phi(r)}\right] = -n\int d\mathbf{r}'\,\left\{\ln\left[g(r')e^{\beta\phi(r')}\right] - h(r')\right\}h(|\mathbf{r} - \mathbf{r}'|), \quad (6.66a)$$

$$\text{PY} \Rightarrow g(r)e^{\beta\phi(r)} - 1 = -n\int d\mathbf{r}'\,\left[g(r')e^{\beta\phi(r')} - 1 - h(r')\right]h(|\mathbf{r} - \mathbf{r}'|). \quad (6.66b)$$

Interestingly, if one *formally* assumes that $y(r) \equiv g(r)e^{\beta\phi(r)} \approx 1$ and applies the linearization property $\ln\left[g(r)e^{\beta\phi(r)}\right] \rightarrow g(r)e^{\beta\phi(r)} - 1$, then the HNC integral equation (6.66a) becomes the PY integral equation (6.66b). On the other hand, the PY approximation stands by itself, even if $y(r)$ is not close to 1.

A few comments are in order. First, the density expansion of $h_{HNC}(r)$ and $y_{HNC}(r)$ can be obtained from the closed integral equation (6.66a) by iteration. It turns out that not only the bridge diagrams disappear, but also *some* chain and open parallel (or bundle) diagrams (of order n^3 and higher) are not retained either [1]. This is because the neglect of $\mathscr{B}(r)$ at the level of (6.61) propagates to other non-bridge diagrams at the level of (6.47). For instance, while (6.52) is an identity, we cannot neglect $\mathscr{B}(r)$ on both sides, i.e., $\mathscr{C} \neq \ln(1 + \mathscr{C} + \mathscr{P})$. A similar comment applies to $h_{PY}(r)$ and $y_{PY}(r)$, in which case some chain diagrams disappear along with *all* the bridge and open parallel diagrams. This is illustrated by comparison between Tables 6.3 and 6.4.

Another interesting feature is that all the diagrams neglected in the density expansion of $y_{HNC}(r)$ are neglected in the density expansion of $y_{PY}(r)$ as well. However, the latter neglects extra diagrams which are retained by $y_{HNC}(r)$ (see Table 6.4). Thus, one could think that the HNC equation is *always* a better approximation than the PY equation. On the other hand, this is not necessarily the case, especially for HS-like systems. In those cases the diagrams neglected in the PY approximation may cancel each other to a reasonable degree, so that adding more diagrams (as HNC does) may actually worsen the result. For instance, the combination of the two diagrams neglected by the PY approximation to first order in density is

(6.67)

where the dotted line on the right-hand side means an e-bond between the field particles 3 and 4, i.e., a pair Boltzmann factor $e(r_{34}) \equiv 1 + f(r_{34}) = e^{-\beta\phi(r_{34})}$. The right-hand side diagram of (6.67) is a simple example of the diagrams introduced

Table 6.4 Diagrams contributing to $\gamma_2(r)$, $\gamma_3(r)$, $\gamma_4(r)$, and $\gamma_5(r)$ in the expansions (6.23)

Density term	Coefficient	Diagrams
n^0	$\gamma_2 = 1$	∘ ∘
n	γ_3	(diagram)
$\dfrac{n^2}{2}$	$2\gamma_4$	2 (diagram) $+4$ (diagram) $+$ (diagram) $+$ (diagram)
$\dfrac{n^3}{6}$	$6\gamma_5$	6 (diagram) $+6$ (diagram) $+12$ (diagram) $+12$ (diagram) $+6$ (diagram) $+12$ (diagram) $+12$ (diagram) $+6$ (diagram) $+6$ (diagram) $+6$ (diagram) $+6$ (diagram) $+6$ (diagram) $+12$ (diagram) $+$ (diagram) $+3$ (diagram) $+12$ (diagram) $+12$ (diagram) $+3$ (diagram) $+3$ (diagram) $+6$ (diagram) $+6$ (diagram) $+6$ (diagram) $+3$ (diagram) $+$ (diagram)

The diagrams framed with a box are neglected by the HNC approximation, while the diagrams enclosed with a box or with an oval are neglected by the PY approximation

Fig. 6.4 William Graham Hoover (b. 1937) (Photograph by Baidurya Bhattacharya, courtesy of Wm. G. Hoover)

by Ree and Hoover (see Fig. 6.4) [12], which are obtained from the standard Mayer diagrams by substituting $1 = e(r) - f(r)$ for all field–field and field–root pairs not connected by an f-bond. For instance,

$$2 \text{(diagram)} + 4 \text{(diagram)} = 2 \text{(diagram)} - 2 \text{(diagram)} . \tag{6.68}$$

For HS-like interactions, the f-bonds force particles to remain close together, while the e-bonds force them to be apart. These competing conditions make some Ree–Hoover diagrams, like the one in (6.67), to be almost negligible.

Fig. 6.5 Plot of $\frac{1}{2}$ 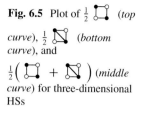 *(top curve)*, $\frac{1}{2}$ 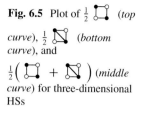 *(bottom curve)*, and $\frac{1}{2}\left(\right.$ 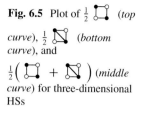 $+$ 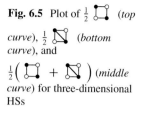 $\left.\right)$ *(middle curve)* for three-dimensional HSs

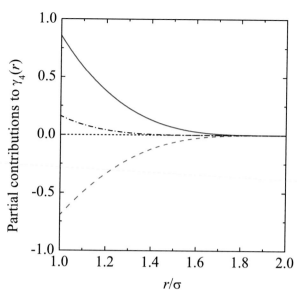

In the HS specific case, the three diagrams in (6.67) vanish if $r_{12} > 2\sigma$ since in that case it is impossible that either particle 3 or particle 4 can be separated from both 1 and 2 a distance smaller than σ. If $r_{12} < 2\sigma$, the only configurations which contribute to the Ree–Hoover diagram on the right-hand side of (6.67) are those where $r_{13} < \sigma$, $r_{23} < \sigma$, $r_{14} < \sigma$, and $r_{24} < \sigma$ *but* $r_{34} > \sigma$. It is obvious that those configurations represent a smaller volume than the ones contributing to any of the two diagrams on the left-hand side of (6.67), especially if $r_{12} > \sigma$. In fact, as can be seen from (6.33c) and (6.33d), the right-hand side of (6.67) vanishes if $r > \sqrt{3}\sigma$ in the three-dimensional case.

The distance dependencies of the three diagrams in (6.67) are plotted in Fig. 6.5 in the range $1 \leq r/\sigma \leq 2$ for three-dimensional HSs. Figure 6.6 displays the exact function $\gamma_4(r)$ (obtained by adding the four contributing diagrams shown in the third row of Table 6.4), together with the HNC approximation $\gamma_4^{HNC}(r)$ (obtained by adding the first three diagrams and neglecting the fourth one) and the PY approximation $\gamma_4^{PY}(r)$ (obtained by adding the first two diagrams and neglecting the third and fourth ones). We observe that the complete star diagram 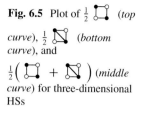 neglected by the HNC approximation is negative, what yields $\gamma_4^{HNC}(r) > \gamma_4(r)$. On the other hand, the diagram 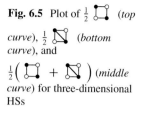 is almost equally positive, so that the sum 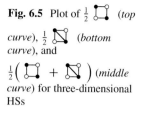 $+$ 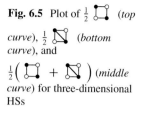 is weakly positive (see Fig. 6.5). As a consequence, $\gamma_4^{PY}(r) < \gamma_4(r)$ but the PY error $|\gamma_4^{PY}(r) - \gamma_4(r)|$ is clearly much smaller than the HNC one $|\gamma_4^{HNC}(r) - \gamma_4(r)|$.

Being approximate, the RDF $g(r)$ obtained from either the PY or the HNC integral equations is *not* thermodynamically consistent, i.e., in general, virial route $\neq$ chemical-potential route $\neq$ compressibility route $\neq$ energy route. However, it can be proved that the virial and energy routes are equivalent in the HNC approximation for *any* interaction potential [13, 14].

Fig. 6.6 Plot of $\gamma_4(r)$ for three-dimensional HSs. The *solid, dashed,* and *dash-dotted curves* correspond to the exact, HNC, and PY functions, respectively

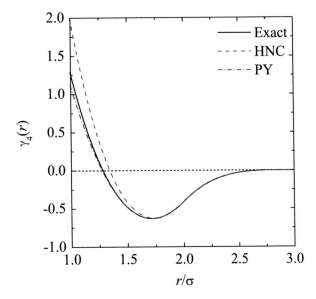

As a simple illustration of the thermodynamic inconsistency problem, let us consider again the fourth virial coefficient of three-dimensional HS fluids. From (6.28b) and (6.33), we can easily obtain

$$y(\sigma) = 1 + \frac{5}{2}\eta + \left(4 + \frac{25}{8}\lambda\right)\eta^2 + \cdots , \quad (d = 3), \quad (6.69a)$$

$$\widetilde{S}(0) = 1 + 4\pi n \int_0^\infty dr\, r^2 h(r)$$

$$= 1 - 8\eta + 34\eta^2 - \left(108 - \frac{2\,357}{105}\lambda\right)\eta^3 + \cdots , \quad (d = 3), \quad (6.69b)$$

where a factor λ has been attached to the diagram , so that the PY and HNC approximations correspond to $\lambda = 0$, and $\lambda = 1$, respectively. Notice the difference between (6.69) and the exact result (6.41). Using again the compressibility and virial routes [see (4.29) and (4.88), respectively], one can obtain

$$\text{Virial route} \Rightarrow b_4 = 16 + \frac{25}{2}\lambda , \quad (d = 3), \quad (6.70a)$$

$$\text{Compressibility route} \Rightarrow b_4 = 19 - \frac{2\,357}{420}\lambda , \quad (d = 3). \quad (6.70b)$$

The values obtained by setting $\lambda = 0$ (PY) and $\lambda = 1$ (HNC) are shown in Table 6.5, which also includes the results obtained through the chemical-potential route [15, 16]. As can be observed, the three PY values are closer to the exact result $b_4 \simeq$

Table 6.5 Fourth virial coefficient b_4 for three-dimensional HSs, according to the virial, chemical-potential, and compressibility routes in the PY and HNC approximations

PY			HNC		
Virial	Chemical-potential	Compressibility	Virial	Chemical-potential	Compressibility
16	$\frac{67}{4} = 16.75$	19	$\frac{57}{2} = 28.5$	$\frac{227}{8} = 28.375$	$\frac{5623}{420} \simeq 13.388$

Table 6.6 Some characteristic closures of the form $\mathscr{B}(r) = \mathfrak{B}_{appr}[\overline{\gamma}(r)]$

Closure	Reference	$\mathfrak{B}_{appr}[\overline{\gamma}(r)]$
HNC	[9, 10]	0
PY	[11]	$\ln[1 + \overline{\gamma}(r)] - \overline{\gamma}(r)$
Verlet (modified)	[31, 32]	$-a_{V,1}\dfrac{[\overline{\gamma}(r)]^2}{1 + a_{V,2}\overline{\gamma}(r)}$, $\quad a_{V,1} = \dfrac{1}{2}$, $\quad a_{V,2} = \dfrac{4}{5}$
Martynov–Sarkisov	[33]	$\sqrt{1 + 2\overline{\gamma}(r)} - \overline{\gamma}(r) - 1$
Rogers–Young	[34]	$\ln\left\{1 + \dfrac{\exp\left[(1 - e^{-a_{RY}r})\overline{\gamma}(r)\right] - 1}{1 - e^{-a_{RY}r}}\right\} - \overline{\gamma}(r)$, $\quad a_{RY} = \dfrac{4}{25}$
Ballone–Pastore–Galli–Gazzillo	[35]	$[1 + a_{BPGG}\overline{\gamma}(r)]^{1/a_{BPGG}} - \overline{\gamma}(r) - 1$, $\quad a_{BPGG} = \dfrac{15}{8}$

18.365 than any of the three HNC values. Note also that the virial and chemical-potential routes yield rather similar values in both approximations, especially in the HNC case.

What makes the PY integral equation particularly appealing is that it admits a non-trivial *exact* solution for three-dimensional HS [17–19] and SHS [20] liquids, their additive mixtures [21–23], and their generalizations to $d = $ odd [24–28]. We will return to this point in Chap. 7.

6.8.2 A Few Other Closures

Apart from the classical PY and HNC approximations, many other ones have been proposed in the literature [13, 29, 30]. Most of them are formulated by closing the formally *exact* relation (6.61) with an *approximation* of the form $\mathscr{B}(r) = \mathfrak{B}_{appr}[\overline{\gamma}(r)]$, where

$$\boxed{\overline{\gamma}(r) \equiv h(r) - c(r)} \tag{6.71}$$

is the *indirect* correlation function, which is made of all the chain diagrams, i.e., $\overline{\gamma}(r) = \mathscr{C}(r)$, as can be seen from (6.50) and (6.58). A few examples of closure relations are displayed in Table 6.6, where, for comparison, the HNC and PY closures are included as well.

The parameters $a_{V,1}$, $a_{V,2}$, a_{RY}, and a_{BPGG} are usually determined by ensuring thermodynamic consistency between the virial and compressibility routes and, therefore, they depend in principle on the thermodynamic state (n, T) of the system. The numerical values given in Table 6.6 are empirical fractional values adequate for HSs.

6.8.3   Linearized Debye–Hückel and Mean Spherical Approximations

Some other closures are proposed in forms different from $c(r) = \mathfrak{c}_{\text{approx}}[h(r)]$ or $\mathscr{B}(r) = \mathfrak{B}_{\text{approx}}[\overline{\gamma}(r)]$. Two simple examples are the linearized Debye–Hückel (LDH) theory (see Figs. 6.7 and 6.8) and the mean spherical approximation (MSA).

Fig. 6.7 Peter Joseph William Debye (1884–1966) (Photograph from Wikimedia Commons, http://en. wikipedia.org/wiki/File: Debije-boerhaave.jpg)

Fig. 6.8 Erich Hückel (1896–1980) (Photograph from Wikimedia Commons, http://en. wikipedia.org/wiki/File: Hueckel.jpg)

The LDH theory consists in retaining only the *linear* chain diagrams in the expansion of the cavity function $y(r)$ [see (6.43)]:

$$w(r) = n \; \circ\!\!-\!\!\bullet\!\!-\!\!\circ + n^2 \; \circ\!\!-\!\!\bullet\!\!-\!\!\bullet\!\!-\!\!\circ + n^3 \; \circ\!\!-\!\!\bullet\!\!-\!\!\bullet\!\!-\!\!\bullet\!\!-\!\!\circ + \cdots, \tag{6.72}$$

where we have denoted by $w(r) \equiv y(r) - 1$ the *shifted* cavity function. This apparently crude approximation is justified in the case of *long-range* interactions (like Coulomb's) since the linear chains are the most divergent diagrams but their sum gives a convergent result [29]. The approximation (6.72) is also valid for *bounded* potentials in the high-temperature limit [36]. For those potentials $|f(r)|$ can be made arbitrarily small by increasing the temperature and thus, at any order in density, the linear chains (having the least number of bonds) are the dominant ones.

In Fourier space, (6.72) becomes

$$\text{LDH} \Rightarrow \tilde{w}(k) = n \left[\tilde{f}(k)\right]^2 + n^2 \left[\tilde{f}(k)\right]^3 + n^3 \left[\tilde{f}(k)\right]^4 + \cdots = \frac{n \left[\tilde{f}(k)\right]^2}{1 - n\tilde{f}(k)} . \tag{6.73}$$

The conventional Debye–Hückel theory is obtained from (6.73) by assuming that (i) $|w(r)| \ll 1$, so that $\ln y(r) \approx w(r)$, and (ii) $f(r) \approx -\beta\phi(r)$. In that case (6.25) yields

$$\beta\tilde{\psi}(k) \approx \beta\tilde{\phi}(k) - \tilde{w}(k) \approx \frac{\beta\tilde{\phi}(k)}{1 + n\beta\tilde{\phi}(k)} \tag{6.74}$$

for the Fourier transform of the potential of mean force.

Another approximation closely related to the LDH one (6.73) is the MSA for *soft* potentials. By a soft potential we refer to any function $\phi(r)$ satisfying the conditions

$$\lim_{r \to 0} r^d \phi(r) = 0 , \quad \lim_{r \to \infty} r^d \phi(r) = 0 . \tag{6.75}$$

While the second condition (meaning that the potential is sufficiently short ranged) is quite general, the first condition actually defines the restricted class of soft potentials. It includes bounded potentials (such as the Gaussian-core model [37] or the PS model [38]), logarithmically diverging potentials [39], or even potentials diverging algebraically as $\phi(r) \sim r^{-s}$ with $s < d$. On the other hand, conventional molecular models (such as HS, SW, SS, or LJ fluids) are excluded from the class of potentials (6.75). A consequence of the conditions (6.75) is that $\tilde{\phi}(0) = $ finite.

To derive the MSA for soft potentials, we first start from the identity $h(r) = f(r)y(r) + y(r) - 1$. Next, in the same spirit as the assumption (i) above, we assume (i') $f(r)y(r) \approx f(r)$, so that $\tilde{h}(k) \approx \tilde{f}(k) + \tilde{w}(k)$. Insertion of (6.73) then yields

$$\tilde{h}(k) \approx \frac{\tilde{f}(k)}{1 - n\tilde{f}(k)} . \tag{6.76}$$

According to the OZ relation (4.27), the approximation (6.76) is equivalent to $\tilde{c}(k) \approx \tilde{f}(k)$. Going back to real space, $c(r) \approx f(r)$. Finally, repeating the assumption (ii) above, i.e., $f(r) \approx -\beta\phi(r)$, we get

$$\text{MSA} \Rightarrow c(r) = -\beta\phi(r) \Rightarrow \tilde{h}(k) = \frac{-\beta\widetilde{\phi}(k)}{1 + n\beta\widetilde{\phi}(k)} . \tag{6.77}$$

It must be noted that, in the MSA, the DCF is independent of density but differs from its correct zero-density limit $c(r) \to f(r)$ [see (6.57)]. The MSA $c(r) = -\beta\phi(r)$ is also known as the random-phase approximation [40].

As said before, the MSA (6.77) has usually been applied to soft potentials [41]. For potentials with a hard core at $r = \sigma$ plus an attractive tail for $r \geq \sigma$, the MSA (6.77) is replaced by the double condition [40]

$$\begin{cases} g(r) = 0 , & r < \sigma , \\ c(r) = -\beta\phi(r) , & r > \sigma . \end{cases} \tag{6.78}$$

This version of the MSA is exactly solvable for Yukawa fluids [42, 43], which is characterized by the interaction potential

$$\phi(r) = \begin{cases} \infty , & r < \sigma , \\ -K\dfrac{e^{-z(r-\sigma)}}{r} , & r > \sigma . \end{cases} \tag{6.79}$$

6.9   Some Thermodynamic Consistency Relations in Approximate Theories

As sketched in Fig. 4.12, an *approximate* RDF $g(r)$ does not guarantee thermodynamic consistency among the different routes. However, there are a few cases where either a partial consistency or a certain simple relationship may exist.

6.9.1   Are the Virial-Route HNC and the Compressibility-Route Percus–Yevick Values of the Fourth Virial Coefficient Related?

As summarized in Fig. 6.1, the knowledge of the coefficients $\gamma_k(r)$ in the density expansion of the cavity function allows one to obtain the virial coefficients B_k. In general, unless the functions $\gamma_k(r)$ are exact, the virial coefficients B_k will depend on the thermodynamic route followed. This is exemplified by Table 6.5 for the fourth virial coefficient predicted by the HNC and PY approximations for a three-dimensional HS fluid.

Here, we will consider an arbitrary potential and focus on the compressibility route [see (4.29)] and the virial route [see (4.41)], denoting the corresponding virial coefficients by $B_k^{(c)}$ and $B_k^{(v)}$, respectively.

From Table 6.5 we can observe the simple relation $B_4^{(HNC,v)}/B_4^{(PY,c)} = \frac{3}{2}$. Is that simple fractional number restricted to three-dimensional HSs? Does the ratio $B_4^{(HNC,v)}/B_4^{(PY,c)}$ depend on the dimensionality d of the system? Does it depend (in general) on the temperature T and on the interaction potential $\phi(r)$? Do the answers to the above questions depend on whether the system is a mixture or not? We will see now that, quite interestingly, the simple relation

$$B_4^{(HNC,v)}(T) = \frac{3}{2}B_4^{(PY,c)}(T)$$ (6.80)

turns out to hold regardless of the value of d, the form of $\phi(r)$, and the detailed composition (one-component or multicomponent) of the system [44].

Let us start by writing again (6.36) for the virial route:

$$B_k^{(v)}(T) = \frac{1}{2d} \int d\mathbf{r}\, \gamma_k(r) r \frac{\partial f(r)}{\partial r}\ .$$ (6.81)

As for the compressibility route, from (4.29) one has

$$\chi_T(n, T) = 1 + n \int d\mathbf{r}\, \{[f(r) + 1]\, y(r) - 1\}$$

$$= 1 + \chi_{T,2}(T)n + \chi_{T,3}(T)n^2 + \chi_{T,4}(T)n^3 + \cdots ,$$ (6.82)

where

$$\chi_{T,2}(T) = \int d\mathbf{r} f(r)\ ,$$ (6.83a)

$$\chi_{T,k}(T) = \int d\mathbf{r}\, [f(r) + 1]\, \gamma_k(r)\ , \quad k \geq 3\ .$$ (6.83b)

Then, taking into account (1.39), we obtain

$$B_2^{(c)}(T) = -\frac{1}{2}\chi_{T,2}(T) , \tag{6.84a}$$

$$B_3^{(c)}(T) = -\frac{1}{3}\left[\chi_{T,3}(T) - \chi_{T,2}^2(T)\right] , \tag{6.84b}$$

$$B_4^{(c)}(T) = -\frac{1}{4}\left[\chi_{T,4}(T) - 2\chi_{T,2}(T)\chi_{T,3}(T) + \chi_{T,2}^3(T)\right] . \tag{6.84c}$$

Let us now particularize to the HNC and PY approximations. Since $\gamma_3(r)$ is exactly retained by those approximations (see second row of Table 6.4), the same holds for the third virial coefficient B_3, regardless of the thermodynamic route. On the other hand, $\gamma_4^{(PY)} \neq \gamma_4^{(HNC)} \neq \gamma_4^{(exact)}$ (see third row of Table 6.4). Therefore, it can be expected that

$$B_4^{(PY,v)} \neq B_4^{(PY,c)} \neq B_4^{(HNC,v)} \neq B_4^{(HNC,c)} \neq B_4^{(exact)} . \tag{6.85}$$

In order to account for all possibilities (exact, HNC, and PY) for the function $\gamma_4(r)$, let us construct a "tunable" function

$$\gamma_4 = \frac{1}{2}\left(2\ \text{[diagram]} + 4\ \text{[diagram]} + \lambda_1\ \text{[diagram]} + \lambda_2\ \text{[diagram]}\right), \tag{6.86}$$

in analogy with what was done in (6.69). The cases $(\lambda_1, \lambda_2) = (1, 1)$, $(1, 0)$, and $(0, 0)$ correspond to $\gamma_4^{(exact)}$, $\gamma_4^{(HNC)}$, and $\gamma_4^{(PY)}$, respectively. Inserting (6.86) into (6.81), one has

$$B_4^{(v)} = \frac{1}{4d}\left(2\ \text{[diagram]} + 4\ \text{[diagram]} + \lambda_1\ \text{[diagram]} + \lambda_2\ \text{[diagram]}\right), \tag{6.87}$$

where a dashed line [not to be confused with the meaning of dotted lines in (6.67) and (6.68)] denotes a factor $r\partial f(r)/\partial r$. By integrating by parts, the following properties can be proved [44]:

$$\text{[diagram]} = -\frac{3d}{4}\ \text{[diagram]} , \tag{6.88a}$$

$$\text{[diagram]} + \frac{1}{4}\ \text{[diagram]} = -\frac{3d}{4}\ \text{[diagram]} , \tag{6.88b}$$

$$\text{[diagram]} = -\frac{d}{2}\ \text{[diagram]} . \tag{6.88c}$$

Consequently,

$$B_4^{(v)} = -\frac{3}{8}\,\square - \frac{3}{4}\,\square\!\!\!\diagup - \frac{\lambda_2}{8}\,\boxtimes + \frac{\lambda_1 - 1}{4d}\,\square\!\!\!\diagup . \tag{6.89}$$

In the case of the compressibility route, (6.83) yields

$$\chi_{T,2} = \circ\!\!-\!\!\bullet , \tag{6.90a}$$

$$\chi_{T,3} = \wedge + \triangle , \tag{6.90b}$$

$$\chi_{T,4} = \square + \frac{2 + \lambda_1}{2}\,\square + 2\,\square\!\!\!\diagup + \frac{4 + \lambda_1 + \lambda_2}{2}\,\square\!\!\!\diagup + \frac{\lambda_2}{2}\,\boxtimes , \tag{6.90c}$$

where in (6.90c) use has been made of the property

$$\square\!\!\!\diagup = \square\!\!\!\diagup = V^{-1}\,\square\!\!\!\diagup . \tag{6.91}$$

Noting that

$$\chi_{T,2}\chi_{T,3} = \square + \square\!\!\!\diagup , \qquad \chi_{T,2}^3 = \square , \tag{6.92}$$

and using (6.84c), we finally obtain

$$B_4^{(c)} = -\frac{2 + \lambda_1}{8}\,\square - \frac{4 + \lambda_1 + \lambda_2}{8}\,\square\!\!\!\diagup - \frac{\lambda_2}{8}\,\boxtimes . \tag{6.93}$$

Comparison between (6.89) and (6.93) shows that

$$\boxed{ B_4^{(v)}\Big|_{\lambda_1=1,\lambda_2=\frac{3\lambda}{2+\lambda}} = \frac{3}{2+\lambda}\,B_4^{(c)}\Big|_{\lambda_1=\lambda_2=\lambda} . } \tag{6.94}$$

In the case of the exact $\gamma_4(r)$ we have $\lambda = 1$ in both sides of (6.94) and therefore $B_4^{(\text{exact},v)} = B_4^{(\text{exact},c)}$, as expected. On the other hand, the choice $\lambda = 0$ makes the left- and right-hand sides correspond to the HNC and PY approximations, respectively, and then (6.94) reduces to the sought result (6.80).

More in general, (6.94) implies that for any approximation of the class $\lambda_1 = \lambda_2$ there exists a specific approximation of the class $\lambda_1 = 1$, such that the compressibility and virial values, respectively, of B_4 are proportional to each other. The connection between both classes is schematically illustrated in Fig. 6.9. Interestingly, the largest deviation of the proportionality factor from 1 occurs in the

Fig. 6.9 The *diagonal* (labeled *c*) and *vertical* (labeled *v*) *lines* represent the classes of approximations $\lambda_1 = \lambda_2$ and $\lambda_1 = 1$, respectively. The *dashed tie lines* connect the pairs of approximations whose respective values of $B_4^{(c)}$ and $B_4^{(v)}$ are related by (6.94)

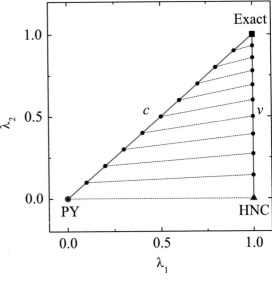

Fig. 6.10 Temperature dependence of the fourth virial coefficient for the three-dimensional PS fluid [8]. The *thick line* represents the exact results, while $B_4^{(PY,v)}$, $B_4^{(PY,c)}$, $B_4^{(PY,e)}$, $B_4^{(HNC,v)}$, and $B_4^{(HNC,c)}$ are represented by the *dash-dotted, dashed, dotted with crosses, dash-dot-dotted, and dotted lines*, respectively. The *circles* represent the values of $\frac{3}{2}B_4^{(PY,c)}$, which fall on the $B_4^{(HNC,v)}$ curve

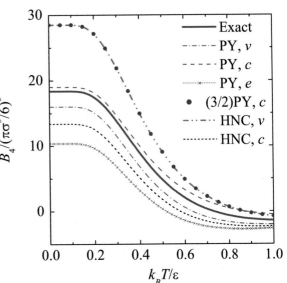

case of the PY and HNC pair. The proof of (6.94) can be easily extended to mixtures [44].

The simple relationship (6.80) is confirmed by analytical results for the three-dimensional PS model [8] and for the one-dimensional PSW model [45]. For instance, the temperature-dependence of $B_4^{(exact)}$, $B_4^{(PY,v)}$, $B_4^{(PY,c)}$, $B_4^{(PY,e)}$ (energy route), $B_4^{(HNC,v)} = B_4^{(HNC,e)}$, and $B_4^{(HNC,c)}$ is displayed in Fig. 6.10 for the PS fluid. As can be observed, the best approximation in the low-temperature regime

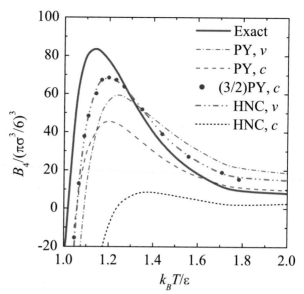

Fig. 6.11 Temperature dependence of the fourth virial coefficient for the three-dimensional SW fluid with $\sigma'/\sigma = \frac{3}{2}$ [5]. The *thick line* represents the exact results, while $B_4^{(PY,v)}$, $B_4^{(PY,c)}$, $B_4^{(HNC,v)}$, and $B_4^{(HNC,c)}$ are represented by the *dash-dotted, dashed, dash-dot-dotted*, and *dotted lines*, respectively. The *circles* represent the values of $\frac{3}{2}B_4^{(PY,c)}$, which fall on the $B_4^{(HNC,v)}$ curve

(up to $T^* \equiv k_B T/\varepsilon \simeq 0.71$) is provided by $B_4^{(PY,c)}$. In the intermediate range ($0.71 \lesssim T^* \lesssim 1.04$), however, $B_4^{(PY,v)}$ presents the best agreement. Finally, for $T^* \gtrsim 1.04$ (not shown in Fig. 6.10) the best performance turns out to correspond to $B_4^{(HNC,v)} = B_4^{(HNC,e)}$. What is especially interesting is that (6.80) is exactly verified.

As for the three-dimensional SW fluid, Fig. 6.11 shows numerical values [5] describing the temperature-dependence of $B_4^{(exact)}$, $B_4^{(PY,v)}$, $B_4^{(PY,c)}$, $B_4^{(HNC,v)}$, and $B_4^{(HNC,c)}$ for the SW system with $\sigma'/\sigma = \frac{3}{2}$. It can be observed that $B_4^{(HNC,v)}$ and $B_4^{(PY,c)}$ give the most accurate results for $1 \lesssim T^* \lesssim 1.5$ and $T^* \gtrsim 1.5$, respectively. Moreover, both approximations are (within numerical uncertainties) clearly consistent with (6.80), as expected.

6.9.2   Energy and Virial Routes in the Linearized Debye–Hückel and Mean Spherical Approximations

As already said in Sect. 6.8.1, the energy and virial routes are fully equivalent in the HNC approximation. Now we will see that the same property holds in the LDH approximation (6.73) [46] and in the MSA (6.77) for soft potentials [47].

6.9.2.1   Linearized Debye–Hückel Approximation

We start by recalling the energy and virial routes (4.33) and (4.41), respectively. In terms of the shifted cavity function $w(r) = y(r) - 1$, they are given by

$$u^{\text{ex}} \equiv \frac{\langle E \rangle^{\text{ex}}}{N} = -\frac{n}{2} \int d\mathbf{r} \, [1 + w(r)] \frac{\partial f(r)}{\partial \beta} \,, \tag{6.95a}$$

$$Z \equiv \frac{\beta p}{n} = 1 + \frac{n}{2d} \int d\mathbf{r} \, [1 + w(r)] \, \mathbf{r} \cdot \nabla f(r) \,. \tag{6.95b}$$

The consistency condition between both routes is provided by the Maxwell relation [see (1.38)]

$$n \frac{\partial u^{\text{ex}}}{\partial n} = \frac{\partial Z}{\partial \beta} \,. \tag{6.96}$$

Given the mathematical identity

$$-\int d\mathbf{r} f(r) = \frac{1}{d} \int d\mathbf{r} \, \mathbf{r} \cdot \nabla f(r) \,, \tag{6.97}$$

the consistency condition (6.96) becomes

$$-\frac{\partial}{\partial n} \left[n \int d\mathbf{r} \, w(r) \frac{\partial f(r)}{\partial \beta} \right] = \frac{1}{d} \frac{\partial}{\partial \beta} \left[\int d\mathbf{r} \, w(r) \mathbf{r} \cdot \nabla f(r) \right] \,. \tag{6.98}$$

Since the LDH approximation (6.73) is formulated in Fourier space, it is convenient to express the spatial integrals in (6.98) as wavevector integrals:

$$\frac{\partial}{\partial n} \left[n \int d\mathbf{k} \, \tilde{w}(k) \frac{\partial \tilde{f}(k)}{\partial \beta} \right] = \frac{1}{d} \frac{\partial}{\partial \beta} \left\{ \int d\mathbf{k} \, \tilde{w}(k) \nabla_{\mathbf{k}} \cdot \left[\mathbf{k} \tilde{f}(k) \right] \right\} \,. \tag{6.99}$$

We now make use of the mathematical identity

$$\frac{\partial}{\partial \beta} \left\{ \tilde{w}(k) \, \nabla_{\mathbf{k}} \cdot \left[\mathbf{k} \tilde{f}(k) \right] \right\} = \mathbf{k} \cdot \left[\frac{\partial \tilde{w}(k)}{\partial \beta} \nabla_{\mathbf{k}} \tilde{f}(k) - \frac{\partial \tilde{f}(k)}{\partial \beta} \nabla_{\mathbf{k}} \tilde{w}(k) \right]$$

$$+ d \frac{\partial \tilde{w}(k)}{\partial \beta} \tilde{f}(k) + \nabla_{\mathbf{k}} \cdot \left[\mathbf{k} \tilde{w}(k) \frac{\partial \tilde{f}(k)}{\partial \beta} \right] \tag{6.100}$$

to rewrite (6.99) as

$$
\frac{\partial}{\partial n}\left[n\int d\mathbf{k}\,\tilde{w}\,(k)\,\frac{\partial \tilde{f}\,(k)}{\partial \beta}\right] = \frac{1}{d}\int d\mathbf{k}\,\mathbf{k}\cdot\left[\frac{\partial \tilde{w}\,(k)}{\partial \beta}\nabla_{\mathbf{k}}\tilde{f}\,(k) - \frac{\partial \tilde{f}\,(k)}{\partial \beta}\nabla_{\mathbf{k}}\tilde{w}\,(k)\right]
$$
$$
+ \int d\mathbf{k}\,\frac{\partial \tilde{w}\,(k)}{\partial \beta}\tilde{f}\,(k) \ . \tag{6.101}
$$

It must be emphasized that no approximations have been carried out so far. There-fore, *any* $\tilde{w}(k)$ satisfying condition (6.101) gives thermodynamically consistent results via the energy and virial routes.

Let us suppose a *closure* relation of the form

$$
\tilde{w}(k) = n^{-1}\mathfrak{w}[n\tilde{f}(k)] \ , \quad \mathfrak{w}[z] = \text{arbitrary} \ . \tag{6.102}
$$

This implies the relations

$$
\frac{\partial}{\partial n}\,[n\tilde{w}\,(k)] = \mathfrak{w}_1[n\tilde{f}\,(k)]\tilde{f}\,(k) \ , \tag{6.103a}
$$

$$
\frac{\partial \tilde{w}\,(k)}{\partial \beta} = \mathfrak{w}_1[n\tilde{f}\,(k)]\frac{\partial \tilde{f}\,(k)}{\partial \beta} \ , \tag{6.103b}
$$

$$
\nabla_{\mathbf{k}}\tilde{w}\,(k) = \mathfrak{w}_1[n\tilde{f}\,(k)]\nabla_{\mathbf{k}}\tilde{f}\,(k) \ , \tag{6.103c}
$$

where $\mathfrak{w}_1[z] \equiv d\mathfrak{w}[z]/dz$. It is then straightforward to check that the energy-virial consistency condition (6.101) is identically satisfied.

As a corollary, the LDH approximation (6.73) belongs to the scaling class (6.102) with the particular choice $\mathfrak{w}[z] = z^2/(1-z)$, what closes the proof.

6.9.2.2 Mean Spherical Approximation for Soft Potentials

The proof in the case of the MSA (6.77) follows along similar lines [47]. Now, instead of (6.95), we start from the energy and virial routes (4.32) and (4.40), respectively, written as

$$
u^{ex} = \frac{n}{2}\int d\mathbf{r}\,[1 + h(r)]\,\frac{\partial\,[\beta\phi(r)]}{\partial \beta} \ , \tag{6.104a}
$$

$$
Z = 1 - \frac{n}{2d}\int d\mathbf{r}\,[1 + h(r)]\,\mathbf{r}\cdot\nabla\,[\beta\phi(r)] \ . \tag{6.104b}
$$

We observe that (6.95) becomes (6.104) with the formal changes $w(r) \to h(r)$ and $f(r) \to -\beta\phi(r)$. Since all the steps leading from (6.96) to (6.101) are purely technical, it is clear that we obtain a consistency condition analogous to (6.101), except for the formal changes $\tilde{w}(k) \rightarrow \tilde{h}(k)$ and $\tilde{f}(k) \rightarrow -\beta\tilde{\phi}(k)$. Consequently, that consistency condition is automatically satisfied by closures of the form

$$\tilde{h}(k) = n^{-1}\mathfrak{h}[-n\beta\tilde{\phi}(k)] \,, \quad \mathfrak{h}[z] = \text{arbitrary} \,. \tag{6.105}$$

As shown in (6.77), the MSA for soft potentials belongs to that class of closures with the particular choice $\mathfrak{h}[z] = z/(1-z)$.

6.9.2.3   The Free-Energy Route

As we saw in Sect. 4.5.5, the free-energy route (4.56) or (4.57) includes the energy and virial routes as just two particular choices of the protocol $\phi^{(\xi)}(r)$: the energy scaling (4.58) for the energy route and the spatial scaling (4.61) for the virial route. Given that these two routes are thermodynamically equivalent in the LDH approximation and in the MSA for soft potentials, one can wonder whether that equivalence extends to any protocol. In other words, is the Helmholtz free energy obtained through (4.56) or (4.57) independent of the protocol choice? As we will see, the answer is indeed affirmative for the classes of approximations (6.102) and (6.105), which include the LDH approximation and the MSA for soft potentials, respectively.

Let us start by rewriting (4.57) in the form

$$\begin{aligned}
\beta a^{\text{ex}} &= -\frac{n}{2}\int_0^1 d\xi \int d\mathbf{r}\, \frac{\partial f^{(\xi)}(r)}{\partial \xi}\left[1 + w^{(\xi)}(r)\right] \\
&= -\frac{n}{2}\tilde{f}(0) - \frac{n}{2}\int \frac{d\mathbf{k}}{(2\pi)^d}\int_0^1 d\xi\, \tilde{w}^{(\xi)}(k)\frac{\partial \tilde{f}^{(\xi)}(k)}{\partial \xi} \,.
\end{aligned} \tag{6.106}$$

Now we assume approximations of the class (6.102), i.e., $\tilde{w}^{(\xi)}(k) = n^{-1}\mathfrak{w}[n\tilde{f}^{(\xi)}(k)]$, and introduce the function

$$\mathfrak{w}_0[z] \equiv \int_0^z dz'\, \mathfrak{w}[z'] \,. \tag{6.107}$$

Then, the formally exact expression (6.106) becomes

$$\beta a^{\text{ex}} = -\frac{n}{2}\tilde{f}(0) - \frac{1}{2n}\int \frac{d\mathbf{k}}{(2\pi)^d}\, \mathfrak{w}_0[n\tilde{f}(k)] \,. \tag{6.108}$$

As anticipated, the final result is independent of the specific protocol $\phi^{(\xi)}(r)$ chosen. In the particular case of the LDH approximation, $\mathfrak{w}[z]=z^2/(1-z) \Rightarrow \mathfrak{w}_0[z]$

$= -z(1 + z/2) - \ln(1 - z)$, so that

$$
\boxed{
\begin{aligned}
\beta a^{\mathrm{ex}} &= -\frac{n}{2}\tilde{f}(0) + \frac{1}{2n}\int \frac{d\mathbf{k}}{(2\pi)^d}\left\{ n\tilde{f}(k)\left[1 + \frac{n\tilde{f}(k)}{2}\right]\right. \\
&\quad \left. + \ln\left[1 - n\tilde{f}(k)\right]\right\}, \qquad\qquad \text{(LDH)} .
\end{aligned}}
\tag{6.109}
$$

From the thermodynamic relations (1.36) it is straightforward to obtain

$$
u^{\mathrm{ex}} = -\frac{n}{2}\frac{\partial \tilde{f}(0)}{\partial \beta} - \frac{1}{2}\int \frac{d\mathbf{k}}{(2\pi)^d}\frac{\left[n\tilde{f}(k)\right]^2}{1 - n\tilde{f}(k)}\frac{\partial \tilde{f}(k)}{\partial \beta} , \quad \text{(LDH)} ,
\tag{6.110a}
$$

$$
Z = 1 - \frac{n}{2}\tilde{f}(0) - \frac{1}{2n}\int \frac{d\mathbf{k}}{(2\pi)^d}\left\{ n\tilde{f}(k)\left[\frac{1}{1 - n\tilde{f}(k)} - \frac{n\tilde{f}(k)}{2}\right] + \ln\left[1 - n\tilde{f}(k)\right]\right\} ,
$$
$$
\text{(LDH)} ,
\tag{6.110b}
$$

$$
\beta\mu^{\mathrm{ex}} = -n\tilde{f}(0) - \frac{1}{2n}\int \frac{d\mathbf{k}}{(2\pi)^d}\frac{\left[n\tilde{f}(k)\right]^3}{1 - n\tilde{f}(k)} , \quad \text{(LDH)} .
\tag{6.110c}
$$

A similar conclusion holds for the class (6.105). First, in analogy with (6.106), we rewrite (4.56) as

$$
\begin{aligned}
\beta a^{\mathrm{ex}} &= \frac{n}{2}\int_0^1 d\xi \int d\mathbf{r}\, \frac{\partial \beta \phi^{(\xi)}(r)}{\partial \xi}\left[1 + h^{(\xi)}(r)\right] \\
&= \frac{n}{2}\beta\tilde{\phi}(0) + \frac{n}{2}\int \frac{d\mathbf{k}}{(2\pi)^d}\int_0^1 d\xi\, \tilde{h}^{(\xi)}(k)\frac{\partial \beta \widetilde{\phi^{(\xi)}}(k)}{\partial \xi} .
\end{aligned}
\tag{6.111}
$$

As before, notice the formal analogy between (6.106) and (6.111) under the replacements $\tilde{w}(k) \leftrightarrow \tilde{h}(k)$ and $\tilde{f}(k) \leftrightarrow -\beta\tilde{\phi}(k)$. Consequently, approximations belonging to the class (6.105) yield, regardless of the protocol,

$$
\beta a^{\mathrm{ex}} = \frac{n}{2}\beta\tilde{\phi}(0) - \frac{1}{2n}\int \frac{d\mathbf{k}}{(2\pi)^d}\, \mathfrak{h}_0[-n\beta\tilde{\phi}(k)] ,
\tag{6.112}
$$

where

$$
\mathfrak{h}_0[z] \equiv \int_0^z dz'\, \mathfrak{h}[z'] .
\tag{6.113}
$$

In the MSA, $\mathfrak{h}[z] = z/(1-z) \Rightarrow \mathfrak{h}_0[z] = -z - \ln(1-z)$, so that

$$\beta a^{\text{ex}} = \frac{n}{2}\beta\widetilde{\phi}(0) - \frac{1}{2n}\int\frac{d\mathbf{k}}{(2\pi)^d}\left\{n\beta\widetilde{\phi}(k) - \ln\left[1 + n\beta\widetilde{\phi}(k)\right]\right\}\,, \quad \text{(MSA)}\,. \tag{6.114}$$

Again, use of (1.36) yields

$$\beta u^{\text{ex}} = \frac{n}{2}\beta\widetilde{\phi}(0) - \frac{1}{2n}\int\frac{d\mathbf{k}}{(2\pi)^d}\frac{\left[n\beta\widetilde{\phi}(k)\right]^2}{1 + n\beta\widetilde{\phi}(k)}\,, \quad \text{(MSA)}\,, \tag{6.115a}$$

$$Z = 1 + \frac{n}{2}\beta\widetilde{\phi}(0) + \frac{1}{2n}\int\frac{d\mathbf{k}}{(2\pi)^d}\left\{\frac{n\beta\widetilde{\phi}(k)}{1 + n\beta\widetilde{\phi}(k)} - \ln\left[1 + n\beta\widetilde{\phi}(k)\right]\right\}\,,$$
$$\text{(MSA)}\,, \tag{6.115b}$$

$$\beta\mu^{\text{ex}} = n\beta\widetilde{\phi}(0) - \frac{1}{2n}\int\frac{d\mathbf{k}}{(2\pi)^d}\frac{\left[n\beta\widetilde{\phi}(k)\right]^2}{1 + n\beta\widetilde{\phi}(k)}\,, \quad \text{(MSA)}\,. \tag{6.115c}$$

6.9.3   *"Energy" Route in Hard-Sphere Liquids*

We saw in (4.87) that the energy route is *useless* for HSs. In fact, the consistency condition (6.96) is *trivially* satisfied since

$$n\frac{\partial u_{\text{HS}}^{\text{ex}}}{\partial n} = 0\,, \qquad \frac{\partial Z_{\text{HS}}}{\partial\beta} = 0\,. \tag{6.116}$$

The last equality expresses the fact that the HS compressibility factor

$$Z_{\text{HS}}(\eta) = 1 + 2^{d-1}\eta y_{\text{HS}}(\sigma;\eta) \tag{6.117}$$

is independent of temperature. Thus, there is no possibility of extracting non-trivial thermodynamic information from the energy route in the case of HSs.

However, a physical meaning can be ascribed if *first* the energy route is applied to a non-HS system that includes the HS one as a special case and *then* the HS limit is taken.

Let us take the SS potential (see Table 3.1) as a convenient choice of a non-HS potential. It has the interesting property of reducing to the HS potential in three independent limits:

$$\lim_{\beta\varepsilon\to 0}\phi_{\text{SS}}(r) = \phi_{\text{HS}}(r)\ (\text{diameter }\sigma)\,, \tag{6.118a}$$

$$\lim_{\beta\varepsilon\to\infty}\phi_{\text{SS}}(r) = \phi_{\text{HS}}(r)\ (\text{diameter }\sigma')\,, \tag{6.118b}$$

$$\lim_{\sigma'\to\sigma}\phi_{\text{SS}}(r) = \phi_{\text{HS}}(r)\ (\text{diameter }\sigma' = \sigma)\,. \tag{6.118c}$$

It also reduces to the PS potential (see again Table 3.1) in the limit $\sigma \to 0$ and, from it, to the HS potential in the limit $\varepsilon \to \infty$:

$$\lim_{\sigma \to 0} \phi_{SS}(r) = \phi_{PS}(r) \text{ (diameter } \sigma') ,$$
(6.119a)

$$\lim_{\beta \varepsilon \to \infty} \phi_{PS}(r) = \phi_{HS}(r) \text{ (diameter } \sigma') .$$
(6.119b)

The limits (6.118) and (6.119) are illustrated in Figs. 6.12 and 6.13. They are represented in Fig. 6.13 by the paths A, B, C, D, and D', respectively.

Suppose now that an *approximate* cavity function $y_{SS}(r; n, \beta)$ is known (for instance, as the solution to an integral equation) for the SS fluid. Then, the energy route (4.33) gives (at a certain point P in Fig. 6.13)

$$u_{SS}^{ex}(n, \beta) = d2^{d-1} v_d n \varepsilon e^{-\beta \varepsilon} \int_{\sigma}^{\sigma'} dr\, r^{d-1} y_{SS}(r; n, \beta) .$$
(6.120)

The associated energy-route compressibility factor is obtained from (6.96) as

$$Z_{SS}(n, \beta) - Z_{HS}(n\sigma^d) = n \frac{\partial}{\partial n} \int_0^\beta d\beta'\, u_{SS}^{ex}(n, \beta')$$

$$= d2^{d-1} v_d n \varepsilon \frac{\partial}{\partial n} n \int_0^\beta d\beta'\, e^{-\beta' \varepsilon} \int_{\sigma}^{\sigma'} dr\, r^{d-1} y_{SS}(r; n, \beta') ,$$
(6.121)

where in the first step the integration constant has been fixed by the physical condition (6.118a), while (6.120) has been used in the second step. Note that

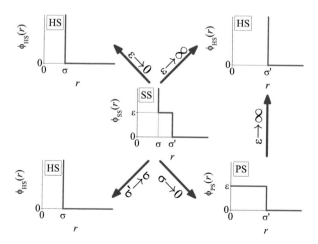

Fig. 6.12 Graphical representation of the limits (6.118) and (6.119)

Fig. 6.13 Parameter space for SS fluids. The plane $\sigma/\sigma' = 0$ represents PS fluids (of diameter σ'), the plane $e^{-\beta\varepsilon} = 0$ corresponds to HS fluids of diameter σ', and the planes $e^{-\beta\varepsilon} = 1$ and $\sigma/\sigma' = 1$ define HS fluids of diameter σ. Starting from a given SS fluid (represented by point P), it is possible to go to the HS fluid at the same density by following different paths

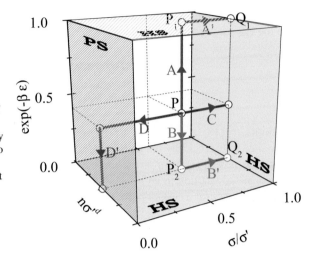

(6.121) gives the difference between the compressibility factors at points P and P_1 of Fig. 6.13 in terms of the SS cavity function along path A.

Next, we take the limit $\beta\varepsilon \to \infty$ on both sides of (6.121), apply (6.118b), and divide both sides of the equation by $n\sigma'^d - n\sigma^d$. The result is

$$\frac{Z_{HS}(n\sigma'^d) - Z_{HS}(n\sigma^d)}{n\sigma'^d - n\sigma^d} = \frac{d2^{d-1}v_d\varepsilon}{\sigma'^d - \sigma^d}\frac{\partial}{\partial n}n\int_0^\infty d\beta\, e^{-\beta\varepsilon}\int_\sigma^{\sigma'} dr\, r^{d-1}y_{SS}(r;n,\beta)\,. \tag{6.122}$$

This corresponds to moving point P to point P_2 along path B.

Finally, we take the limit $\sigma' \to \sigma$, i.e., we move points P_1 and P_2 to points Q_1 and Q_2 along paths A' and B', respectively. The left-hand side of (6.122) becomes

$$\lim_{\sigma'\to\sigma}\frac{Z_{HS}(n\sigma'^d) - Z_{HS}(n\sigma^d)}{n\sigma'^d - n\sigma^d} = \sigma^{-d}\frac{\partial}{\partial n}Z_{HS}(n\sigma^d)\,. \tag{6.123}$$

Moreover, the spatial integral on the right-hand side of (6.122) reduces to

$$\lim_{\sigma'\to\sigma}\frac{1}{\sigma'^d - \sigma^d}\int_\sigma^{\sigma'} dr\, r^{d-1}y_{SS}(r;n,\beta) = \frac{1}{d}y_{HS}(\sigma;n\sigma^d)\,, \tag{6.124}$$

where the third limit (6.118c) has been used (path C for any point P). Taking into account (6.123) and (6.124) in (6.122), one gets

$$\frac{\partial}{\partial n}Z_{HS}(n\sigma^d) = 2^{d-1}v_d\frac{\partial}{\partial n}n\sigma^d y_{HS}(\sigma;n\sigma^d)\,. \tag{6.125}$$

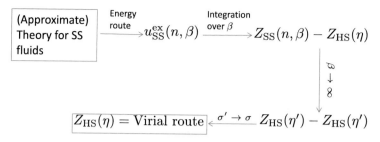

Fig. 6.14 Scheme of the steps followed to derive (6.117) starting from (6.120)

Integration over density and application of the ideal-gas boundary condition $Z_{HS}(0) = 1$ yields (6.117), which is not but the virial EoS! The generalization to mixtures follows essentially the same steps [48].

In summary, the ill definition of the energy route to the EoS of HSs can be avoided by first considering a SS fluid and then taking the limit of a vanishing shoulder width. The resulting EoS coincides exactly with the one obtained through the virial route. This applies regardless of the approximation followed to describe the SS and HS fluids. *From that point of view*, the energy and virial routes to the EoS of HS fluids can be considered as equivalent. Figure 6.14 presents a scheme of the energy route $\rightarrow$ virial route path.

It must be emphasized that the application of the three limits (6.118) is *essential* to derive (6.117) from (6.120) [49]. For instance, if the limit $\sigma \rightarrow 0$ (instead of $\sigma' \rightarrow \sigma$) is taken in (6.122) (i.e., points P_1 and P_2 are moved along paths parallel to path D of Fig. 6.13), the result is

$$Z_{HS}(n\sigma'^d) = 1 + d2^{d-1}nv_d\varepsilon\frac{\partial}{\partial n}n\int_0^\infty d\beta\, e^{-\beta\varepsilon}\int_0^{\sigma'} dr\, r^{d-1}y_{PS}(r;n,\beta)\,, \qquad (6.126)$$

where the change $y_{SS} \rightarrow y_{PS}$ is a consequence of (6.119a). Equation (6.126) is an alternative recipe to obtain the HS EoS from the energy route applied to PSs. In general, it gives a result different from (6.117) when an *approximate* cavity function y_{PS} is used. For instance, in the PY approximation for three-dimensional systems, (6.117) gives a (rescaled) fourth virial coefficient $b_4^{HS} = 16$ (see Table 6.5), while (6.126) gives the rather poor result $b_4^{HS} = 1\,814/175 \simeq 10.37$ [8, 49].

Exercises

6.1 Derive (6.6) and (6.7).

6.2 Check the correctness of (6.11) and (6.14)

6.3 Check (6.16).

6.4 Check (6.18).

6.5 Check (6.21).

6.6 Check (6.29).

6.7 Derive (6.31).

6.8 Check that (6.32) results from (6.31) and (3.81a).

6.9 Obtain (6.33c).

6.10 Obtain (6.41).

6.11 Make use of the property (6.35) and of the HS expressions (6.28b) and (6.33) to find $y(\sigma)$ and $\widetilde{S}(0)$ for the PS model to second order in density. Next, obtain the first four virial coefficients from the virial and compressibility routes. (In case of need, consult [8].)

6.12 Check that (6.42) is obtained from (6.41) by application of the compressibility [see (4.29)] and virial [see (4.88)] routes.

6.13 Check (6.51).

6.14 Taking into account (6.51), prove that (6.52) is indeed satisfied to order n^3.

6.15 Check (6.56).

6.16 Check that the HNC closure (6.64) and the PY closure (6.65) are equivalent to

$$c(r) = y(r)f(r) + y(r) - 1 - \ln y(r) , \quad \text{(HNC)} ,$$
$$c(r) = y(r)f(r) , \quad \text{(PY)} ,$$

when expressed in terms of the DCF $c(r)$ and the cavity function $y(r)$.

6.17 Check that the HNC closure (6.64) and the PY closure (6.65) are equivalent to

$$y(r) = e^{\bar{\gamma}(r)} , \quad \text{(HNC)} ,$$
$$y(r) = 1 + \bar{\gamma}(r) , \quad \text{(PY)} ,$$

when expressed in terms of the indirect correlation function $\bar{\gamma}(r)$ [see (6.71)] and the cavity function $y(r)$.

6.18 Check (6.68).

6.19 Obtain (6.69).

6.20 Check that (6.70) is obtained from (6.69) by application of the compressibility [see (4.29)] and virial [see (4.88)] routes.

6.21 Find the value of the factor λ that makes both routes in (6.70) be consistent to each other. What is the value of the "consistent" coefficient b_4? How does it compare with the exact value of b_4 (see Table 3.11)?

6.22 Using (6.61), check that a closure relation of the form $\mathscr{B}(r) = \mathfrak{B}_{appr}[\overline{\gamma}(r)]$ is equivalent to

$$\ln y(r) = \overline{\gamma}(r) + \mathfrak{B}_{appr}[\overline{\gamma}(r)] .$$

6.23 Check that all the closure relations of Table 6.6 give $\mathscr{B}(r) \approx -A[\overline{\gamma}(r)]^2$ in the limit $\overline{\gamma}(r) \to 0$ and find the value of the coefficient A in each case. Taking into account that $\overline{\gamma}(r) = \mathscr{O}(n)$ [compare (6.54) and (6.57)], this implies that $\mathscr{B}(r) = \mathscr{O}(n^2)$, in agreement with (6.46).

6.24 Check that the Rogers–Young closure [see Table 6.6] reduces to the PY and HNC closures in the limits $a_{RY} \to 0$ and $a_{RY} \to \infty$, respectively.

6.25 Check that the Ballone–Pastore–Galli–Gazzillo closure [see Table 6.6] reduces to the HNC and Martynov–Sarkisov closures in the limits $a_{BPGG} \to 1$ and $a_{BPGG} \to 2$, respectively.

6.26 Derive (6.74).

6.27 Check (6.83) and (6.84).

6.28 With the help of the Appendix of [44], prove (6.88).

6.29 Derive (6.90).

6.30 Check (6.92)–(6.94).

6.31 Following the steps described in [44], generalize (6.94) to mixtures.

6.32 Derive (6.97) by integrating by parts the right-hand side.

6.33 Derive (6.98) and (6.99).

6.34 Check (6.100).

6.35 Check (6.103).

6.36 Derive (6.106) and (6.108).

6.37 Derive (6.110) from (6.109) and (6.115) from (6.114)

6.38 Check that (6.110) and (6.115) are consistent with the Maxwell relation (6.96).

6.39 Derive (6.120).

6.40 Check (6.123) and (6.124).

6.41 Check (6.126).

References

1. R. Balescu, *Equilibrium and Nonequilibrium Statistical Mechanics* (Wiley, New York, 1974)
2. M. Baus, J.L. Colot, Phys. Rev. A **36**, 3912 (1987)
3. M. Abramowitz, I.A. Stegun (eds.), *Handbook of Mathematical Functions* (Dover, New York, 1972)
4. F.W.J. Olver, D.W. Lozier, R.F. Boisvert, C.W. Clark (eds.), *NIST Handbook of Mathematical Functions* (Cambridge University Press, New York, 2010)
5. J.A. Barker, D. Henderson, Can. J. Phys. **44**, 3959 (1967)
6. B.R.A. Nijboer, L. van Hove, Phys. Rev. **85**, 777 (1952)
7. F.H. Ree, N. Keeler, S.L. McCarthy, J. Chem. Phys. **44**, 3407 (1966)
8. A. Santos, A. Malijevský, Phys. Rev. E **75**, 021201 (2007)
9. T. Morita, Prog. Theor. Phys. **20**, 920 (1958)
10. J.M.J. van Leeuwen, J. Groeneveld, J. de Boer, Physica **25**, 792 (1959)
11. J.K. Percus, G.J. Yevick, Phys. Rev. **110**, 1 (1958)
12. F.H. Ree, W.G. Hoover, J. Chem. Phys. **41**, 1635 (1964)
13. J.A. Barker, D. Henderson, Rev. Mod. Phys. **48**, 587 (1976)
14. T. Morita, Prog. Theor. Phys. **23**, 829 (1960)
15. A. Santos, Phys. Rev. Lett. **109**, 120601 (2012)
16. E. Beltrán-Heredia, A. Santos, J. Chem. Phys. **140**, 134507 (2014)
17. M.S. Wertheim, Phys. Rev. Lett. **10**, 321 (1963)
18. E. Thiele, J. Chem. Phys. **39**, 474 (1963)
19. M.S. Wertheim, J. Math. Phys. **5**, 643 (1964)
20. R.J. Baxter, J. Chem. Phys. **49**, 2770 (1968)
21. J.L. Lebowitz, Phys. Rev. **133**, A895 (1964)
22. J.W. Perram, E.R. Smith, Chem. Phys. Lett. **35**, 138 (1975)
23. B. Barboy, Chem. Phys. **11**, 357 (1975)
24. C. Freasier, D.J. Isbister, Mol. Phys. **42**, 927 (1981)
25. E. Leutheusser, Physica A **127**, 667 (1984)
26. R.D. Rohrmann, A. Santos, Phys. Rev. E **76**, 051202 (2007)
27. R.D. Rohrmann, A. Santos, Phys. Rev. E **83**, 011201 (2011)
28. R.D. Rohrmann, A. Santos, Phys. Rev. E **84**, 041203 (2011)
29. J.P. Hansen, I.R. McDonald, *Theory of Simple Liquids*, 3rd edn. (Academic, London, 2006)
30. J.R. Solana, *Perturbation Theories for the Thermodynamic Properties of Fluids and Solids* (CRC Press, Boca Raton, 2013)
31. L. Verlet, Mol. Phys. **41**, 183 (1980)
32. L. Verlet, Mol. Phys. **42**, 1291 (1981)
33. G.A. Martynov, G.N. Sarkisov, Mol. Phys. **49**, 1495 (1983)
34. F.J. Rogers, D.A. Young, Phys. Rev. A **30**, 999 (1984)
35. P. Ballone, G. Pastore, G. Galli, D. Gazzillo, Mol. Phys. **59**, 275 (1986)
36. L. Acedo, A. Santos, Phys. Lett. A **323**, 427 (2004). Erratum: **376**, 2274–2275 (2012)
37. F.H. Stillinger, D.K. Stillinger, Physica A **244**, 358 (1997)
38. C. Marquest, T.A. Witten, J. Phys. France **50**, 1267 (1989)
39. C.N. Likos, H. Löwen, M. Watzlawek, B. Abbas, O. Jucknischke, J. Allgaier, D. Richter, Phys. Rev. Lett. **80**, 4450 (1998)
40. W. Schirmacher, *Theory of Liquids and Other Disordered Media. A Short Introduction*. Lecture Notes in Physics, vol. 887 (Springer, Cham, 2014)
41. B.M. Mladek, G. Kahl, M. Neuman, J. Chem. Phys. **124**, 064503 (2006)
42. E. Waisman, Mol. Phys. **25**, 45 (1973)
43. Y. Tang, J. Chem. Phys. **118**, 4140 (2003)
44. A. Santos, G. Manzano, J. Chem. Phys. **132**, 144508 (2010)

45. A. Santos, R. Fantoni, A. Giacometti, Phys. Rev. E **77**, 051206 (2008)
46. A. Santos, R. Fantoni, A. Giacometti, J. Chem. Phys. **131**, 181105 (2009)
47. A. Santos, J. Chem. Phys. **126**, 116101 (2007)
48. A. Santos, J. Chem. Phys. **123**, 104102 (2005)
49. A. Santos, Mol. Phys. **104**, 3411 (2006)

Chapter 7
Exact Solution of the Percus–Yevick
Approximation for Hard Spheres ...and Beyond

One of the milestones of the statistical-mechanical theory of liquids in equilibrium was the exact analytical solution of the Percus–Yevick integral equation for three-dimensional hard spheres by, independently, Wertheim and Thiele in 1963 [1–4]. The solution was later extended to mixtures [5, 6], hard hyperspheres in odd dimensions [7, 8], and sticky hard spheres [9–12]. This chapter presents some of those solutions within the framework of the rational-function approximation [13–15], which lends itself to generalizations beyond the Percus–Yevick approximation.

7.1 Introduction

Particularized to $d = 3$, the OZ relation (4.26) can be written as

$$rh(r) = rc(r) + 2\pi n \int_0^\infty dr'\, r'c(r') \int_{|r-r'|}^{r+r'} dr''\, r''h(r'') \,, \tag{7.1}$$

where bipolar coordinates have been used. In the HS case, one necessarily has $g(r) = 0$ for $r < \sigma$. Moreover, the PY closure (6.65) implies that $c(r) = 0$ for $r > \sigma$. Thus, the mathematical problem consists in solving (7.1) subject to the boundary conditions

$$\begin{cases} g(r) = 0 \,, & r < \sigma \text{ (exact hard-core condition)} \,, \\ c(r) = 0 \,, & r > \sigma \text{ (PY approximation for HSs)} \,. \end{cases} \tag{7.2}$$

Despite the apparent formidable difficulty of (7.1) with the boundary conditions (7.2), Wertheim (see Fig. 7.1) [2, 4] and, independently, Thiele [3] were able to find an exact solution by analytical means. The solution relies on the use of Laplace

© Springer International Publishing Switzerland 2016
A. Santos, *A Concise Course on the Theory of Classical Liquids*,
Lecture Notes in Physics 923, DOI 10.1007/978-3-319-29668-5_7

Fig. 7.1 Michael Stephen
Wertheim (b. 1931)
(Photograph courtesy of M.S.
Wertheim)

transforms, as suggested by the structure of (7.1), and stringent analytical properties
of *entire functions* of complex variable. The interested reader is referred to [1–4] for
further details.

In this chapter, however, we will follow an alternative method [13–16] that does
not make explicit use of (7.2) and admits extensions and generalizations.

7.2   An Alternative Approach: The Rational-Function Approximation

The main steps we will follow are the following ones:

1. Introduce the Laplace transform $\widehat{G}(s)$ of $rg(r)$.
2. Define an auxiliary function $\widehat{F}(s)$ directly related to $\widehat{G}(s)$.
3. Find the exact properties of $\widehat{F}(s)$ for small s and for large s.
4. Propose a rational-function *approximation* (RFA) for $\widehat{F}(s)$ satisfying the previous exact properties.

As will be seen, the *simplest* approximation (i.e., the one with the least
number of parameters) directly yields the PY solution. Furthermore, the next-order
approximation contains two free parameters which can be determined by prescribing
a given EoS and thermodynamic consistency between the virial and compressibility
routes.

The same approach can be extended to mixtures, to other related systems with
piecewise-constant potentials, and to higher dimensionalities with $d = $ odd.

We now proceed with the four steps described above.

7.2.1   Introduction of $\widehat{G}(s)$

Let us introduce the Laplace transform of $rg(r)$:

$$\boxed{\widehat{G}(s) = \mathscr{L}\left[rg(r)\right](s) = \int_0^\infty dr\, e^{-sr} rg(r) \, .} \tag{7.3}$$

Note that in the one-dimensional case $\widehat{G}(s)$ was defined in (5.8) as $\mathscr{L}\left[g(r)\right](s)$. In the three-dimensional case, the choice of $rg(r)$ instead of $g(r)$ as the function to be Laplace transformed is suggested by the structure of (7.1) and also by the link of $\widehat{G}(s)$ to the *Fourier* transform $\tilde{h}(k)$ of the total correlation function $h(r) = g(r) - 1$ and hence to the structure factor $\widetilde{S}(k) = 1 + n\tilde{h}(k)$:

$$\tilde{h}(k) = -2\pi \left[\frac{\widehat{H}(s) - \widehat{H}(-s)}{s}\right]_{s=ik} = -2\pi \left[\frac{\widehat{G}(s) - \widehat{G}(-s)}{s}\right]_{s=ik} , \tag{7.4}$$

where

$$\widehat{H}(s) = \widehat{G}(s) - s^{-2} \tag{7.5}$$

is the Laplace transform of $rh(r)$. Had we defined $\widehat{G}(s)$ as the Laplace transform of $g(r)$, (7.4) would have involved the derivative $\widehat{G}'(s)$, what would be far less convenient.

In the more general case of $d =$ odd ≥ 3, it can be seen that the right choice for $\widehat{G}(s)$ is [17]

$$\widehat{G}(s) = \int_0^\infty dr\, e^{-sr} \theta_{\frac{d-3}{2}}(sr) rg(r) \, , \tag{7.6}$$

where

$$\theta_k(x) = \sum_{j=0}^k \frac{(2k-j)!}{2^{k-j}(k-j)!j!} x^j \tag{7.7}$$

is a so-called *reverse Bessel polynomial* [18]. It is related to the spherical Bessel function of the first kind, $j_k(x)$, by

$$j_k(x) \equiv \sqrt{\frac{\pi}{2x}} J_{k+1/2}(x) = \frac{\theta_k(-ix)e^{ix} - \theta_k(ix)e^{-ix}}{2ix^{k+1}} . \tag{7.8}$$

In this case $d = \text{odd} \geq 3$, (7.4) becomes

$$\tilde{h}(k) = (-2\pi)^{(d-1)/2} \left[\frac{\widehat{H}(s) - \widehat{H}(-s)}{s^{d-2}} \right]_{s=ik} = (-2\pi)^{(d-1)/2} \left[\frac{\widehat{G}(s) - \widehat{G}(-s)}{s^{d-2}} \right]_{s=ik},$$

$$(7.9)$$

where

$$\widehat{H}(s) = \widehat{G}(s) - (d-2)!! s^{-2} \qquad (7.10)$$

is defined as in (7.6), except for the replacement $g(r) \to h(r)$, and, as usual, the double factorial is defined as $k!! = \prod_{i=0}^{[k/2]-1} (k - 2i)$ with the convention $0!! = 1$, $[k/2]$ denoting the integer part of $k/2$.

7.2.2 Definition of $\widehat{F}(s)$

Henceforth we return to the three-dimensional case ($d = 3$) and, for simplicity, we take $\sigma = 1$ as the unit length. Taking (6.23a) and (6.28b) into account, the HS RDF to first order in density is

$$g(r) = \Theta(r - 1) \left[1 + \Theta(2 - r)(r - 2)^2 \left(\frac{r}{2} + 2 \right) \eta + \cdots \right], \qquad (7.11)$$

where we recall that $\eta = \frac{\pi}{6} n \sigma^3$ is the packing fraction. To that order, the Laplace transform of $rg(r)$ is given by

$$s^{-1} \widehat{G}(s) = \left[\widehat{F}_0(s) + \widehat{F}_1(s) \eta \right] e^{-s} - 12\eta \left[\widehat{F}_0(s) \right]^2 e^{-2s} + \cdots, \qquad (7.12)$$

where

$$\widehat{F}_0(s) = s^{-2} + s^{-3}, \qquad (7.13a)$$

$$\widehat{F}_1(s) = \frac{5}{2} s^{-2} - 2s^{-3} - 6s^{-4} + 12s^{-5} + 12s^{-6}. \qquad (7.13b)$$

The exact form (7.12) of $\widehat{G}(s)$ to order η suggests the *definition* of an auxiliary function $\widehat{F}(s)$ through

$$s^{-1} \widehat{G}(s) = \widehat{F}(s) e^{-s} - 12\eta \left[\widehat{F}(s) \right]^2 e^{-2s} + (12\eta)^2 \left[\widehat{F}(s) \right]^3 e^{-3s} - \cdots$$

$$= \frac{\widehat{F}(s) e^{-s}}{1 + 12\eta \widehat{F}(s) e^{-s}}. \qquad (7.14)$$

Equivalently,

$$\widehat{F}(s) \equiv e^s \frac{s^{-1}\widehat{G}(s)}{1 - 12\eta s^{-1}\widehat{G}(s)} \,. \tag{7.15}$$

Of course, $\widehat{F}(s)$ depends on η. To first order,

$$\widehat{F}(s) = \widehat{F}_0(s) + \widehat{F}_1(s)\eta + \cdots , \tag{7.16}$$

with $\widehat{F}_0(s)$ and $\widehat{F}_1(s)$ given by (7.13).

In analogy with the one-dimensional case [see (5.61)], the introduction of $\widehat{F}(s)$ allows one to express $g(r)$ as a succession of *shells* ($1 < r < 2, 2 < r < 3, \ldots$) in a natural way. First, according to (7.14),

$$\widehat{G}(s) = \sum_{\ell=1}^{\infty} (-12\eta)^{\ell-1} s \left[\widehat{F}(s)\right]^{\ell} e^{-\ell s} \,. \tag{7.17}$$

Then, Laplace inversion term by term gives

$$g(r) = \frac{1}{r} \sum_{\ell=1}^{\infty} (-12\eta)^{\ell-1} \bar{\Psi}_\ell(r - \ell)\Theta(r - \ell) \,, \tag{7.18}$$

where

$$\bar{\Psi}_\ell(r) = \mathcal{L}^{-1}\left[s\left[\widehat{F}(s)\right]^{\ell}\right](r) \,. \tag{7.19}$$

7.2.3   Exact Properties of $\widehat{F}(s)$ for Small s and Large s

In order to derive the exact behavior of $\widehat{G}(s)$ for large s, we need to start from the behavior of $g(r)$ for $r \gtrsim 1$:

$$g(r) = \Theta(r - 1)\left[g(1^+) + g'(1^+)(r - 1) + \frac{1}{2}g''(1^+)(r - 1)^2 + \cdots\right] \,. \tag{7.20}$$

In Laplace space,

$$se^s\widehat{G}(s) = g(1^+) + \left[g(1^+) + g'(1^+)\right]s^{-1} + \mathcal{O}(s^{-2}) \,. \tag{7.21}$$

Therefore, according to (7.15),

$$\widehat{F}(s) = g(1^+)s^{-2} + \left[g(1^+) + g'(1^+)\right]s^{-3} + \mathcal{O}(s^{-4}) . \tag{7.22}$$

In particular,

$$\boxed{\lim_{s \to \infty} s^2 \widehat{F}(s) = g(1^+) = \text{finite} .} \tag{7.23}$$

Thus, we see that $\widehat{F}(s)$ must necessarily behave as s^{-2} for large s.

Now we turn to the small-s behavior. Let us expand the Laplace transform of $rh(r)$ in powers of s:

$$\widehat{H}(s) = \widehat{H}^{(0)} + \widehat{H}^{(1)}s + \cdots , \tag{7.24}$$

where

$$\widehat{H}^{(k)} \equiv (-1)^k \int_0^\infty dr \, r^{k+1} h(r) . \tag{7.25}$$

In particular, $\widehat{H}^{(1)}$ is directly related to the isothermal compressibility [see (4.29)]:

$$\chi_T = 1 + n\tilde{h}(0) = 1 - 24\eta \widehat{H}^{(1)} . \tag{7.26}$$

Since χ_T must be finite, and recalling (7.5), we find

$$s^2 \widehat{G}(s) = 1 + 0 \times s + \widehat{H}^{(0)}s^2 + \widehat{H}^{(1)}s^3 + \mathcal{O}(s^4) . \tag{7.27}$$

Therefore, from (7.15) the small-s behavior of $\widehat{F}(s)$ is found to be

$$\frac{e^s}{\widehat{F}(s)} = -12\eta + \frac{s}{\widehat{G}(s)}$$
$$= -12\eta + 0 \times s + 0 \times s^2 + 1 \times s^3 + 0 \times s^4 - \widehat{H}^{(0)}s^5 - \widehat{H}^{(1)}s^6 + \mathcal{O}(s^7) . \tag{7.28}$$

Thus, just the condition $\chi_T = $ finite univocally fixes the first *five* coefficients in the power series expansion of $\widehat{F}(s)$. More specifically,

$$\boxed{\widehat{F}(s) = -\frac{1}{12\eta}\left[1 + s + \frac{s^2}{2} + \frac{1 + 2\eta}{12\eta}s^3 + \frac{1 + \eta/2}{12\eta}s^4\right] + \mathcal{O}(s^5) .} \tag{7.29}$$

7.2.4   Construction of the Approximation: Percus–Yevick Solution

Thus far, all the preceding results are formally exact. To recapitulate, we have defined the Laplace transform $\widehat{G}(s)$ in (7.3) and the auxiliary function $\widehat{F}(s)$ in (7.15). This latter function must comply with the two basic requirements (7.23) and (7.29).

A simple way of satisfying both conditions is by means of a *rational-function* form:

$$\widehat{F}(s) = \frac{\text{Polynomial in } s \text{ of degree } k}{\text{Polynomial in } s \text{ of degree } k + 2} \tag{7.30}$$

with $2k + 3 \geq 5 \Rightarrow k \geq 1$. The *simplest* RFA corresponds to $k = 1$:

$$\boxed{\widehat{F}(s) = -\frac{1}{12\eta} \frac{1 + L^{(1)}s}{1 + S^{(1)}s + S^{(2)}s^2 + S^{(3)}s^3}} \tag{7.31}$$

where the coefficients are determined from (7.29). They are

$$L^{(1)} = \frac{1 + \eta/2}{1 + 2\eta}, \tag{7.32a}$$

$$S^{(1)} = -\frac{3}{2}\frac{\eta}{1 + 2\eta}, \quad S^{(2)} = -\frac{1}{2}\frac{1 - \eta}{1 + 2\eta}, \quad S^{(3)} = -\frac{1}{12\eta}\frac{(1 - \eta)^2}{1 + 2\eta}. \tag{7.32b}$$

The rational function (7.31) can also be rewritten as

$$\widehat{F}(s) = \frac{\Lambda^{(0)} + \Lambda^{(1)}s}{\sigma^3 s^3 - 12\eta\left[\left(1 - \sigma s + \frac{1}{2}\sigma^2 s^2\right)\Lambda^{(0)} + s\left(1 - \sigma s\right)\Lambda^{(1)}\right]}, \tag{7.33}$$

where

$$\Lambda^{(0)} = \frac{1 + 2\eta}{(1 - \eta)^2}, \quad \Lambda^{(1)} = \sigma\frac{1 + \eta/2}{(1 - \eta)^2}, \tag{7.34}$$

and, for later convenience, we have considered an arbitrary length unit (i.e., not necessarily $\sigma = 1$). Inserting (7.33) into (7.14), it is possible to express $\widehat{G}(s)$ as

$$\boxed{\widehat{G}(s) = \frac{e^{-\sigma s}}{s^2}\frac{\Lambda^{(0)} + \Lambda^{(1)}s}{1 - 2\pi n\left[\varphi_2(\sigma s)\sigma^3\Lambda^{(0)} + \varphi_1(\sigma s)\sigma^2\Lambda^{(1)}\right]}, \tag{7.35}}$$

where

$$\varphi_k(x) \equiv x^{-(k+1)} \left(\sum_{\ell=0}^{k} \frac{(-x)^\ell}{\ell!} - e^{-x} \right) . \tag{7.36}$$

Note that $\lim_{x \to 0} \varphi_k(x) = (-1)^k/(k+1)!$.

7.2.4.1   Structural Properties

Once $\widehat{F}(s)$ and hence $\widehat{G}(s)$ have been completely determined by the approximation (7.31), it is easy to go back to real space (here again we assume the length unit $\sigma = 1$) and obtain the corresponding RDF $g(r)$. Three alternative ways are possible. First, one can invert numerically the Laplace transform $\widehat{G}(s)$ by means of efficient algorithms [19]. A second method consists in obtaining $\tilde{h}(k)$ from (7.4) and then performing a numerical Fourier inversion. The third method is purely analytical and is based on (7.18) and (7.19). From a practical point of view, one is interested in determining $g(r)$ up to a certain distance $r_{\max}$ since $g(r) \to 1$ for large distances. In that case, the summation in (7.18) can be truncated for $\ell > r_{\max}$ [20]. To obtain $\bar{\Psi}_\ell(r)$ from (7.19) and (7.31) one only needs the three roots $\{s_i, i = 1, 2, 3\}$ of the cubic equation $1 + S^{(1)}s + S^{(2)}s^2 + S^{(3)}s^3 = 0$ and to apply the residue theorem, i.e.,

$$\bar{\Psi}_\ell(r) = \sum_{j=1}^{\ell} \frac{\sum_{i=1}^{3} a_{\ell j}^{(i)} e^{s_i r}}{(\ell-j)!(j-1)!} r^{\ell-j} , \tag{7.37a}$$

$$a_{\ell j}^{(i)} = \lim_{s \to s_i} \left(\frac{\partial}{\partial s} \right)^{j-1} \left\{ s \left[(s - s_i)\widehat{F}(s) \right]^\ell \right\} . \tag{7.37b}$$

This method is analogous to the one followed for the one-dimensional systems in (5.53), (5.61), and (5.73). The values at contact are simply obtained from (7.22) as

$$g(1^+) = -\frac{1}{12\eta} \frac{L^{(1)}}{S^{(3)}} = \frac{1 + \eta/2}{(1 - \eta)^2} , \tag{7.38a}$$

$$g'(1^+) = -\frac{1}{12\eta S^{(3)}} \left[1 - L^{(1)} \left(1 + \frac{S^{(2)}}{S^{(3)}} \right) \right] = -\frac{9}{2} \frac{\eta(1 + \eta)}{(1 - \eta)^3} . \tag{7.38b}$$

As for the structure factor, application of (4.24) and (7.4) yields the explicit expression

$$
\begin{aligned}
\frac{1}{\widetilde{S}(k)} = 1 &+ \frac{72\eta^2(2+\eta)^2}{(1-\eta)^4}k^{-4} + \frac{288\eta^2(1+2\eta)^2}{(1-\eta)^4}k^{-6} - \left[\frac{12\eta(2+\eta)}{(1-\eta)^2}k^{-2}\right. \\
&+ \frac{72\eta^2(2-4\eta-7\eta^2)}{(1-\eta)^4}k^{-4} + \left.\frac{288\eta^2(1+2\eta)^2}{(1-\eta)^4}k^{-6}\right]\cos k \\
&+ \left[\frac{24\eta(1-5\eta-5\eta^2)}{(1-\eta)^3}k^{-3} - \frac{288\eta^2(1+2\eta)^2}{(1-\eta)^4}k^{-5}\right]\sin k .
\end{aligned} \tag{7.39}
$$

To complete the description of the structural properties stemming from the approximation (7.31), let us consider the DCF $c(r)$. Its Fourier transform is obtained from $\widetilde{h}(k)$ via the OZ relation (4.27). The inverse Fourier transform can be performed analytically with the result

$$
c(r) = \begin{cases} -\dfrac{(1+2\eta)^2}{(1-\eta)^4} + \dfrac{6\eta(1+\eta/2)^2}{(1-\eta)^4}r - \dfrac{\eta(1+2\eta)^2}{2(1-\eta)^4}r^3 , & r < 1 , \\[4mm] 0 , & r > 1 . \end{cases} \tag{7.40}
$$

We observe that $c(r) = 0$ for $r > 1$. But this is precisely the *signature* of the PY approximation for HSs [see (7.2)]. This shows that the *simplest* realization (7.31) of the RFA (7.30) turns out to coincide with the exact PY solution.

Figures 7.2, 7.3, and 7.4 display the PY functions $g(r)$, $c(r)$, and $\widetilde{S}(k)$, respectively, at several densities.

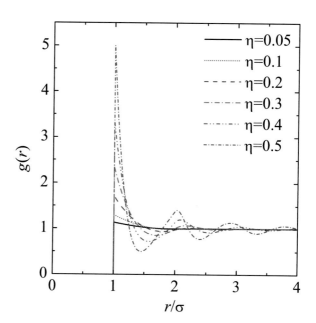

Fig. 7.2 RDF of a three-dimensional HS fluid, as obtained from the PY approximation, at several values of the packing fraction: $\eta \equiv (\pi/6)n\sigma^3 = 0.05, 0.1, 0.2, 0.3, 0.4,$ and 0.5

$\eta=0.05$
$\eta=0.1$
$\eta=0.2$
$\eta=0.3$
$\eta=0.4$
$\eta=0.5$

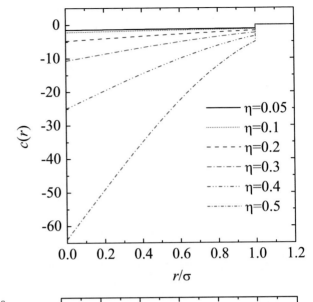

Fig. 7.3 DCF of a three-dimensional HS fluid, as obtained from the PY approximation, at several values of the packing fraction: $\eta \equiv (\pi/6)n\sigma^3 = 0.05, 0.1, 0.2, 0.3, 0.4,$ and 0.5

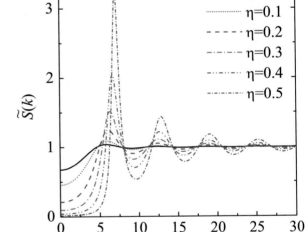

Fig. 7.4 Structure factor of a three-dimensional HS fluid, as obtained from the PY approximation, at several values of the packing fraction: $\eta \equiv (\pi/6)n\sigma^3 = 0.05, 0.1, 0.2, 0.3, 0.4,$ and 0.5

7.2.4.2 Equation of State

Once $\widehat{G}(s)$ is fully determined, one can easily obtain the EoS. As expected, the result depends on the thermodynamic route employed. Let us start with the virial route. According to (6.117), the virial route in the three-dimensional case is

$$Z^{(v)} = 1 + 4\eta g(1^+) . \tag{7.41}$$

From the contact value (7.38a) one immediately obtains

$$\boxed{Z_{\mathrm{PY}}^{(v)} = \frac{1 + 2\eta + 3\eta^2}{(1 - \eta)^2}.}$$
(7.42)

In the case of the compressibility route, (7.26) shows that we need to determine $\widehat{H}^{(1)}$. This quantity is evaluated from the coefficient of s^6 in the Taylor expansion of $e^s/\widehat{F}(s)$, as shown in (7.28). The result is

$$\widehat{H}^{(1)} = \frac{8 - 2\eta + 4\eta^2 - \eta^3}{24(1 + 2\eta)^2}.$$
(7.43)

Insertion into (7.26) yields

$$\boxed{\chi_{T,\mathrm{PY}} = \frac{(1 - \eta)^4}{(1 + 2\eta)^2} \quad \text{(compressibility route)}.}$$
(7.44)

The associated compressibility factor is obtained upon integration [see (1.39)] as

$$\boxed{Z_{\mathrm{PY}}^{(c)} = \int_0^1 \frac{dt}{\chi_{T,\mathrm{PY}}(\eta t)} = \frac{1 + \eta + \eta^2}{(1 - \eta)^3}.}$$
(7.45)

This expression turns out to coincide with the SPT EoS [21–23] [see (3.115)].

Finally, we consider the chemical-potential EoS. In the three-dimensional one-component case, (4.92) gives

$$\beta\mu^{\mathrm{ex}} = -\ln(1 - \eta) + 24\eta \int_{\frac{1}{2}}^1 d\sigma_{01}\sigma_{01}^2 g_{01}(\sigma_{01}^+) \quad (\mu \text{ route}).$$
(7.46)

We see that the contact value (7.38a) is not enough to compute μ^{ex}. We need to "borrow" the solute–solvent contact value $g_{01}(\sigma_{01}^+)$ from the PY solution for mixtures [5]:

$$g_{01}(\sigma_{01}^+) = \frac{1}{1 - \eta} + \frac{3}{2}\frac{\eta}{(1 - \eta)^2}\left(2 - \frac{1}{\sigma_{01}}\right).$$
(7.47)

This expression is exact if $\sigma_{01} = \frac{1}{2}$ [24] and reduces to (7.38a) if $\sigma_{01} = 1$. Performing the integration in (7.46), one finds

$$\boxed{\beta\mu_{\mathrm{PY}}^{\mathrm{ex}} = \eta\frac{7 + \eta/2}{(1 - \eta)^2} - \ln(1 - \eta) \quad (\mu \text{ route}).}$$
(7.48)

The excess free energy per particle consistent with (7.48) is obtained, taking into account the thermodynamic relation (1.36c), as

$$\beta a_{\mathrm{PY}}^{\mathrm{ex}} = \int_0^1 dt\, \beta \mu_{\mathrm{PY}}^{\mathrm{ex}}(\eta t) = \frac{3}{2}\frac{6-\eta}{1-\eta} + \frac{9-\eta}{\eta}\ln(1-\eta) \quad (\mu \text{ route}) . \qquad (7.49)$$

Then, the EoS is derived from the thermodynamic relation (1.36b) with the result

$$Z_{\mathrm{PY}}^{(\mu)} = -8\frac{1-31\eta/16}{(1-\eta)^2} - \frac{9}{\eta}\ln(1-\eta) . \qquad (7.50)$$

While the virial and compressibility EoS (7.42) and (7.45), respectively, are known since 1963 [3], the chemical-potential EoS (7.50) seems to have remained hidden until much later [24].

Interestingly, the CS EoS [see (3.113)] can be recovered as an *interpolation* between the PY virial and compressibility equations:

$$Z_{\mathrm{CS}} = \frac{1}{3}Z_{\mathrm{PY}}^{(v)} + \frac{2}{3}Z_{\mathrm{PY}}^{(c)} . \qquad (7.51)$$

The most relevant thermodynamic quantities corresponding to each route as obtained from the PY solution, as well as from the CS EoS, are summarized in Table 7.1.

Table 7.1 Main thermodynamic quantities as obtained from the PY solution for HS fluids via different routes

Route	βa^{ex}	$\beta\mu^{\mathrm{ex}}$	Z	χ_T^{-1}
v	$\dfrac{6\eta}{1-\eta}$ $+2\ln(1-\eta)$	$\dfrac{2\eta(5-2\eta)}{(1-\eta)^2}$ $+2\ln(1-\eta)$	$\dfrac{1+2\eta+3\eta^2}{(1-\eta)^2}$	$\dfrac{1+5\eta+9\eta^2-3\eta^3}{(1-\eta)^3}$
c	$\dfrac{3\eta(2-\eta)}{2(1-\eta)^2}$ $-\ln(1-\eta)$	$\dfrac{\eta(14-13\eta+5\eta^2)}{2(1-\eta)^3}$ $-\ln(1-\eta)$	$\dfrac{1+\eta+\eta^2}{(1-\eta)^3}$	$\dfrac{(1+2\eta)^2}{(1-\eta)^4}$
μ	$\dfrac{3(6-\eta)}{2(1-\eta)}$ $+\dfrac{(9-\eta)}{\eta}\ln(1-\eta)$	$\dfrac{\eta(14+\eta)}{2(1-\eta)^2}$ $-\ln(1-\eta)$	$-\dfrac{16-31\eta}{2(1-\eta)^2}$ $-\dfrac{9}{\eta}\ln(1-\eta)$	$\dfrac{1+5\eta+9\eta^2}{(1-\eta)^3}$
CS	$\dfrac{\eta(4-3\eta)}{(1-\eta)^2}$	$\dfrac{\eta(8-9\eta+3\eta^2)}{(1-\eta)^3}$	$\dfrac{1+\eta+\eta^2-\eta^3}{(1-\eta)^3}$	$\dfrac{1+4\eta+4\eta^2-4\eta^3+\eta^4}{(1-\eta)^4}$

Table 7.2 First twelve (reduced) virial coefficients b_k as obtained exactly and from several EoS related to the PY approximation

k	Exact	$Z_{PY}^{(v)}$	$Z_{PY}^{(c)}$	$Z_{PY}^{(\mu)}$	Z_{CS}	$Z^{(\mu c,1)}$	$Z^{(\mu c,2)}$
2	4	4	4	4	4	4	4
3	10	10	10	10	10	10	10
4	$18.364\,768\cdots$	16	19	$\dfrac{67}{4}=16.75$	18	$\dfrac{181}{10}=18.1$	$\dfrac{145}{8}=18.125$
5	$28.224\,376(15)$	22	31	$\dfrac{119}{5}=23.8$	28	$\dfrac{703}{25}=28.12$	$\dfrac{141}{5}=28.2$
6	$39.815\,23(10)$	28	46	31	40	40	$\dfrac{241}{6}\simeq 40.2$
7	$53.342\,1(5)$	34	64	$\dfrac{268}{7}\simeq 38.3$	54	$\dfrac{376}{7}\simeq 53.7$	54
8	$68.529(3)$	40	85	$\dfrac{365}{8}=45.6$	70	$\dfrac{277}{4}=69.25$	$\dfrac{1115}{16}\simeq 69.7$
9	$85.83(2)$	46	109	53	88	$\dfrac{433}{5}=86.6$	$\dfrac{785}{9}\simeq 87.2$
10	$105.70(10)$	52	136	$\dfrac{302}{5}=60.4$	108	$\dfrac{2\,644}{25}=105.76$	$\dfrac{533}{5}=106.6$
11	$126.5(6)$	58	166	$\dfrac{746}{11}\simeq 67.8$	130	$\dfrac{1\,394}{11}\simeq 126.7$	$\dfrac{1\,406}{11}\simeq 127.8$
12	$130(25)$	64	199	$\dfrac{301}{4}=75.25$	154	$\dfrac{299}{2}=149.5$	$\dfrac{1\,207}{8}\simeq 150.9$

The reduced virial coefficients b_k [see (3.104)] predicted by the three EoS (7.42), (7.45), and (7.50) are

$$b_k^{(\mathrm{PY},v)} = 2(3k-4)\,, \quad b_k^{(\mathrm{PY},c)} = \frac{3k^2 - 3k + 2}{2}\,, \quad b_k^{(\mathrm{PY},\mu)} = \frac{15k^2 - 31k + 18}{2k}\,. \tag{7.52}$$

Those virial coefficients are compared with the exact values [25–30] (see Table 3.11) in Table 7.2, which also includes the CS coefficients. We observe that (7.45) overestimates the known coefficients, while (7.42) and (7.50) underestimate them, the chemical-potential route being slightly more accurate than the virial one. Those trends agree with what can be observed from Fig. 7.5, where the deviations of the three PY EoS from MD simulation results [31] are shown.

As we already saw in Fig. 3.22, Z_{CS} is an excellent EoS. On the other hand, since $Z_{PY}^{(\mu)}$ is more reliable than $Z_{PY}^{(v)}$, one may wonder whether an interpolation formula

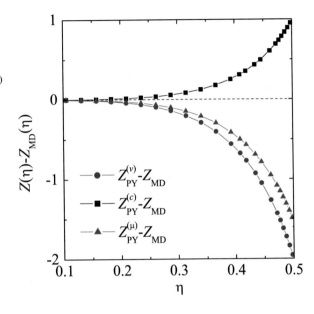

Fig. 7.5 Plot of the deviations $Z_{\mathrm{PY}}^{(v)}(\eta) - Z_{\mathrm{MD}}(\eta)$ (*circles*), $Z_{\mathrm{PY}}^{(c)}(\eta) - Z_{\mathrm{MD}}(\eta)$ (*squares*), and $Z_{\mathrm{PY}}^{(\mu)}(\eta) - Z_{\mathrm{MD}}(\eta)$ (*triangles*) for three-dimensional HS fluids

similar to (7.51), but this time between $Z_{\mathrm{PY}}^{(\mu)}$ and $Z_{\mathrm{PY}}^{(c)}$, i.e.,

$$Z^{(\mu c)} = \lambda Z_{\mathrm{PY}}^{(\mu)} + (1 - \lambda) Z_{\mathrm{PY}}^{(c)} , \qquad (7.53)$$

might be even more accurate. From an analysis of the virial coefficients one can check that the optimal value of the interpolation parameter is $\lambda \approx 0.4$. In particular, the two choices

$$\lambda = \frac{2}{5} \Rightarrow Z^{(\mu c,1)} , \quad \lambda = \frac{7}{18} \Rightarrow Z^{(\mu c,2)} \qquad (7.54)$$

are analyzed in Table 7.2 at the level of the virial coefficients, where

$$b_k^{(\mu c,1)} = \frac{9k^3 + 21k^2 - 56k + 36}{10k} , \quad b_k^{(\mu c,2)} = \frac{11k^3 + 24k^2 - 65k + 42}{12k} . \qquad (7.55)$$

A better general performance than that of the CS coefficients is clearly observed.

The good quality of the compressibility factors $Z^{(\mu c,1)}$ and $Z^{(\mu c,2)}$ is confirmed by Fig. 7.6, where again deviations from MD simulation values (Z_{MD}) [31] are plotted as functions of the packing fraction.

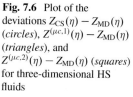

Fig. 7.6 Plot of the deviations $Z_{CS}(\eta) - Z_{MD}(\eta)$ (*circles*), $Z^{(\mu c,1)}(\eta) - Z_{MD}(\eta)$ (*triangles*), and $Z^{(\mu c,2)}(\eta) - Z_{MD}(\eta)$ (*squares*) for three-dimensional HS fluids

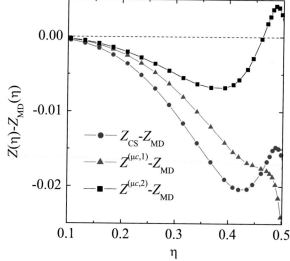

7.3 Percus–Yevick Approximation for Hard-Sphere and Sticky-Hard-Sphere Mixtures

Just 1 year after Wertheim [2] and Thiele [4] found the exact solution of the (three-dimensional) OZ equation (7.1) with the PY closure (7.2) for a HS fluid, Joel L. Lebowitz (see Fig. 7.7) extended the solution to AHS mixtures [5]. Not much later, in the same paper [9] where Baxter (see Fig. 3.4) had introduced the SHS interaction model [see (3.4), (3.5), and Table 3.1], he solved it with the PY approximation. The generalization to (additive) SHS mixtures was subsequently carried out by Perram and Smith [10] and, independently, by Barboy [11, 12].

Since the SHS multicomponent fluid includes as particular cases both the AHS mixture and the SHS one-component fluid, we will focus here on the former system. Moreover, as in Sect. 7.2.4, we will obtain the PY solution by applying the RFA methodology [15, 32–35].

7.3.1 Sticky-Hard-Sphere Mixtures

7.3.1.1 General Relations

Let us consider a (three-dimensional) multicomponent fluid with a general interaction potential $\phi_{\alpha\gamma}(r)$ for particles of species α and γ. Similarly to what we did in the one-component case [see (7.3)], we introduce the Laplace transform of

Fig. 7.7 Joel L. Lebowitz
(b. 1930) (Photograph
courtesy of P. Cvitanović,
https://www.flickr.com/
photos/birdtracks/
3106543914)

$rg_{\alpha\gamma}(r)$:

$$\widehat{G}_{\alpha\gamma}(s) = \int_0^\infty dr\, e^{-sr} rg_{\alpha\gamma}(r) \ . \tag{7.56}$$

In analogy to (7.4), the Fourier transform of the total correlation function $h_{\alpha\gamma}(r)$ is given by

$$\tilde{h}_{\alpha\gamma}(k) = -2\pi \left[\frac{\widehat{H}_{\alpha\gamma}(s) - \widehat{H}_{\alpha\gamma}(-s)}{s} \right]_{s=ik} = -2\pi \left[\frac{\widehat{G}_{\alpha\gamma}(s) - \widehat{G}_{\alpha\gamma}(-s)}{s} \right]_{s=ik} , \tag{7.57}$$

where

$$\widehat{H}_{\alpha\gamma}(s) = \widehat{G}_{\alpha\gamma}(s) - s^{-2} \tag{7.58}$$

is the Laplace transform of $rh_{\alpha\gamma}(r)$. The condition of a finite isothermal compressibility [see (4.74) and (4.79)] implies that $\tilde{h}_{\alpha\gamma}(0) = $ finite. As a consequence,

$$s^2 \widehat{G}_{\alpha\gamma}(s) = 1 + \widehat{H}_{\alpha\gamma}^{(0)} s^2 + \widehat{H}_{\alpha\gamma}^{(1)} s^3 + \cdots \tag{7.59}$$

with $\widehat{H}_{\alpha\gamma}^{(0)} = $ finite and $\widehat{H}_{\alpha\gamma}^{(1)} = -\tilde{h}_{\alpha\gamma}(0)/4\pi = $ finite, where

$$\widehat{H}_{\alpha\gamma}^{(k)} \equiv (-1)^k \int_0^\infty dr\, r^{k+1} h_{\alpha\gamma}(r) \ . \tag{7.60}$$

The behavior of $\widehat{G}_{\alpha\gamma}(s)$ for small s is therefore constrained by (7.59).

Thus far, the interaction potential $\phi_{\alpha\gamma}(r)$ is arbitrary. Now, we first assume that $\phi_{\alpha\gamma}(r)$ has a SW form characterized by a hard-core diameter $\sigma_{\alpha\gamma}$, a well depth $\varepsilon_{\alpha\gamma}$, and a well width $\sigma'_{\alpha\gamma} - \sigma_{\alpha\gamma}$ (see Table 3.1). The degree of "stickiness" of the (α, γ) pair interaction is measured by the parameter [see (3.4)]

$$\tau_{\alpha\gamma}^{-1} \equiv 4 \left[\left(\frac{\sigma_{\alpha\gamma}'}{\sigma_{\alpha\gamma}} \right)^3 - 1 \right] \left(e^{\beta \varepsilon_{\alpha\gamma}} - 1 \right) . \tag{7.61}$$

The SHS fluid corresponds to the combined limits $\sigma_{\alpha\gamma}' \to \sigma_{\alpha\gamma}$ and $\varepsilon_{\alpha\gamma} \to \infty$ with $\tau_{\alpha\gamma}^{-1} = $ finite. In that case, the Mayer function becomes [see (3.5)]

$$f_{\alpha\gamma}(r) = -\Theta(\sigma_{\alpha\gamma} - r) + \frac{\sigma_{\alpha\gamma}}{12\tau_{\alpha\gamma}} \delta(r - \sigma_{\alpha\gamma}) , \tag{7.62}$$

so that the relationship between the RDF $g_{\alpha\gamma}(r)$ and the cavity function $y_{\alpha\gamma}(r)$ is

$$g_{\alpha\gamma}(r) = y_{\alpha\gamma}(r) \left[\Theta(r - \sigma_{\alpha\gamma}) + \frac{\sigma_{\alpha\gamma}}{12\tau_{\alpha\gamma}} \delta(r - \sigma_{\alpha\gamma}) \right] . \tag{7.63}$$

In particular, in the region near contact,

$$\begin{aligned} g_{\alpha\gamma}(r) = {} & \frac{\sigma_{\alpha\gamma}}{12\tau_{\alpha\gamma}} y_{\alpha\gamma}(\sigma_{\alpha\gamma}) \delta(r - \sigma_{\alpha\gamma}) \\ & + \Theta(r - \sigma_{\alpha\gamma}) \left[y_{\alpha\gamma}(\sigma_{\alpha\gamma}) + y_{\alpha\gamma}'(\sigma_{\alpha\gamma})(r - \sigma_{\alpha\gamma}) + \cdots \right] . \end{aligned} \tag{7.64}$$

The condition of finite $y_{\alpha\gamma}(\sigma_{\alpha\gamma})$ translates into the following behavior of the Laplace transform $\widehat{G}_{\alpha\gamma}(s)$ for large s:

$$\begin{aligned} e^{\sigma_{\alpha\gamma} s} \widehat{G}_{\alpha\gamma}(s) = {} & \frac{\sigma_{\alpha\gamma}^2}{12\tau_{\alpha\gamma}} y_{\alpha\gamma}(\sigma_{\alpha\gamma}) + \sigma_{\alpha\gamma} y_{\alpha\gamma}(\sigma_{\alpha\gamma}) s^{-1} \\ & + \left[y_{\alpha\gamma}(\sigma_{\alpha\gamma}) + \sigma_{\alpha\gamma} y_{\alpha\gamma}'(\sigma_{\alpha\gamma}) \right] s^{-2} + \mathcal{O}(s^{-3}) . \end{aligned} \tag{7.65}$$

Moreover, from (4.84) we see that the virial EoS for the SHS mixture is given by

$$Z^{(v)} = 1 + \frac{2\pi}{3} n \sum_{\alpha,\gamma} x_\alpha x_\gamma \sigma_{\alpha\gamma}^3 y_{\alpha\gamma}(\sigma_{\alpha\gamma}) \left\{ 1 - \frac{\tau_{\alpha\gamma}^{-1}}{12} \left[3 + \frac{\sigma_{\alpha\gamma} y_{\alpha\gamma}'(\sigma_{\alpha\gamma})}{y_{\alpha\gamma}(\sigma_{\alpha\gamma})} \right] \right\} , \tag{7.66}$$

where $y_{\alpha\gamma}'(r) \equiv \partial y_{\alpha\gamma}(r)/\partial r$. Analogously, the energy route becomes (see Table 4.3)

$$\begin{aligned} u^{\text{ex}} = {} & -\frac{\pi}{6} n \sum_{\alpha,\gamma} x_\alpha x_\gamma \sigma_{\alpha\gamma}^3 y_{\alpha\gamma}(\sigma_{\alpha\gamma}) \frac{\varepsilon_{\alpha\gamma}}{\tau_{\alpha\gamma}} \\ = {} & -\frac{\pi}{6} n \sum_{\alpha,\gamma} x_\alpha x_\gamma \sigma_{\alpha\gamma}^3 y_{\alpha\gamma}(\sigma_{\alpha\gamma}) \frac{\partial \tau_{\alpha\gamma}^{-1}}{\partial \beta} \quad \text{(energy route)} , \end{aligned} \tag{7.67}$$

where (7.63) has been applied in the first step and in the second step we have taken into account that $\partial \tau_{\alpha\gamma}^{-1}/\partial \beta = \varepsilon_{\alpha\gamma} \tau_{\alpha\gamma}^{-1}$ in the SHS limit [see (7.61)].

The case of a HS system is recovered by taking the limit of vanishing stickiness, $\tau_{\alpha\gamma}^{-1} \to 0$, in (7.62)–(7.67). It is important to notice that, while $e^{\sigma_{\alpha\gamma}s}\widehat{G}_{\alpha\gamma}(s) \sim s^{-1}$ for large s in the case of HSs [see also (7.21)], one has $e^{\sigma_{\alpha\gamma}s}\widehat{G}_{\alpha\gamma}(s) \sim s^0$ in the case of SHSs. However, the behavior for small s is still given by (7.59), regardless of the interaction model.

7.3.1.2   Percus–Yevick Approximation

Now we are in conditions of constructing an analytical approximation for $\widehat{G}_{\alpha\gamma}(s)$ (within the spirit of the RFA) consistent with the requirements (7.59) and (7.65). To that end, we keep a structure similar to that of (7.35), except that we need to deal with matrices [see (5.31) for the exact one-dimensional solution]. Taking all of this into account, we construct the approximation

$$\widehat{G}_{\alpha\gamma}(s) = \frac{e^{-\sigma_{\alpha\gamma}s}}{s^2} \left\{ \Lambda(s) \cdot [\mathbb{I} - \Sigma(s)]^{-1} \right\}_{\alpha\gamma} , \qquad (7.68)$$

where $\Lambda(s)$ and $\Sigma(s)$ are the matrices

$$\Lambda_{\alpha\gamma}(s) = \Lambda_{\alpha\gamma}^{(0)} + \Lambda_{\alpha\gamma}^{(1)}s + \Lambda_{\alpha\gamma}^{(2)}s^2 , \qquad (7.69a)$$

$$\Sigma_{\alpha\gamma}(s) = 2\pi n x_\alpha \left[\varphi_2(\sigma_\alpha s)\sigma_\alpha^3 \Lambda_{\alpha\gamma}^{(0)} + \varphi_1(\sigma_\alpha s)\sigma_\alpha^2 \Lambda_{\alpha\gamma}^{(1)} + \varphi_0(\sigma_\alpha s)\sigma_\alpha \Lambda_{\alpha\gamma}^{(2)} \right] . \qquad (7.69b)$$

Here we have restricted ourselves to *additive* mixtures, i.e., $\sigma_{\alpha\gamma} = \frac{1}{2}(\sigma_\alpha + \sigma_\gamma)$. The presence of the terms $\Lambda_{\alpha\gamma}^{(2)}$ in (7.69) is directly related to the SHS interaction since $\Lambda_{\alpha\gamma}^{(2)} = 0$ in the HS case. In fact, since $\lim_{x\to\infty} \varphi_k(x) = 0$, one simply has $\lim_{s\to\infty} e^{\sigma_{\alpha\gamma}s}\widehat{G}_{\alpha\gamma}(s) = \Lambda_{\alpha\gamma}^{(2)}/2\pi$ and then (7.65) implies

$$\Lambda_{\alpha\gamma}^{(2)} = \frac{\sigma_{\alpha\gamma}^2}{12\tau_{\alpha\gamma}} y_{\alpha\gamma}(\sigma_{\alpha\gamma}) . \qquad (7.70)$$

But (7.65) also imposes the condition that the ratio between the first and second terms in the expansion of $e^{\sigma_{\alpha\gamma}s}\widehat{G}_{\alpha\gamma}(s)$ in powers of s^{-1} must be *exactly* equal to $\sigma_{\alpha\gamma}/12\tau_{\alpha\gamma}$. Let us see what constraints on the coefficients $\Lambda_{\alpha\gamma}^{(k)}$ are implied by this condition.

The property $\varphi_k(x) = (-1)^k x^{-1}/k! + \mathcal{O}(x^{-2})$ leads to

$$\Sigma(s) = s^{-1}\bar{\Sigma} + \mathcal{O}(s^{-2}) \Rightarrow [\mathbb{I} - \Sigma(s)]^{-1} = \mathbb{I} + s^{-1}\bar{\Sigma} + \mathcal{O}(s^{-2}) , \qquad (7.71a)$$

$$\bar{\Sigma}_{\alpha\gamma} = 2\pi n x_\alpha \left(\frac{1}{2}\sigma_\alpha^2 \Lambda_{\alpha\gamma}^{(0)} - \sigma_\alpha \Lambda_{\alpha\gamma}^{(1)} + \Lambda_{\alpha\gamma}^{(2)} \right) . \tag{7.71b}$$

Consequently, the behavior for large s that follows from (7.68) is

$$e^{\sigma_{\alpha\gamma}s}\widehat{G}_{\alpha\gamma}(s) = \Lambda_{\alpha\gamma}^{(2)} + \left[\Lambda_{\alpha\gamma}^{(1)} + \left(\Lambda^{(2)} \cdot \bar{\Sigma} \right)_{\alpha\gamma} \right] s^{-1} + \mathcal{O}(s^{-2}) . \tag{7.72}$$

Comparison with (7.65) yields, apart from (7.70),

$$\sigma_{\alpha\gamma} y_{\alpha\gamma}(\sigma_{\alpha\gamma}) = \Lambda_{\alpha\gamma}^{(1)} + \sum_\delta \Lambda_{\alpha\delta}^{(2)} \bar{\Sigma}_{\delta\gamma} . \tag{7.73}$$

Elimination of the contact value $y_{\alpha\gamma}(\sigma_{\alpha\gamma})$ between (7.70) and (7.73) gives a relationship between the matrices $\Lambda^{(0)}$, $\Lambda^{(1)}$, and $\Lambda^{(2)}$. In order to close the problem, we need two more relations. They are provided by the small-s behavior (7.59), according to which the coefficients of s^0 and s in the power series expansion of $s^2 \widehat{G}_{\alpha\gamma}(s)$ must be 1 and 0, respectively. After some algebra, this yields [35]

$$\Lambda_{\alpha\gamma}^{(0)} = \frac{1}{1-\eta} + 3\frac{\eta}{(1-\eta)^2}\frac{M_2}{M_3}\sigma_\gamma - \frac{\eta}{1-\eta}\frac{12}{M_3}\sum_\delta x_\delta \sigma_\delta \Lambda_{\delta\gamma}^{(2)} , \tag{7.74a}$$

$$\Lambda_{\alpha\gamma}^{(1)} = \frac{\sigma_{\alpha\gamma}}{1-\eta} + \frac{3}{2}\frac{\eta}{(1-\eta)^2}\frac{M_2}{M_3}\sigma_\alpha\sigma_\gamma - \frac{\eta}{1-\eta}\frac{6\sigma_\alpha}{M_3}\sum_\delta x_\delta \sigma_\delta \Lambda_{\delta\gamma}^{(2)} , \tag{7.74b}$$

where the moments M_k of the size distribution and the total packing fraction η are defined by (3.70) and (3.116), respectively. Note that $\Lambda_{\alpha\gamma}^{(0)}$ does not actually depend on the row index α. Finally, inserting (7.74) into (7.73) and combining the result with (7.70), the following closed *bilinear* equation for $\Lambda^{(2)}$ is obtained:

$$\frac{12\tau_{\alpha\gamma}\Lambda_{\alpha\gamma}^{(2)}}{\sigma_{\alpha\gamma}} = \frac{\sigma_{\alpha\gamma}}{1-\eta} + \frac{3}{2}\frac{\eta}{(1-\eta)^2}\frac{M_2}{M_3}\sigma_\alpha\sigma_\gamma + \eta\frac{12}{M_3}\sum_\delta x_\delta \Lambda_{\alpha\delta}^{(2)}\Lambda_{\delta\gamma}^{(2)}$$
$$- \frac{\eta}{1-\eta}\frac{6}{M_3}\sum_\delta x_\delta\sigma_\delta\left(\Lambda_{\alpha\delta}^{(2)}\sigma_\gamma + \Lambda_{\delta\gamma}^{(2)}\sigma_\alpha\right) . \tag{7.75}$$

The physical solution is obtained by the condition $\lim_{\tau_{\alpha\gamma}\to\infty}\Lambda_{\alpha\gamma}^{(2)} = 0$, in consistency with (7.70).

Equations (7.68), (7.69), (7.74), and (7.75) fully determine the RDFs (in Laplace space) for an *arbitrary* (additive) SHS mixture, as obtained from the simplest implementation of the RFA [15, 35]. Although not derived explicitly here as the solution of the OZ relation with the PY closure, it turns out that (7.68), (7.69), (7.74), and (7.75) actually coincide with such a solution [10–12].

Fig. 7.8 Cavity functions $y_{11}(r) = y_{22}(r)$ and $y_{12}(r)$ at $\eta = 0.4$ for an equimolar binary mixture $(x_1 = x_2 = \frac{1}{2})$ with $\sigma_1 = \sigma_2 = \sigma_{12} = \sigma$. The *solid lines* correspond to MC simulations [36] of a SW mixture with $\sigma_1' = \sigma_2' = \sigma$, $\sigma_{12}' = 1.01\sigma$, $\varepsilon_{11} = \varepsilon_{22} = 0$, and $k_B T/\varepsilon_{12} = 0.266$. The *dashed lines* represent the analytical solution of the PY approximation for a SHS mixture with $\tau_{11}^{-1} = \tau_{22}^{-1} = 0$ and $\tau_{12} = 0.2$

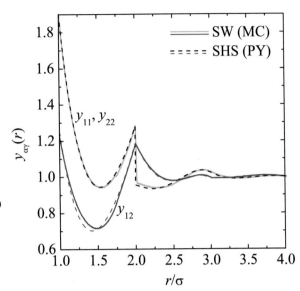

As a test of the practical usefulness of the PY solution for SHS mixtures, Fig. 7.8 compares the cavity functions predicted by the PY analytical solution with those obtained from MC simulations of a true short-range SW mixture [36]. A general good agreement is found, especially in the case of the pair correlation function $y_{11}(r) = y_{22}(r)$ of the particles interacting via the HS potential.

In what concerns the thermodynamic properties, note that the contact values $y_{\alpha\gamma}(\sigma_{\alpha\gamma})$ are readily obtained from either (7.70) or (7.73). Analogously, the coefficient of s^{-2} in the series expansion of $e^{\sigma_{\alpha\gamma}}\widehat{G}_{\alpha\gamma}(s)$ in powers of s^{-1} allows us to identify $y_{\alpha\gamma}'(\sigma_{\alpha\gamma})$ from (7.65). Then, the virial EoS can be obtained via (7.66). Furthermore, the coefficient of s^3 in the series expansion of $s^2\widehat{G}_{\alpha\gamma}(s)$ in powers of s provides $\tilde{h}_{\alpha\gamma}(0) = -4\pi\widehat{H}_{\alpha\gamma}^{(1)}$ and, hence, the isothermal susceptibility [see (4.74) and (4.79)]. The expressions are too cumbersome to be reproduced here, so we particularize now to AHS mixtures and to one-component SHS fluids.

7.3.2 Additive Hard-Sphere Mixtures

The AHS case [5, 37] is obtained from the more general SHS system by just setting $\tau_{\alpha\gamma}^{-1} = 0$ for every pair (α, γ). In view of (7.70), this implies that $\Lambda_{\alpha\gamma}^{(2)} = 0$. Also, (7.73) gives $y_{\alpha\gamma}(\sigma_{\alpha\gamma}) = \Lambda_{\alpha\gamma}^{(1)}/\sigma_{\alpha\gamma}$, i.e.,

$$y_{\alpha\gamma}(\sigma_{\alpha\gamma}) = g_{\alpha\gamma}(\sigma_{\alpha\gamma}^+) = \frac{1}{1-\eta} + \frac{3}{2}\frac{\eta}{(1-\eta)^2}\frac{\sigma_\alpha\sigma_\gamma}{\sigma_{\alpha\gamma}}\frac{M_2}{M_3}, \qquad (7.76)$$

where use has been made of (7.74b). Insertion of (7.76) into (7.66) (with $\tau_{\alpha\gamma}^{-1} = 0$) yields

$$Z_{PY}^{(v)} = \frac{1}{1-\eta} + \frac{3\eta}{(1-\eta)^2} \frac{m_2}{m_3} + \frac{3\eta^2}{(1-\eta)^2} \frac{m_2^3}{m_3^2},$$

(7.77)

where the reduced moments m_k are defined in (3.123).

The evaluation of $\widehat{H}_{\alpha\beta}^{(1)}$ from (7.59) is more involved but, however, the associated expression for the isothermal susceptibility is rather simple:

$$\chi_{T,PY}^{-1} = \frac{1}{(1-\eta)^2} + \frac{6\eta}{(1-\eta)^3} \frac{m_2}{m_3} + \frac{9\eta^2}{(1-\eta)^4} \frac{m_2^3}{m_3^2} \quad \text{(compressibility route)}.$$

(7.78)

Integration over density [see (1.39)] gives

$$Z_{PY}^{(c)} = \frac{1}{1-\eta} + \frac{3\eta}{(1-\eta)^2} \frac{m_2}{m_3} + \frac{3\eta^2}{(1-\eta)^3} \frac{m_2^3}{m_3^2}.$$

(7.79)

Note that the virial and compressibility expressions of Z_{PY} differ only in the numerical value of the negative power of $1 - \eta$ in the third term.

If an impurity particle of diameter σ_0 is added to the AHS mixture, the impurity–fluid contact value $y_{0\alpha}(\sigma_{0\alpha})$ is simply given by (7.76) with $\sigma_\gamma \to \sigma_0$ since a single impurity particle does not modify the size moments of the mixture. Then, the chemical-potential route (4.92) yields [37]

$$\beta\mu_{v,PY}^{ex} = -\ln(1-\eta) + \frac{3\eta}{1-\eta} \frac{m_2}{m_3} \frac{\sigma_v}{M_1} + \frac{3\eta}{1-\eta} \left(\frac{m_2}{m_3} + \frac{3}{2} \frac{\eta}{1-\eta} \frac{m_2^3}{m_3^2} \right) \frac{\sigma_v^2}{M_2}$$

$$+ \frac{\eta}{1-\eta} \left(1 + 3 \frac{\eta}{1-\eta} \frac{m_2}{m_3} \right) \frac{\sigma_v^3}{M_3} \quad (\mu \text{ route}).$$

(7.80)

The associated (excess) Helmholtz free energy per particle can then be derived from the fundamental equation of thermodynamics as expressed by the first equality in (1.37) as

$$\beta a_{PY}^{ex} = -\ln(1-\eta) + \frac{3\eta}{1-\eta} \frac{m_2}{m_3} + \frac{3\eta^2}{2(1-\eta)^2} \frac{m_2^3}{m_3^2}$$

$$+ \frac{3}{2} \left[\frac{6 - 9\eta + 2\eta^2}{(1-\eta)^2} + 6\frac{\ln(1-\eta)}{\eta} \right] \frac{m_2^3}{m_3^2} \quad (\mu \text{ route}).$$

(7.81)

Applying now the second equality in (1.37) we can obtain the EoS in the chemical-potential route:

$$Z_{PY}^{(\mu)} = \frac{1}{1-\eta} + \frac{3\eta}{(1-\eta)^2}\frac{m_2}{m_3} - 9\left[\frac{1-\frac{3}{2}\eta}{(1-\eta)^2} + \frac{\ln(1-\eta)}{\eta}\right]\frac{m_2^3}{m_3^2}. \qquad (7.82)$$

Equations (7.76)–(7.82) represent the extensions to mixtures of (7.38a), (7.42), (7.44), (7.45), and (7.48)–(7.50), respectively.

The PY expression (7.76) for the contact values is similar to the one derived from the SPT [21, 23, 38–40]:

$$y_{\alpha\gamma}(\sigma_{\alpha\gamma}) = \frac{1}{1-\eta} + \frac{3}{2}\frac{\eta}{(1-\eta)^2}\frac{\sigma_\alpha\sigma_\gamma}{\sigma_{\alpha\gamma}}\frac{M_2}{M_3} + \frac{3}{4}\frac{\eta^2}{(1-\eta)^3}\left(\frac{\sigma_\alpha\sigma_\gamma}{\sigma_{\alpha\gamma}}\frac{M_2}{M_3}\right)^2, \quad \text{(SPT)}.$$

$$\qquad (7.83)$$

Interestingly, when the above expression is inserted into the virial EoS (7.66) (again with $\tau_{\alpha\gamma}^{-1} = 0$), the result coincides with the PY compressibility-route EoS (7.79) [see also (3.147)], as happened in the one-component case [see (3.115) and (7.45)]. A simple, and yet accurate, prescription for the contact values is obtained, in analogy with (7.51), by a linear interpolation (with respective weights $\frac{1}{3}$ and $\frac{2}{3}$) between the PY and SPT results (7.76) and (7.83), respectively. This was proposed, independently, by Boublík [41], Grundke and Henderson [42], and Lee and Levesque [43] (BGHLL):

$$y_{\alpha\gamma}(\sigma_{\alpha\gamma}) = \frac{1}{1-\eta} + \frac{3}{2}\frac{\eta}{(1-\eta)^2}\frac{\sigma_\alpha\sigma_\gamma}{\sigma_{\alpha\gamma}}\frac{M_2}{M_3} + \frac{1}{2}\frac{\eta^2}{(1-\eta)^3}\left(\frac{\sigma_\alpha\sigma_\gamma}{\sigma_{\alpha\gamma}}\frac{M_2}{M_3}\right)^2, \quad \text{(BGHLL)}.$$

$$\qquad (7.84)$$

Obviously, the compressibility factor corresponding to the contact values (7.84) is equal to $\frac{1}{3}Z_{PY}^{(v)} + \frac{2}{3}Z_{PY}^{(c)}$. This EoS was first obtained by Boublík [41] and, independently, Mansoori, Carnahan, Starling, and Leland [44] (BMCSL):

$$Z_{BMCSL} = \frac{1}{1-\eta} + \frac{3\eta}{(1-\eta)^2}\frac{m_2}{m_3} + \frac{\eta^2(3-\eta)}{(1-\eta)^3}\frac{m_2^3}{m_3^2}. \qquad (7.85)$$

The expressions for the excess Helmholtz free energy per particle and for the compressibility factor, as predicted by the PY and BMCSL approximations, are displayed in Table 7.3.

As an assessment of the performance of the PY compressibility factors (7.77), (7.79), and (7.82), as well as of the BMCSL prescription (7.85), Figs. 7.9 and 7.10 compare them against computer simulations [45] for binary mixtures at $\eta = 0.49$ and, respectively, $\sigma_2/\sigma_1 = 0.6$ and $\sigma_2/\sigma_1 = 0.3$. It is observed that, as expected

Table 7.3 Main thermodynamic quantities as obtained from the PY solution for AHS fluids via different routes

Route	βa^{ex}	Z
v	$\ln(1-\eta)\left(3\dfrac{m_2^3}{m_3^2}-1\right)+\dfrac{3\eta}{1-\eta}$ $\times\left(\dfrac{m_2}{m_3}+\dfrac{m_2^3}{m_3^2}\right)$	$\dfrac{1}{1-\eta}+\dfrac{3\eta}{(1-\eta)^2}\dfrac{m_2}{m_3}+\dfrac{3\eta^2}{(1-\eta)^2}\dfrac{m_2^3}{m_3^2}$
c	$-\ln(1-\eta)+\dfrac{3\eta}{1-\eta}\dfrac{m_2}{m_3}+\dfrac{3\eta^2}{2(1-\eta)^2}\dfrac{m_2^3}{m_3^2}$	$\dfrac{1}{1-\eta}+\dfrac{3\eta}{(1-\eta)^2}\dfrac{m_2}{m_3}+\dfrac{3\eta^2}{(1-\eta)^3}\dfrac{m_2^3}{m_3^2}$
μ	$-\ln(1-\eta)+\dfrac{3\eta}{1-\eta}\dfrac{m_2}{m_3}+\dfrac{3\eta^2}{2(1-\eta)^2}\dfrac{m_2^3}{m_3^2}$ $+\dfrac{3}{2}\left[\dfrac{6-9\eta+2\eta^2}{(1-\eta)^2}+6\dfrac{\ln(1-\eta)}{\eta}\right]\dfrac{m_2^3}{m_3^2}$	$\dfrac{1}{1-\eta}+\dfrac{3\eta}{(1-\eta)^2}\dfrac{m_2}{m_3}$ $-9\left[\dfrac{1-\frac{3}{2}\eta}{(1-\eta)^2}+\dfrac{\ln(1-\eta)}{\eta}\right]\dfrac{m_2^3}{m_3^2}$
BMCSL	$\ln(1-\eta)\left(\dfrac{m_2^3}{m_3^2}-1\right)+\dfrac{3\eta}{1-\eta}\dfrac{m_2}{m_3}$ $+\dfrac{\eta}{(1-\eta)^2}\dfrac{m_2^3}{m_3^2}$	$\dfrac{1}{1-\eta}+\dfrac{3\eta}{(1-\eta)^2}\dfrac{m_2}{m_3}+\dfrac{\eta^2(3-\eta)}{(1-\eta)^3}\dfrac{m_2^3}{m_3^2}$

from the one-component case (see Fig. 7.5), $Z_{\text{PY}}^{(v)}$ underestimates the simulation values, while $Z_{\text{PY}}^{(c)}$ overestimates them. The chemical-potential route $Z_{\text{PY}}^{(\mu)}$ lies below the simulation data, but it exhibits a better behavior than the virial route $Z_{\text{PY}}^{(v)}$. The weighted average between $Z_{\text{PY}}^{(v)}$ and $Z_{\text{PY}}^{(c)}$ made in the construction of the BMCSL EoS (7.85) does a very good job. A slightly better agreement is obtained from the weighted average between $Z_{\text{PY}}^{(\mu)}$ and $Z_{\text{PY}}^{(c)}$ [see (7.53)] with $\lambda=\frac{7}{18}$.

7.3.3 One-Component Sticky Hard Spheres

7.3.3.1 Structural Properties

The special case of a one-component SHS fluid [9, 32, 33] can be obtained from the multicomponent one by taking $\sigma_{\alpha\gamma}=\sigma$ and $\tau_{\alpha\gamma}=\tau$. Thus, from (7.68) and (7.69), the Laplace transform of $rg(r)$ becomes

$$\widehat{G}(s)=\frac{\text{e}^{-s}}{s^2}\frac{\Lambda^{(0)}+\Lambda^{(1)}s+\Lambda^{(2)}s^2}{1-12\eta\left[\varphi_2(s)\Lambda^{(0)}+\varphi_1(s)\Lambda^{(1)}+\varphi_0(s)\Lambda^{(2)}\right]},\qquad(7.86)$$

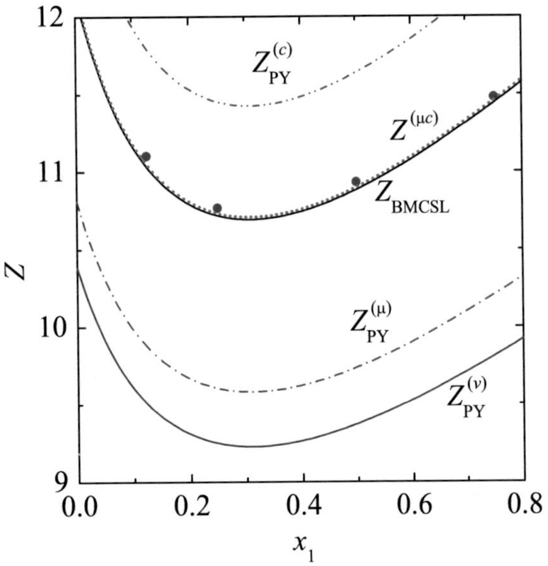

Fig. 7.9 Plot of the compressibility factor Z as a function of the mole fraction x_1 for an AHS binary mixture with a packing fraction $\eta = 0.49$ and a size ratio $\sigma_2/\sigma_1 = 0.6$. The *symbols* are computer simulation values [45], while the *lines* stand for (from *top* to *bottom*) (7.79), (7.53) with $\lambda = \frac{7}{18}$, (7.85), (7.82), and (7.77), respectively. Note that $Z^{(\mu c)}$ and Z_{BMCSL} are hardly distinguishable

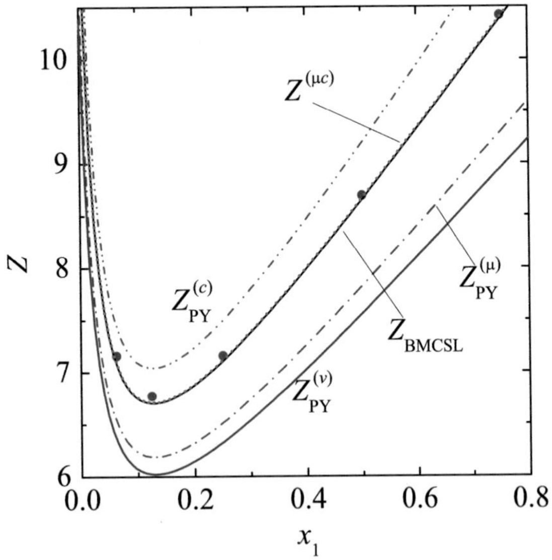

Fig. 7.10 Plot of the compressibility factor Z as a function of the mole fraction x_1 for an AHS binary mixture with a packing fraction $\eta = 0.49$ and a size ratio $\sigma_2/\sigma_1 = 0.3$. The *symbols* are computer simulation values [45], while the *lines* stand for (from *top* to *bottom*) (7.79), (7.53) with $\lambda = \frac{7}{18}$, (7.85), (7.82), and (7.77), respectively. Note that $Z^{(\mu c)}$ and Z_{BMCSL} are hardly distinguishable

where we have taken $\sigma = 1$. Moreover, (7.74) reduces to

$$\Lambda^{(0)} = \frac{1 + 2\eta}{(1 - \eta)^2} - \frac{12\eta}{1 - \eta}\Lambda^{(2)} , \qquad \Lambda^{(1)} = \frac{1 + \frac{1}{2}\eta}{(1 - \eta)^2} - \frac{6\eta}{1 - \eta}\Lambda^{(2)} , \qquad (7.87)$$

and (7.75) becomes the quadratic equation

$$12\tau\Lambda^{(2)} = \frac{1 + \frac{1}{2}\eta}{(1 - \eta)^2} - \frac{12\eta}{1 - \eta}\Lambda^{(2)} + 12\eta\Lambda^{(2)2} , \qquad (7.88)$$

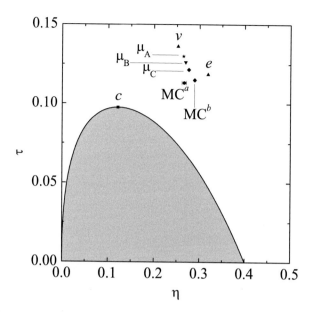

Fig. 7.11 Plane τ vs η showing the (*gray*) region inside which the PY approximation does not possess any real solution for the three-dimensional SHS fluid. The *symbols* denote the locations of the critical points (see Sect. 7.3.3.3 below) predicted by the virial (v), compressibility (c), energy (e), and chemical-potential (μ, protocols A, B, and C) routes. In addition, the *symbols* denoted as MC^a and MC^b represent the critical points obtained by MC simulations for the SHS fluid [46, 47] and by extrapolation to $\sigma' \to \sigma$ of MC simulations for the SW fluid, respectively [48]

whose physical root is

$$\Lambda^{(2)} = \frac{1 - (1 - \tau^{-1})\eta - w}{2\tau^{-1}(1 - \eta)\eta}, \tag{7.89}$$

where

$$w \equiv \sqrt{(1-\eta)\left[1 - \eta\left(1 - 2\tau^{-1} + \frac{\tau^{-2}}{3}\right)\right] + \frac{\tau^{-2}}{2}\eta^2}. \tag{7.90}$$

It must be remarked that the quantity w ceases to be real if $\tau < \tau_{\text{th}} = \frac{2-\sqrt{2}}{6} \simeq 0.097631$ and $\eta_-(\tau) < \eta < \eta_+(\tau)$, where

$$\eta_\pm(\tau) = \frac{1 - \tau^{-1} + \frac{1}{6}\tau^{-2} \pm \frac{1}{2}\tau^{-1}\sqrt{2 - \frac{4}{3}\tau^{-1}(1 - \frac{1}{12}\tau^{-1})}}{1 - 2\tau^{-1} + \frac{5}{6}\tau^{-2}}. \tag{7.91}$$

In the limit $\tau \to \tau_{\text{th}}$ one has $\eta_\pm(\tau) \to \eta_{\text{th}} = (3\sqrt{2} - 4)/2 \simeq 0.12132$, while $\lim_{\tau \to 0}\eta_-(\tau) = 0$ and $\lim_{\tau \to 0}\eta_+(\tau) = \frac{2}{5}$. The dome-shaped region where the PY approximation does not have a real solution for SHS fluids is shown in Fig. 7.11.

Fig. 7.12 Cavity function at $\eta = 0.32$. The *solid* and *dotted lines* correspond to MC simulations of a SW fluid (with $\sigma'/\sigma = 1.01$ and $k_B T/\varepsilon = 0.266$) [36] and of a SHS fluid with $\tau = 0.2$ [49], respectively. The *dashed line* represents the analytical solution of the PY approximation for a SHS fluid with $\tau = 0.2$

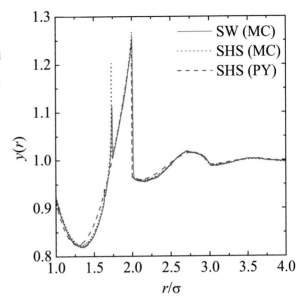

Fig. 7.13 Daan Frenkel (b. 1948) (Photograph courtesy of D. Frenkel)

In analogy to the HS case in (7.33) and (7.35), combination of (7.15) and (7.86) allows us to find a rational-function form for $\widehat{F}(s)$, namely

$$\widehat{F}(s) = \frac{\Lambda^{(0)} + \Lambda^{(1)}s + \Lambda^{(2)}s^2}{s^3 - 12\eta\left[\left(1 - s + \frac{1}{2}s^2\right)\Lambda^{(0)} + s\left(1 - s\right)\Lambda^{(1)} + s^2\Lambda^{(2)}\right]} . \tag{7.92}$$

The RDF is subsequently obtained from (7.18) and (7.19), although some care is needed to isolate the singular term $\Lambda^{(2)}\delta(r)$ from $\bar{\Psi}_1(r)$ [see (7.63)].

Figure 7.12 compares the PY cavity function at $\eta = 0.32$ and $\tau = 0.2$ with MC simulation results by Miller and Frenkel (see Fig. 7.13) [49]. MC data for a SW fluid at a reduced temperature $T^* \equiv k_B T/\varepsilon = 0.266$ [36] are also included. Similarly to the case of $y_{12}(r)$ in Fig. 7.8, the main deviations of the theoretical curve with respect to the simulation results are located around the first minimum of $y(r)$. On the other hand, the PY approximation does not account for the singularities (delta-peaks and/or discontinuities) of the SHS cavity function $y(r)$ at $r = \sqrt{8/3}, 5/3, \sqrt{3}, 2, \ldots$ [47, 49–53].

7.3.3.2 Equation of State

Let us consider now the thermodynamic properties. The one-component version of (7.70) is

$$y(1) = 12\tau \Lambda^{(2)} . \tag{7.93}$$

Moreover, the first derivative $y'(1)$ can be identified from the one-component version of (7.65) with the result

$$y'(1) = -\frac{9\eta(1+\eta)}{2(1-\eta)^3} - (12\tau - 1)\frac{1+\eta/2}{(1-\eta)^2}$$
$$-6\left[\eta\frac{2-11\eta}{(1-\eta)^2} - 2\tau\left(12\tau - \frac{1-13\eta}{1-\eta}\right)\right]\Lambda^{(2)} . \tag{7.94}$$

Insertion of (7.93) and (7.94) into (7.66) gives the virial EoS

$$\boxed{\begin{aligned} Z_{\text{PY}}^{(v)} &= \frac{1+2\eta+3\eta^2}{(1-\eta)^2} - \frac{\eta}{3}\frac{1-5\eta-5\eta^2}{(1-\eta)^3}\tau^{-1} \\ &\quad - \frac{8\eta}{1-\eta}\left(1+5\eta - \frac{\eta}{2}\frac{1-\frac{11}{2}\eta}{1-\eta}\tau^{-1}\right)\Lambda^{(2)} . \end{aligned}} \tag{7.95}$$

As for the energy route, combination of the one-component version of (7.67) and the thermodynamic relation (1.36a) shows that

$$\frac{\partial(\beta a^{\text{ex}})}{\partial \tau^{-1}} = -\eta y(1) \Rightarrow \beta a^{\text{ex}} = \beta a_{\text{HS}}^{\text{ex}} - \eta\tau^{-1}\int_0^1 dt\, y(1; \eta, t\tau^{-1}) \quad \text{(energy route)} ,$$
$$\tag{7.96}$$

where $a_{\text{HS}}^{\text{ex}}$ is the excess Helmholtz free energy per particle of the HS fluid at the same packing fraction and the contact function in the integrand is evaluated at a scaled stickiness $t\tau^{-1}$. Then, application of (1.36b) gives

$$Z = Z_{\text{HS}} - \eta\tau^{-1}\int_0^1 dt\,\frac{\partial[\eta y(1; \eta, t\tau^{-1})]}{\partial \eta} . \tag{7.97}$$

Use of (7.89), (7.90), and (7.93) allows one to get the following explicit expressions:

$$\beta a^{\text{ex}} = \beta a_{\text{HS}}^{\text{ex}} + 12\eta \Lambda^{(2)} + \frac{6\eta}{1-\eta} \ln\left[24\tau \Lambda^{(2)} \frac{(1-\eta)^2}{2+\eta}\right]$$

$$+ \frac{\sqrt{6\eta|2-5\eta|}}{1-\eta} \mathscr{W} \quad \text{(energy route)}, \tag{7.98a}$$

$$Z_{\text{PY}}^{(e)} = Z_{\text{HS}} + \frac{6\eta}{(1-\eta)^2} \ln\left[24\tau \Lambda^{(2)} \frac{(1-\eta)^2}{2+\eta}\right]$$

$$+ \frac{1-4\eta}{(1-\eta)^2} \frac{\sqrt{6\eta|2-5\eta|}}{2-5\eta} \mathscr{W}, \tag{7.98b}$$

where

$$\mathscr{W} \equiv \cos^{-1}\sqrt{\frac{6\eta}{2+\eta}} - \cos^{-1}\left[\sqrt{\frac{6\eta}{2+\eta}}\left(1 - \frac{2-5\eta}{6\tau(1-\eta)}\right)\right] \tag{7.99}$$

if $\eta < \frac{2}{5}$ but $\eta \notin (\eta_-, \eta_+)$, and

$$\mathscr{W} \equiv \cosh^{-1}\left[\sqrt{\frac{6\eta}{2+\eta}}\left(1 + \frac{5\eta-2}{6\tau(1-\eta)}\right)\right] - \cosh^{-1}\sqrt{\frac{6\eta}{2+\eta}} \tag{7.100}$$

if $\eta > \frac{2}{5}$. We recall that the "constant of integration" Z_{HS} remains undetermined in the energy route.

As before, the moment $\widehat{H}^{(1)}$ of the total correlation function [see (7.25)] can be obtained from the small-s behavior (7.27). Inserting the resulting expression into (7.26), one obtains

$$\chi_{T,\text{PY}}^{-1} = \frac{\left[1 + 2\eta - 12\eta(1-\eta)\Lambda^{(2)}\right]^2}{(1-\eta)^4} \quad \text{(compressibility route)}. \tag{7.101}$$

The associated compressibility factor is

$$Z_{\text{PY}}^{(c)} = \frac{1-3\eta-\eta^2}{(1-\eta)^3} - 2\tau \frac{2+\eta}{(1-\eta)^2} + 8\left[6\tau^2 + \frac{12\eta\tau}{1-\eta} - \eta\frac{2-5\eta}{(1-\eta)^2}\right]\Lambda^{(2)}. \tag{7.102}$$

From here, a further integration over density [see (1.36b)] gives

$$\beta a_{\text{PY}}^{\text{ex}} = \frac{9 + 4w\tau(5-8\eta)}{2(1-\eta)^2} + \frac{6(9\tau - 3 - 10w\tau^2)}{1-\eta} - \frac{24\tau^3(1-w)}{\eta} - \ln(1-\eta) + 24\tau^3$$

$$+36\tau^2 - 60\tau + \frac{27}{2} + 12\tau(6\tau - 1)\ln\left|\frac{1 + w - \eta(1 - \tau^{-1} + \frac{1}{6}\tau^{-2})}{2(1 - \eta)}\right|$$

$$-\sqrt{2}\left(54\tau^2 - 12\tau + 1\right)\ln\left|\frac{6\tau - 1 - 2\eta(3\tau - 2) + 3\sqrt{2}w\tau}{(1 - \eta)(\tau/\tau_{\text{th}} - 1)}\right|$$

$$\text{(compressibility route)}. \qquad (7.103)$$

It is remarkable that, despite the complex density dependence of $\chi_{T,\text{PY}}^{-1}$ in (7.101), explicit expressions for $Z_{\text{PY}}^{(c)}$ and $\beta a_{\text{PY}}^{\text{ex}}$ can be found upon integration over packing fraction from $\eta = 0$ to the value of interest. Strictly speaking, in view of Fig. 7.11, this is allowed only if either $\tau > \tau_{\text{th}}$ or $\tau < \tau_{\text{th}}$ and $\eta < \eta_-(\tau)$. On the other hand, (7.102) and (7.103) can be analytically continued to the region $\tau < \tau_{\text{th}}$ and $\eta > \eta_+(\tau)$.

In order to apply the chemical-potential route, we consider a solute particle of (relative) diameter $\sigma_0 = 2\xi - 1$ (with $\frac{1}{2} \leq \xi \leq 1$) that interacts with the solvent particles via an interaction potential $\phi^{(\xi)}(r)$ such that [see (7.62)]

$$e^{-\beta\phi^{(\xi)}(r)} = \Theta(r - \xi) + \frac{\xi}{12\tau_\xi}\delta(r - \xi), \qquad (7.104)$$

where $\tau_{01}^{-1} = \tau_\xi^{-1}$ measures the degree of stickiness of the solute–solvent interaction. Consequently, (4.52) yields [54]

$$\beta\mu^{\text{ex}}(\eta, \tau) = -\ln(1 - \eta) - 2\eta\int_{\frac{1}{2}}^{1} d\xi\,\xi^2\,\mathscr{I}(\xi, \eta, \tau) \quad (\mu \text{ route}), \qquad (7.105)$$

where

$$\mathscr{I}(\xi, \eta, \tau) \equiv \left(\xi\frac{\partial\tau_\xi^{-1}}{\partial\xi} + 3\tau_\xi^{-1} - 12\right)y_{01}^{(\xi)}(\xi) + \tau_\xi^{-1}\xi\left.\frac{\partial y_{01}^{(\xi)}(r)}{\partial r}\right|_{r=\xi}. \qquad (7.106)$$

As in previous cases, the Helmholtz free energy per particle and the compressibility factor can be obtained by application of (1.36c) and (1.37), i.e.,

$$\beta a^{\text{ex}}(\eta, \tau) = 1 + \frac{1 - \eta}{\eta}\ln(1 - \eta) - 2\eta\int_{\frac{1}{2}}^{1} d\xi\,\xi^2\int_{0}^{1} dt\,t\mathscr{I}(\xi, t\eta, \tau), \qquad (7.107a)$$

$$Z^{(\mu)}(\eta, \tau) = -\frac{\ln(1 - \eta)}{\eta} - 2\eta\int_{\frac{1}{2}}^{1} d\xi\,\xi^2\left[\mathscr{I}(\xi, \eta, \tau) - \int_{0}^{1} dt\,t\mathscr{I}(\xi, t\eta, \tau)\right].$$

$$(7.107b)$$

Within the PY approximation, use of (7.65), (7.70), (7.74), and (7.75) allows us to get [54]

$$y_{01}^{(\xi)}(\xi) = \frac{12\tau_\xi}{\xi^2} \Lambda_{01}^{(2)} , \qquad (7.108a)$$

$$\xi \left. \frac{\partial y_{01}^{(\xi)}(r)}{\partial r} \right|_{r=\xi} + y_{01}^{(\xi)}(\xi) = 12\eta \Lambda_{01}^{(2)} \left[3\eta \left(\Lambda^{(0)} - 2\Lambda^{(1)} + 2\Lambda^{(2)} \right)^2 - \Lambda^{(0)} + \Lambda^{(1)} \right]$$

$$+ \Lambda^{(0)} + 6\eta \Lambda_{01}^{(1)} \left(\Lambda^{(0)} - 2\Lambda^{(1)} + 2\Lambda^{(2)} \right) , \qquad (7.108b)$$

with

$$\Lambda_{01}^{(1)} = \frac{\xi + \eta(2\xi - 3/2)}{(1 - \eta)^2} - \frac{6\eta(2\xi - 1)}{1 - \eta} \Lambda^{(2)} , \qquad (7.109a)$$

$$\Lambda_{01}^{(2)} = \left(\frac{12\tau_\xi}{\xi} + \frac{6\eta}{1 - \eta} - 12\eta \Lambda^{(2)} \right)^{-1} \Lambda_{01}^{(1)} . \qquad (7.109b)$$

In order to close the PY determination of the function $\mathscr{I}(\xi, \eta, \tau)$ defined by (7.106) and hence the chemical-potential route (7.105)–(7.107), we need to fix the ξ-dependence of τ_ξ^{-1}, i.e., how the solute–solvent stickiness τ_ξ^{-1} changes from zero to the system value τ^{-1} as the diameter $2\xi - 1$ increases from zero to unity. Were the function $\mathscr{I}(\xi)$ exact, the thermodynamic quantities obtained from (7.105)–(7.107) would be of course independent of the choice for the protocol τ_ξ^{-1}. As expected, this is not what happens with the PY approximation [54]. As representative examples, let us consider the following three prototype protocols:

$$\tau_\xi^{-1} = (2\xi - 1)^q \tau^{-1} , \quad q = \begin{cases} \frac{1}{2} , & \text{Protocol A} , \\ 1 , & \text{Protocol B} , \\ 2 , & \text{Protocol C} . \end{cases} \qquad (7.110)$$

In all of them, the solute–solvent stickiness monotonically grows from zero to the solvent–solvent value as the solute diameter $(\sigma_0/\sigma = 2\xi - 1)$ grows from zero to the solvent diameter $(\sigma = 1)$. At a given solute diameter, the strength of the solute–solvent attraction decreases when going from A to B and from B to C. It can be verified that the influence of the protocol is practically negligible for all densities if $\tau^{-1} \lesssim 6$ [54]. For higher stickiness, however, the values of the compressibility factor $Z_{\text{PY}}^{(\mu)}$ from the chemical-potential route become increasingly sensitive to the protocol chosen. In fact, as might be expected on physical grounds, the stronger the relative stickiness τ_ξ^{-1}/τ^{-1} the smaller the pressure.

Fig. 7.14 Compressibility factor Z as a function of the packing fraction η for SHS fluids at $\tau = 0.15$. The *curves* correspond to PY results from various routes, as indicated on the plot, while *solid circles* (joined by a line as a guide to the eye) represent MC results [47]

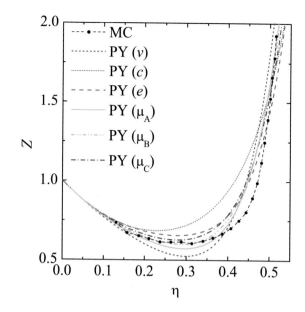

Figure 7.14 compares MC simulations of the SHS fluid at $\tau = 0.15$ [47] with PY predictions from the different routes. Since, as discussed before, the energy route leaves Z_{HS} undetermined, it is common to use the CS EoS (3.113) for Z_{HS} despite the fact that, as discussed in Sect. 6.9.3, the choice (7.42) for Z_{HS} would be more consistent. It can be observed that in the low-density range ($\eta \lesssim 0.15$) all PY routes provide consistent results. For higher densities, the curves corresponding to the three different protocols of the chemical-potential route remain rather close in comparison with those from the virial, energy, and compressibility routes, which show a larger spread. In the range $0.2 \lesssim \eta \lesssim 0.4$, the chemical-potential route gives the best agreement with the simulation data. In the same region, $Z_{PY}^{(e)}$ and $Z_{PY}^{(c)}$ overestimate the simulation values, while $Z_{PY}^{(v)}$ underestimates them. More specifically, up to $\eta \approx 0.4$, one has $Z_{PY}^{(v)} < Z_{PY}^{(\mu_A)} < Z_{PY}^{(\mu_B)} < Z_{PY}^{(\mu_C)} < Z_{PY}^{(e)} < Z_{PY}^{(c)}$. Finally, there is a rather strong disagreement of all the PY routes at high densities, $0.4 \lesssim \eta \lesssim 0.5$, where the simulation data exhibit lower pressure values than the theoretical ones. Interestingly, in contrast to the HS case (see Fig. 7.5), the compressibility route shows the largest deviations from MC results on the whole range of studied densities.

7.3.3.3 Vapor–Liquid Transition

Since the SHS interaction potential possesses an attractive part (although a singular one), a vapor–liquid critical point (η_c, τ_c) and a coexistence (or binodal) line for the (metastable) vapor–liquid phase transition must exist [9, 46, 47, 55, 56]. Given

Table 7.4 Comparison of the SHS vapor–liquid critical values τ_c, η_c, and $\beta_c \mu_c$ from MC simulations [46–48] and from the PY solution in the virial (v), energy (e), compressibility (c), and chemical-potential (μ_A–μ_C) routes

	MC[a]	MC[b]	v	e	c	μ_A	μ_B	μ_C
τ_c	0.1133(5)	0.1150(1)	0.1361	0.1185	0.0976	0.1300	0.1262	0.1215
η_c	0.266(5)	0.2890(5)	0.2524	0.3187	0.1213	0.2645	0.2691	0.2761
$\beta_c \mu_c$	−2.438(1)	−2.394(1)	−2.0806	−2.1672	−2.7869	−2.1323	−2.1728	−2.2230

[a] Miller and Frenkel [46, 47]
[b] Largo et al. [48]

a compressibility factor $Z(\eta, \tau)$, the critical point is obtained by the conditions [1, 57, 58]

$$\left(\frac{\partial(\eta Z)}{\partial \eta} \right)_\tau = \left(\frac{\partial^2(\eta Z)}{\partial \eta^2} \right)_\tau = 0 . \qquad (7.111)$$

As expected, each thermodynamic route (and each protocol τ_ξ^{-1} in the case of the chemical-potential route) applied to the PY approximation yields a different prediction for the critical point. In the special case of the compressibility route, it can be easily checked from (7.101) (see Exercise 7.49) that both $\chi_{T,\mathrm{PY}}^{-1}$ and its first derivative with respect to density vanish at the threshold point $(\eta, \tau) = (\eta_{\mathrm{th}}, \tau_{\mathrm{th}})$. In view of (7.111), this implies that $(\eta_{\mathrm{th}}, \tau_{\mathrm{th}})$ coincides with the critical point in the compressibility route.

The coordinates of the critical point predicted by each PY route, as well as of the ones obtained from MC simulations for the SHS fluid [46, 47] and from extrapolation to $\sigma' \to \sigma$ of MC simulations for the SW fluid [48], were already included in Fig. 7.11. The numerical values of τ_c, η_c, and the chemical potential at the critical point, $\beta_c \mu_c$ (with the convention $\Lambda = \sigma$ for the thermal de Broglie wavelength) are given in Table 7.4. Note that the estimates for the critical parameters proposed in [48] are found to be different (beyond the statistical error bars) from those reported in [46]. This discrepancy might be related to an incomplete mapping of the configuration space in the method employed in [46], which manifests itself in a less complete sampling of the dense region [48]. In what concerns the PY predictions, it can be observed that the compressibility route produces a gross underestimation of the critical density. The latter quantity is much better approximated by the virial route and, especially, the chemical-potential route for the three protocols considered. On the other hand, the critical value of the stickiness parameter evaluated from the virial and the compressibility routes differ significantly from the MC values. For this parameter, the energy route and the chemical-potential route (especially in the case of protocol C) give the best results. It must be remarked, however, that the critical point obtained from $Z_{\mathrm{PY}}^{(e)}$ is quite sensitive to the choice of Z_{HS}. If, instead of the CS EoS (3.113), the more consistent choice (7.42) is used, then no critical point is predicted at all by the energy route.

Fig. 7.15 Dependence on the protocol exponent q [see (7.110)] of the critical parameters τ_c, η_c, and $\beta_c \mu_c$ as obtained from the PY approximation in the chemical-potential route. The *horizontal lines* represent the MC simulation values from [48], while the *vertical line* marks the value $q = 5$ (protocol D)

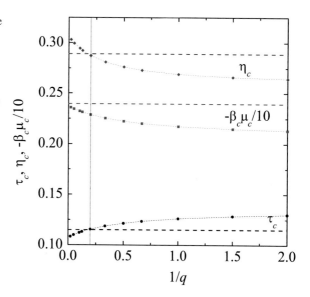

An interesting feature from Fig. 7.11 and Table 7.4 is that the location of the critical point estimated in [48] seems to agree with an extrapolation of the results obtained via the chemical-potential route as one goes from protocol A to protocol B and from B to C, i.e., as one increases the value of the exponent q in (7.110). This is confirmed by Fig. 7.15, which shows that an exponent $q \approx 5$ provides an excellent agreement with the MC data for τ_c and η_c. Even the value of $\beta_c \mu_c$ is relatively well captured by $q \approx 5$. We will refer to $q = 5$ as protocol D, in which case one has $\tau_c = 0.1155$, $\eta_c = 0.2872$, and $\beta_c \mu_c = -2.2873$. Protocol D corresponds to a solute with a very weak (relative) stickiness ($\tau_\xi^{-1}/\tau^{-1} \lesssim 0.03$) for sizes $\sigma_0/\sigma \lesssim 0.5$, followed by a rather rapid increase thereafter.

Below the critical "temperature" τ_c, a vapor–liquid coexistence line signals the locus of mechanical and chemical equilibrium between a vapor phase at a packing fraction $\eta_{\text{vap}}(\tau)$ and a liquid phase at a higher packing fraction $\eta_{\text{liq}}(\tau)$. The conditions of equal pressure and chemical potential between both phases read

$$\eta_{\text{vap}}(\tau) Z(\eta_{\text{vap}}(\tau), \tau) = \eta_{\text{liq}}(\tau) Z(\eta_{\text{liq}}(\tau), \tau) , \qquad (7.112a)$$

$$\ln \eta_{\text{vap}}(\tau) + \beta \mu^{\text{ex}}(\eta_{\text{vap}}(\tau), \tau) = \ln \eta_{\text{liq}}(\tau) + \beta \mu^{\text{ex}}(\eta_{\text{liq}}(\tau), \tau) . \qquad (7.112b)$$

Figure 7.16 displays the coexistence line and the location of the critical point derived from each PY route and from MC computer simulations for the SHS fluid [47]. As mentioned in connection with Table 7.4, it should be noted that the MC data derived in [47] may be affected by an incomplete sampling of the configuration space for the liquid phase [48]. In any case, it may be concluded that the coexistence curves obtained from the virial and compressibility routes differ substantially from

Fig. 7.16 Phase diagram of the SHS fluid showing the vapor–liquid coexistence lines from the PY solution in the virial (v), energy (e), compressibility (c), and chemical-potential (μ_A–μ_D) routes. MC simulation data taken from [47] are shown with error bars. The critical points are indicated with *symbols*

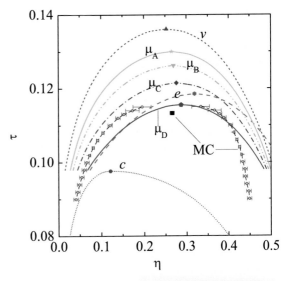

Fig. 7.17 Michael Ellis Fisher (b. 1931) (Photograph courtesy of M.E. Fisher)

computer evaluations. On the contrary, the agreement is reasonably good for the energy (provided that $Z_{HS} = Z_{CS}$) and chemical-potential (protocol D) routes.

The peculiar properties of the critical region predicted by the PY approximation with the compressibility route were analyzed in detail by Fishman and Fisher (see Fig. 7.17) in [59]. If $\tau \approx \tau_{th}$ and $\eta \approx \eta_{th}$, it can be shown that (7.101) becomes

$$\chi_{T,PY}^{-1} \approx c_1 \left(\sqrt{c_2 t + 2x^2} - x \right)^2 , \quad \text{(compressibility route)} , \tag{7.113}$$

with $c_1 = \frac{4}{81}(3 + 2\sqrt{2})$ and $c_2 = 18(\sqrt{2} - 1)$, and where

$$t \equiv \frac{\tau}{\tau_{th}} - 1 , \quad x \equiv \frac{\eta}{\eta_{th}} - 1 . \tag{7.114}$$

While the critical exponents [57] derived from (7.113) take the so-called classical or mean-field values, the functional dependence of $\chi_{T,\text{PY}}^{-1}$ on t and x differs from the mean-field one $\chi_T^{-1} \approx c_1(c_2 t + x^2)$. This gives rise to a number of anomalies [59]. For instance, the critical isotherm ($t = 0$) is highly asymmetric since $\chi_{T,\text{PY}}^{-1} \approx c_1(\sqrt{2} \mp 1)^2 x^2$, where the upper (lower) sign corresponds to $x > 0$ ($x < 0$). A similar asymmetry exists between the vapor and liquid branches of the coexistence line near the critical point, as can be observed in Fig. 7.16. On the other hand, the remaining thermodynamic routes are free from those anomalies and present a fully mean-field critical behavior.

7.4 Beyond the Percus–Yevick Approximation

In Sects. 7.2.4 and 7.3.1.2 we have found the exact solutions of the PY integral equation for HSs and (more generally) for SHS mixtures, respectively, without explicitly addressing the mathematical problem of solving the OZ relation [see (4.26) and (4.76)] with the PY closure [see (6.65)]. Instead, we started from the ansätze (7.31) and (7.68) and then univocally determined the parameters by imposing physical conditions reflected in the exact behavior of the Laplace transforms $\widehat{G}(s)$ and $\widehat{G}_{\alpha\gamma}(s)$ in the regimes of small s and large s. This defines the RFA methodology, as described on p. 204, in its simplest implementation, i.e., when the number of unknowns equals the number of constraints.

Apart from recovering the PY solution by an alternative procedure, the RFA approach can be applied beyond the PY approximation to some model systems. Those systems can be classified into two categories: (1) those that admit an exact solution of the PY approximation and (2) those that are not exactly solvable within the PY approximation. In the first class of systems, the RFA method recovers the PY solution as the *simplest* possible approach and, furthermore, the next-order approach allows one to make contact with a given empirical EoS in a thermodynamically consistent way, thus contributing to a general improvement of the predictions. This method has been applied to three-dimensional one-component HS and SHS fluids [13–15, 33], their additive mixtures [15, 34, 35, 60–62], and hard hyperspheres [17, 63, 64].

The application of the RFA to systems of the second class (i.e., those lacking an exact PY solution) includes the PS model [65, 66], the PSW model [67], the SW potential [68–70], the SS potential [71], piecewise-constant potentials with more than one step [72, 73], NAHS mixtures [74–76], and Janus particles with constrained orientations [77]. In those cases, the *simplest* RFA is already quite accurate, generally improving on the (numerical) solution of the PY approximation.

Let us finish this chapter by considering the HS and SW fluids as representative examples of systems of the first and second class, respectively.

7.4.1   Hard Spheres

In the spirit of the RFA (7.30) for HSs, the next-order approximation is obtained with $k = 2$, i.e.,

$$\widehat{F}(s) = -\frac{1}{12\eta} \frac{1 + L^{(1)}s + L^{(2)}s^2}{1 + S^{(1)}s + S^{(2)}s^2 + S^{(3)}s^3 + S^{(4)}s^4} . \tag{7.115}$$

From the exact series expansion (7.29) one now has

$$L^{(1)} = L^{(1)}_{\text{PY}} + \frac{12\eta}{1 + 2\eta} \left[\frac{1}{2}L^{(2)} - S^{(4)} \right] , \tag{7.116a}$$

$$S^{(1)} = S^{(1)}_{\text{PY}} + \frac{12\eta}{1 + 2\eta} \left[\frac{1}{2}L^{(2)} - S^{(4)} \right] , \tag{7.116b}$$

$$S^{(2)} = S^{(2)}_{\text{PY}} + \frac{12\eta}{1 + 2\eta} \left[\frac{1 - 4\eta}{12\eta}L^{(2)} + S^{(4)} \right] , \tag{7.116c}$$

$$S^{(3)} = S^{(3)}_{\text{PY}} - \frac{12\eta}{1 + 2\eta} \left[\frac{1 - \eta}{12\eta}L^{(2)} + \frac{1}{2}S^{(4)} \right] , \tag{7.116d}$$

where $L^{(1)}_{\text{PY}}, S^{(1)}_{\text{PY}}, S^{(2)}_{\text{PY}}$, and $S^{(3)}_{\text{PY}}$ are given by (7.32).

So far, the two *new* coefficients $L^{(2)}$ and $S^{(4)}$ remain free. They can be conveniently fixed by imposing any desired contact value $g(1^+)$ (or compressibility factor Z) and the corresponding consistent isothermal susceptibility $\chi_T = [\partial(\eta Z)/\partial\eta]^{-1}$. First, the exact condition (7.23) fixes the ratio $L^{(2)}/S^{(4)}$, so that

$$L^{(2)} = -3(Z - 1)S^{(4)} . \tag{7.117}$$

Besides, from (7.22) we can obtain the first derivative at contact as

$$g'(1^+) = -\frac{1}{12\eta S^{(4)}} \left[L^{(1)} - L^{(2)} \left(1 + \frac{S^{(3)}}{S^{(4)}} \right) \right] . \tag{7.118}$$

Higher-order derivatives can be obtained in a similar way [78].

Next, the expansion (7.28) allows us to identify $\widehat{H}^{(1)}$ and, by means of (7.26), relate χ_T, $L^{(2)}$, and $S^{(4)}$. Using (7.117), one gets a quadratic equation for $S^{(4)}$ [14], whose physical solution is

$$S^{(4)} = \frac{1 - \eta}{36\eta(Z - \frac{1}{3})} \left[1 - \sqrt{1 + \frac{Z - \frac{1}{3}}{Z - Z^{(v)}_{\text{PY}}} \left(\frac{\chi_T}{\chi_{T,\text{PY}}} - 1 \right)} \right] , \tag{7.119}$$

where $Z^{(v)}_{\text{PY}}$ and $\chi_{T,\text{PY}}$ are given by (7.42) and (7.44), respectively. In real space, the RDF is still given by (7.18) and (7.37), except that in (7.37a) the summation

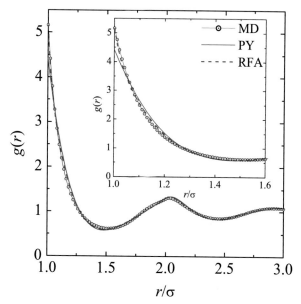

Fig. 7.18 RDF of a three-dimensional HS fluid at a reduced density $n^* = 0.9$ ($\eta = 0.471$) as obtained by MD simulations [31] and from the PY and RFA approaches

$\sum_{i=1}^{3} \rightarrow \sum_{i=1}^{4}$ extends over the *four* roots of the quartic equation $1 + S^{(1)}s + S^{(2)}s^2 + S^{(3)}s^3 + S^{(4)}s^4 = 0$. Explicit expressions of $g(r)$ up to the second coordination shell $\sigma \leq r \leq 3$ can be found in [79].

Figure 7.18 compares MD results of $g(r)$ at $n^* = 0.9$ (see Fig. 4.4) [31] with the predictions obtained from the PY solution (7.31) and from the next-order RFA (7.115). In the latter, Z and χ_T have been chosen as given by the CS EoS [see (3.113) and (3.114a)]. We observe that both theories describe quite well the behavior of $g(r)$ but the PY approximation underestimates the contact value and then decays by crossing the simulation data. Both features are satisfactorily corrected by the next-order RFA.

Once $\widehat{G}(s)$ is fully determined, the Fourier transforms $\tilde{h}(k)$ and $\tilde{c}(k)$ and the static structure factor $\widetilde{S}(k)$ can be explicitly obtained from (7.4), (4.27), and (4.24), respectively. By Fourier inversion, it is possible to find an analytical expression of the DCF in real space [15, 80]. Its functional form is

$$c(r) = \begin{cases} K_+ \dfrac{e^{\kappa r} - 1}{r} - (K_- + K) \dfrac{1 - e^{-\kappa r}}{r} + K_0 \left(1 + \dfrac{\eta}{2}r^3\right) + K_1 r , & r < 1 , \\[3mm] K \dfrac{e^{-\kappa r}}{r} , & r > 1 , \end{cases}$$

(7.120)

with

$$K = g(1^+) + K_+ (e^\kappa - 1) - K_- (1 - e^{-\kappa}) + K_0 \left(1 + \frac{\eta}{2}\right) + K_1 .$$

(7.121)

The expressions for the amplitudes K_+, K_-, K_0, and K_1 and for the damping coefficient κ can be found elsewhere [15]. Note that $c(0) =$ finite and that (7.121) implies $c(1^+) - c(1^-) = g(1^+)$, what is consistent with the expected continuity of the indirect correlation function $\overline{\gamma}(r) = h(r) - c(r)$ at $r = 1$.

In contrast to the PY result (7.40), now the DCF $c(r)$ does not vanish outside the hard core ($r > 1$) but has a Yukawa form in that region. In fact, the RFA (7.115) coincides with the solution of the so-called generalized mean spherical approximation (GMSA) [6, 81, 82], where the OZ relation is solved with the Yukawa closure $c(r)\Theta(r-1) = Kr^{-1}e^{-\kappa r}\Theta(r-1)$. The RFA starting point (7.115), however, is mathematically much more economical and open to applications to other systems.

As said before, a similar RFA scheme to construct an *augmented* PY approximation has also been implemented for AHS [15, 34, 60–62], SHS fluids [13–15, 33, 35], and hard hyperspheres in odd dimensions [17, 63, 64].

7.4.2   Square-Well and Square-Shoulder Fluids

Now we consider the SW interaction potential (see Table 3.1). Since no exact solution of the PY approximation for this potential is known (except in the special SHS limit analyzed in Sect. 7.3.3), the application of the RFA method is more challenging in this case than for HS and SHS fluids.

As in previous cases, the key quantity is the Laplace transform of $rg(r)$ defined by (7.3). It is again convenient to introduce the auxiliary function $\widehat{F}(s)$ through (7.15), where the choice $\sigma = 1$ as length unit is made. As before, the conditions $g(r) =$ finite and $\chi_T =$ finite imply (7.23) and (7.29), respectively. On the other hand, the key difference between the HS and SW cases is that in the latter case $\widehat{G}(s)$ and $\widehat{F}(s)$ must reflect the fact that the RDF $g(r)$ is discontinuous at the well range $r = \sigma'$ as a consequence of the discontinuity of the potential $\phi_{SW}(r)$ and the continuity of the cavity function $y(r)$. This implies that $\widehat{G}(s)$, and hence $\widehat{F}(s)$, must contain the exponential term $e^{-(\sigma'-\sigma)s}$. This is already obvious in the low-density limit, where the condition $\lim_{\eta\to 0} y(r) = 1$ yields

$$\lim_{\eta\to 0} \widehat{F}(s) \equiv \widehat{F}_0(s) = s^{-3}\left[e^{1/T^*}(1+s) - e^{-(\sigma'-1)s}(e^{1/T^*}-1)(1+\sigma's)\right],$$

$$(7.122)$$

where, as usually, $T^* \equiv k_B T/\varepsilon$.

Therefore, any minimal approximation for the auxiliary function $\widehat{F}(s)$ must contain at least five tunable parameters in order to accommodate for the exact form (7.29) for small s, behave as $\widehat{F}(s) \sim s^{-2}$ for large s, and be consistent with the functional structure (7.122). Within the framework of the RFA methodology, the *simplest* possible form that complies with all these requirements [compare with

(7.31)] seems to be [68]

$$\widehat{F}(s) = -\frac{1}{12\eta}\frac{1 + \bar{L}^{(0)} + L^{(1)}s - e^{-(\sigma'-1)s}\left(\bar{L}^{(0)} + \bar{L}^{(1)}s\right)}{1 + S^{(1)}s + S^{(2)}s^2 + S^{(3)}s^3} , \qquad (7.123)$$

where it is already guaranteed that $\lim_{s\to 0}\widehat{F}(s) = -1/12\eta$ but the six coefficients $\bar{L}^{(0)}, L^{(1)}, \bar{L}^{(1)}, S^{(1)}, S^{(2)}$, and $S^{(3)}$ are functions of η, T^*, and σ' yet to be determined. Condition (7.29) allows one to express the parameters $L^{(1)}, S^{(1)}, S^{(2)}$, and $S^{(3)}$ as linear functions of $\bar{L}^{(0)}$ and $\bar{L}^{(1)}$ [68, 69], i.e.,

$$L^{(1)} = L_{\mathrm{PY}}^{(1)} - \left(\sigma' - 1\right)\frac{1 + \frac{1}{2}\eta(\sigma'^3 + \sigma'^2 + \sigma' + 1)}{1 + 2\eta}\bar{L}^{(0)} + \frac{1 + 2\eta\sigma'^3}{1 + 2\eta}\bar{L}^{(1)} , \qquad (7.124a)$$

$$S^{(1)} = S_{\mathrm{PY}}^{(1)} - \frac{\eta(\sigma' - 1)^2}{2}\frac{\sigma'^2 + 2\sigma' + 3}{1 + 2\eta}\bar{L}^{(0)} + \frac{2\eta(\sigma'^3 - 1)}{1 + 2\eta}\bar{L}^{(1)} , \qquad (7.124b)$$

$$S^{(2)} = S_{\mathrm{PY}}^{(2)} - \frac{(\sigma' - 1)^2}{2}\frac{1 - \eta(\sigma' + 1)^2}{1 + 2\eta}\bar{L}^{(0)} + (\sigma' - 1)\frac{1 - 2\eta\sigma'(\sigma' + 1)}{1 + 2\eta}\bar{L}^{(1)} , \qquad (7.124c)$$

$$S^{(3)} = S_{\mathrm{PY}}^{(3)} + \frac{(\sigma' - 1)^2}{6}\frac{\sigma' + 2 - \eta(\frac{3}{2}\sigma'^2 + \sigma' + \frac{1}{2})}{1 + 2\eta}\bar{L}^{(0)} - \frac{\sigma' - 1}{2}\frac{\sigma' + 1 - 2\eta\sigma'^2}{1 + 2\eta}\bar{L}^{(1)} , \qquad (7.124d)$$

where, as in (7.116), the HS quantities $L_{\mathrm{PY}}^{(1)}, S_{\mathrm{PY}}^{(1)}, S_{\mathrm{PY}}^{(2)}$, and $S_{\mathrm{PY}}^{(3)}$ are given by (7.32). Note that consistency of (7.123) with (7.122) requires

$$\lim_{\eta\to 0}\bar{L}^{(0)} = e^{1/T^*} - 1 , \qquad (7.125a)$$

$$\lim_{\eta\to 0}\bar{L}^{(1)} = \sigma'\left(e^{1/T^*} - 1\right) . \qquad (7.125b)$$

Again, we can apply (7.22) to get the contact values [compare with (7.38)]

$$g(1^+) = -\frac{L^{(1)}}{12\eta S^{(3)}} , \qquad (7.126a)$$

$$g'(1^+) = -\frac{1}{12\eta S^{(3)}}\left[1 + \bar{L}^{(0)} - L^{(1)}\left(1 + \frac{S^{(2)}}{S^{(3)}}\right)\right] . \qquad (7.126b)$$

The full RDF can be obtained from (7.18) and (7.19). Since, in contrast to the HS and SHS cases [see (7.31) and (7.92), respectively], the auxiliary function $\widehat{F}(s)$ in (7.123) includes an exponential term, we need to expand $[\widehat{F}(s)]^\ell$ as

$$\left[\widehat{F}(s)\right]^\ell = \sum_{k=0}^{\ell}\binom{\ell}{k}e^{-k(\sigma'-1)s}\widehat{F}_{\ell k}(s) , \qquad (7.127a)$$

$$\widehat{F}_{\ell k}(s) \equiv \frac{(-1)^{\ell-k}}{(12\eta)^{\ell}} \frac{\left(1 + \bar{L}^{(0)} + L^{(1)}s\right)^{\ell-k}\left(\bar{L}^{(0)} + \bar{L}^{(1)}s\right)^{k}}{\left(1 + S^{(1)}s + S^{(2)}s^{2} + S^{(3)}s^{3}\right)^{\ell}} . \qquad (7.127b)$$

Therefore,

$$\bar{\Psi}_{\ell}(r) = \sum_{k=0}^{\ell} \binom{\ell}{k} \bar{\Psi}_{\ell k}(r - k\sigma' + k)\Theta(r - k\sigma' + k) , \qquad (7.128)$$

where

$$\bar{\Psi}_{\ell k}(r) = \mathscr{L}^{-1}\left[s\widehat{F}_{\ell k}(s)\right](r) = \sum_{j=1}^{\ell} \frac{\sum_{i=1}^{3} a_{\ell k j}^{(i)} e^{s_i r}}{(\ell - j)!(j - 1)!} r^{\ell-j} , \qquad (7.129a)$$

$$a_{\ell k j}^{(i)} = \lim_{s \to s_i} \left(\frac{\partial}{\partial s}\right)^{j-1}\left[s(s - s_i)^{\ell}\widehat{F}_{\ell k}(s)\right] . \qquad (7.129b)$$

As in (7.37), $\{s_i, i = 1, 2, 3\}$ are the three roots of the cubic equation $1 + S^{(1)}s + S^{(2)}s^2 + S^{(3)}s^3 = 0$. In particular, in the first coordination shell $(1 < r < 2)$, and assuming $\sigma' < 2$, one has

$$g(r) = -\frac{1}{12\eta r}\sum_{i=1}^{3} \frac{s_i e^{s_i(r-1)}}{S^{(1)} + 2S^{(2)}s_i + 3S^{(3)}s_i^2}\left[1 + \bar{L}^{(0)} + L^{(1)}s_i\right.$$

$$\left. - \left(\bar{L}^{(0)} + \bar{L}^{(1)}s_i\right)e^{-s_i(\sigma'-1)}\Theta(r - \sigma')\right], \quad (1 < r < 2) . \qquad (7.130)$$

To complete the construction of the proposal, we need to determine the parameters $\bar{L}^{(0)}$ and $\bar{L}^{(1)}$ by imposing two new conditions. An obvious condition is the continuity of the cavity function $y(r)$ at $r = \sigma'$, what implies

$$g(\sigma'^+) = e^{-1/T^*} g(\sigma'^-) . \qquad (7.131)$$

If $\sigma' < 2$ this yields $\left(1 - e^{-1/T^*}\right)\bar{\Psi}_{10}(\sigma' - 1) = -\bar{\Psi}_{11}(0)$, i.e.,

$$\frac{\bar{L}^{(1)}}{S^{(3)}} = \left(1 - e^{-1/T^*}\right)\sum_{i=1}^{3} \frac{s_i e^{s_i(\sigma'-1)}\left(1 + \bar{L}^{(0)} + L^{(1)}s_i\right)}{S^{(1)} + 2S^{(2)}s_i + 3S^{(3)}s_i^2} , \qquad (7.132)$$

where we have made use of the Laplace property $\bar{\Psi}_{11}(0) = \lim_{s\to\infty} s^2\widehat{F}_{11}(s) = \bar{L}^{(1)}/12\eta S^{(3)}$. We still need an extra condition. As a convenient compromise between simplicity and accuracy, we can fix the parameter $\bar{L}^{(0)}$ at its exact zero-density limit value (7.125a), i.e., [68]

$$\bar{L}^{(0)} = e^{1/T^*} - 1 . \qquad (7.133)$$

This closes the determination of the six parameters entering in (7.123). Notice that (7.132) becomes a closed transcendental equation for $\bar{L}^{(1)}$ that needs to be solved numerically. Once solved, and together with (7.133), the remaining parameters are simply obtained from (7.124). Therefore, the RFA provides a semi-analytical description of the SW fluid.

It can be proved [68, 69] that the RFA proposal (7.123) reduces to the exact solutions of the PY approximation both in the HS limit ($\varepsilon \to 0$ or $\sigma' \to 1$) and in the SHS limit ($\varepsilon \to \infty$ and $\sigma' \to 1$ with $\tau =$ finite) [see (7.31) and (7.92), respectively]. In fact, the latter limit has been exploited to replace the exact condition (7.131) by a simpler one in the case of *narrow* SW potentials [69]. This allows one to replace the transcendental equation (7.132) by an algebraic one from which $\bar{L}^{(1)}$ can be obtained analytically, which is especially useful for determining the thermodynamic properties [69, 83].

Comparison with computer simulations [68–70, 83] shows that the RFA for SW fluids is rather accurate at any fluid density if the potential well is sufficiently narrow ($\sigma'/\sigma \lesssim 1.2$), as well as for any width if the density is small enough ($n^* = n\sigma^3 \lesssim 0.4$). This is illustrated in Figs. 7.19, 7.20, and 7.21 for three representative cases, where it can be observed that the semi-analytical RFA (which otherwise tends to be less accurate near the discontinuity at $r = \sigma'$) competes favorably well with the numerical solution of the OZ relation complemented by the PY closure. On the

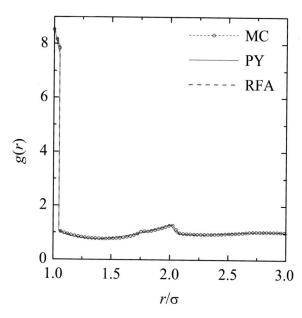

Fig. 7.19 RDF of a three-dimensional SW fluid for $\sigma'/\sigma = 1.05$, $T^* = 0.5$, and $n^* = 0.8$. The *circles* represent MC simulation data [70], the *solid line* refers to the results obtained from a numerical solution of the PY approximation, and the *dashed line* corresponds to the (semi-analytical) prediction from the RFA

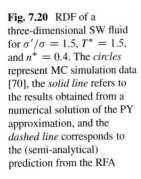

Fig. 7.20 RDF of a three-dimensional SW fluid for $\sigma'/\sigma = 1.5, T^* = 1.5$, and $n^* = 0.4$. The *circles* represent MC simulation data [70], the *solid line* refers to the results obtained from a numerical solution of the PY approximation, and the *dashed line* corresponds to the (semi-analytical) prediction from the RFA

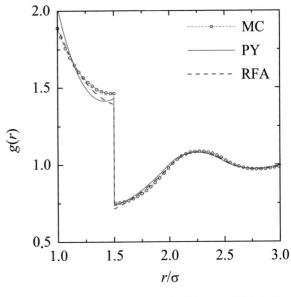

Fig. 7.21 RDF of a three-dimensional SW fluid for $\sigma'/\sigma = 2, T^* = 3$, and $n^* = 0.4$. The *circles* represent MC simulation data [70], the *solid line* refers to the results obtained from a numerical solution of the PY approximation, and the *dashed line* corresponds to the (semi-analytical) prediction from the RFA

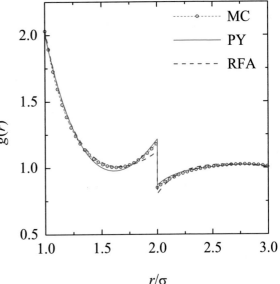

other hand, as the width and/or the density increase, the RFA predictions worsen, especially at low temperatures [70].

From Table 3.1 we can see that the purely repulsive SS potential can be formally defined from the well-known SW potential by the replacement $\varepsilon \rightarrow -\varepsilon$. As a consequence, even though both interaction potentials are physically very different, the implementation of the RFA for SS fluids is as simple as formally setting

Fig. 7.22 RDF of a three-dimensional SS fluid for $\sigma'/\sigma = 1.2$, $T^* = 0.5$, and $\eta = 0.4$. The *circles* represent MC simulation data [84], the *solid line* refers to the results obtained from a numerical solution of the PY approximation, and the *dashed line* corresponds to the (semi-analytical) prediction from the RFA

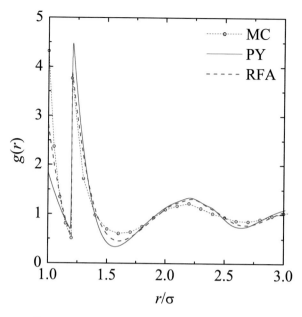

Fig. 7.23 RDF of a three-dimensional SS fluid for $\sigma'/\sigma = 1.5$, $T^* = 0.5$, and $n^* = 0.4$. The *circles* represent MC simulation data [85], the *solid line* refers to the results obtained from a numerical solution of the PY approximation, and the *dashed line* corresponds to the (semi-analytical) prediction from the RFA

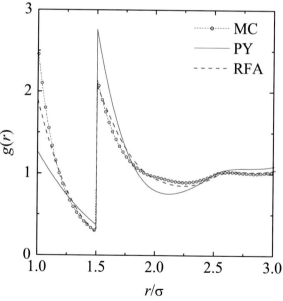

$T^* \to -T^*$ in (7.125a) and (7.132), the remaining equations, in particular (7.123), (7.124), and (7.126)–(7.130), being the same [71]. The RFA results for two SS systems are compared with numerical solutions of the PY approximation and MC simulations [84, 85] in Figs. 7.22 and 7.23. It can be seen that, despite being much simpler, the RFA turns out to be more accurate than the PY approximation.

7.4.2.1   General Piecewise-Constant Potentials: Continuum Limit

Let us conclude this chapter by showing that the RFA structure of the auxiliary function $\widehat{F}(s)$ in (7.123) can be extended to more general J-step piecewise-constant potentials of the form

$$
\phi(r) = \begin{cases}
\infty\,, & r < \sigma\,, \\
\varepsilon_1\,, & \sigma < r < \sigma_1\,, \\
\varepsilon_2\,, & \sigma_1 < r < \sigma_2\,, \\
\;\vdots & \quad\vdots \\
\varepsilon_J\,, & \sigma_{J-1} < r < \sigma_J\,, \\
\varepsilon_{J+1} = 0, r > \sigma_J\,.
\end{cases}
\tag{7.134}
$$

In such a case, (7.123) can be generalized as [72, 73]

$$
\widehat{F}(s) = -\frac{1}{12\eta}\,\frac{1 + \sum_{j=1}^{J}\bar{L}_j^{(0)} + L^{(1)}s - \sum_{j=1}^{J} e^{-(\sigma_j-1)s}\left(\bar{L}_j^{(0)} + \bar{L}_j^{(1)}s\right)}{1 + S^{(1)}s + S^{(2)}s^2 + S^{(3)}s^3}\,,
\tag{7.135}
$$

where, as before, we have taken $\sigma = 1$ as the length unit. The proposal (7.134) contains $2J + 4$ parameters, so we need the same numbers of conditions. Four of them are provided by (7.29), while J additional ones are obtained from the requirement of continuity of the cavity function $y(r)$ at $r = \sigma_j, j = 1, \ldots, J$. This yields (assuming $\sigma_J < 2$) [72, 73]

$$
\begin{aligned}
\frac{\bar{L}_j^{(1)}}{S^{(3)}} = \sum_{i=1}^{3} \frac{\left[e^{\beta(\varepsilon_j - \varepsilon_{j+1})} - 1\right]s_i}{S^{(1)} + 2S^{(2)}s_i + 3S^{(3)}s_i^2} & \left[\sum_{j_1=1}^{j-1}\left(\bar{L}_{j_1}^{(0)} + \bar{L}_{j_1}^{(1)}s_i\right) e^{(\sigma_j - \sigma_{j_1})s_i}\right. \\
& \left. - \left(1 + \sum_{j_1=1}^{J}\bar{L}_{j_1}^{(0)} + L^{(1)}s_i\right) e^{(\sigma_j - 1)s_i}\right]\,, \quad j = 1, \ldots, J\,,
\end{aligned}
\tag{7.136}
$$

where, as before, $\{s_i, i = 1, 2, 3\}$ are the three roots of the cubic equation $1 + S^{(1)}s + S^{(2)}s^2 + S^{(3)}s^3 = 0$. It is straightforward to check that (7.135) and (7.136) reduce to (7.123) and (7.132), respectively, if $J = 1$. As done in (7.133), the remaining J conditions can be provided by the assumption that the parameters $\bar{L}_j^{(0)}$ are independent of density, so that they are determined by their exact zero-density limits, namely [72, 73]

$$
\bar{L}_j^{(0)} = e^{-\beta\varepsilon_j} - e^{-\beta\varepsilon_{j+1}}\,, \quad j = 1, \ldots, J\,.
\tag{7.137}
$$

Comparisons of the predictions from the RFA (7.135) against computer simulations for two-step potentials show a general good agreement [73].

It is tempting to take the *continuum* limit $J \to \infty$ with a *finite* cutoff $\sigma_J \equiv \sigma'$. Before taking the limit, let us first assume $J =$ finite and equispaced values of σ_j, i.e., $\sigma_j = 1 + j\Delta\lambda \equiv \lambda_j$, with $\Delta\lambda \equiv (\sigma' - 1)/J$. We also define $\bar{L}_{\lambda_j}^{(0)} \equiv \bar{L}_j^{(0)}/\Delta\lambda$, $\bar{L}_{\lambda_j}^{(1)} \equiv \bar{L}_j^{(1)}/\Delta\lambda$, and $\phi(\lambda_j) \equiv \varepsilon_j$. Next, taking the limit $J \to \infty$, (7.135)–(7.137) become

$$\widehat{F}(s) = -\frac{1 + \int_1^{\sigma'} d\lambda\, \bar{L}_\lambda^{(0)} + L^{(1)}s - \int_1^{\sigma'} d\lambda\, e^{-(\lambda-1)s}\left(\bar{L}_\lambda^{(0)} + \bar{L}_\lambda^{(1)}s\right)}{12\eta\left(1 + S^{(1)}s + S^{(2)}s^2 + S^{(3)}s^3\right)}, \quad (7.138a)$$

$$\frac{\bar{L}_\lambda^{(1)}}{S^{(3)}} = -\sum_{i=1}^{3} \frac{\beta\phi'(\lambda)s_i}{S^{(1)} + 2S^{(2)}s_i + 3S^{(3)}s_i^2}\left[\int_1^\lambda d\lambda_1\left(\bar{L}_{\lambda_1}^{(0)} + \bar{L}_{\lambda_1}^{(1)}s_i\right)e^{(\lambda-\lambda_1)s_i}\right.$$
$$\left. -\left(1 + \int_1^{\sigma'} d\lambda_1\, \bar{L}_{\lambda_1}^{(0)} + L^{(1)}s_i\right)e^{(\lambda-1)s_i}\right], \quad 1 \le \lambda \le \sigma', \quad (7.138b)$$

$$\bar{L}_\lambda^{(0)} = e^{-\beta\phi(\lambda)}\beta\phi'(\lambda). \quad (7.138c)$$

Equation (7.138b) defines a nonlinear integral equation for the continuous function $\bar{L}_\lambda^{(1)}$ that needs to be solved in the finite interval $1 \le \lambda \le \sigma'$. Once (numerically) solved, (7.138a), together with (7.14), allows us to obtain the RDF $g(r)$ for all distances, including $r > \sigma'$.

Exercises

7.1 Derive (7.1).

7.2 Check (7.4).

7.3 Derive (7.9) from (3.72) and (7.8).

7.4 Making use of the mathematical property (check it!)

$$\int_0^\infty dx\, x\theta_k(x)e^{-x} = \theta_{k+1}(0) = (2k+1)!!\,,$$

derive (7.10).

7.5 Check (7.12) and (7.13).

7.6 Derive (7.17)–(7.19).

7.7 Derive (7.21) and (7.23).

7.8 Check (7.28) and (7.29).

7.9 Check (7.32).

7.10 Take the limit $\eta \to 0$ in (7.32) to prove that (7.31) is consistent with the exact expressions (7.13).

7.11 Check (7.33)–(7.35).

7.12 Taking into account that the complex residue of a function $f(z)$ at a pole z_0 of order ℓ is

$$\frac{1}{(\ell-1)!} \lim_{z\to z_0} \left(\frac{d}{dz}\right)^{\ell-1} \left[(z-z_0)^\ell f(z)\right]$$

and that

$$\left(\frac{d}{dz}\right)^{\ell-1} [f_1(z)f_2(z)] = \sum_{j=1}^{\ell} \binom{\ell-1}{j-1} \left[\left(\frac{d}{dz}\right)^{\ell-j} f_1(z)\right]\left[\left(\frac{d}{dz}\right)^{j-1} f_2(z)\right],$$

prove (7.37) from (7.19).

7.13 Derive (7.38)–(7.40).

7.14 Check from (7.40) that $c(1^-) = -(1 + \eta/2)/(1 - \eta)^2$. Then, taking into account (7.38a), check that the indirect correlation function $\overline{\gamma}(r) = h(r) - c(r)$ is continuous at $r = 1$, as expected.

7.15 Download and install the Wolfram CDF Player (http://www.wolfram.com/cdf-player) in your computer. Play with the Demonstration of reference [20] to explore how the RDF, the structure factor, the DCF, and the bridge function of the three-dimensional HS fluid (as described by the PY approximation) change with the packing fraction η.

7.16 Derive (7.42)–(7.45).

7.17 Reobtain (7.44) as $\lim_{k\to 0} \widetilde{S}(k)$ from (7.39).

7.18 Obtain (7.48)–(7.50).

7.19 Prove that the (excess) Helmholtz free energy per particle corresponding to the PY virial and compressibility EoS (7.42) and (7.45) are

$$\beta a_{PY,v}^{ex} = \frac{6\eta}{1-\eta} + 2\ln(1-\eta),$$

$$\beta a_{PY,c}^{ex} = \frac{3}{2}\frac{\eta(2-\eta)}{(1-\eta)^2} - \ln(1-\eta),$$

respectively. Compare with (7.49). Check that the three routes are consistent with the exact behavior $\beta a^{ex} = 4\eta + 5\eta^2 + \mathcal{O}(\eta^3)$. Reobtain the CS expression (3.114b) as $\frac{1}{3}\beta a_{PY,v}^{ex} + \frac{2}{3}\beta a_{PY,c}^{ex}$.

7.20 Check the entries of Table 7.1 not derived explicitly in Chaps. 3 or 7.

7.21 Derive (7.66) and (7.67).

7.22 Check (7.71)–(7.73).

7.23 Prove (7.74).

7.24 Derive (7.75).

7.25 Prove that (7.47) follows from (7.76) in the special case of a binary HS mixture with $x_0 \to 0$ and $\sigma_1 = 1$.

7.26 Derive (7.77) and (7.79) from (7.76) and (7.78), respectively.

7.27 Derive (7.80)–(7.82).

7.28 Check that (7.76)–(7.82) reduce to (7.38a), (7.42), (7.44), (7.45), and (7.48)–(7.50), respectively, in the special case of a common diameter $\sigma_\alpha = \sigma$, regardless of the mole fractions.

7.29 Express the (excess) chemical potential $\beta\mu_{v,\mathrm{PY}}^{\mathrm{ex}}$ in (7.80) as a function of the set of number densities $\{n_\gamma\}$. Then, check the following violation of a Maxwell relation:

$$\left(\frac{\partial \beta\mu_{v,\mathrm{PY}}^{\mathrm{ex}}}{\partial n_\alpha}\right)_{\{n_{\gamma\neq\alpha}\}} - \left(\frac{\partial \beta\mu_{\alpha,\mathrm{PY}}^{\mathrm{ex}}}{\partial n_v}\right)_{\{n_{\gamma\neq v}\}} = \frac{3\pi}{2}\frac{\eta^2}{(1-\eta)^3}\frac{M_2^2}{M_3^2}\sigma_\alpha^2\sigma_v^2(\sigma_\alpha-\sigma_v)\,.$$

7.30 Obtain (7.79) from (7.83) by application of (7.66) with $\tau_{\alpha\gamma}^{-1}=0$.

7.31 Starting from the SPT contact values (7.83), obtain the excess chemical potential $\mu_{v,\mathrm{SPT}}$ from (4.92). Then, derive the excess Helmholtz free energy per particle a^{ex} and the compressibility factor Z by application of (1.37). Does the resulting Z coincide with (7.79)? In other words, are the virial and chemical-potential thermodynamic routes mutually consistent in the SPT? Hint: Consult [37]

7.32 Repeat Exercise 7.29 but for the (excess) chemical potential $\beta\mu_{v,\mathrm{SPT}}^{\mathrm{ex}}$ derived in Exercise 7.31. Check that in the SPT one consistently has

$$\left(\frac{\partial \beta\mu_{v,\mathrm{SPT}}^{\mathrm{ex}}}{\partial n_\alpha}\right)_{\{n_{\gamma\neq\alpha}\}} = \left(\frac{\partial \beta\mu_{\alpha,\mathrm{SPT}}^{\mathrm{ex}}}{\partial n_v}\right)_{\{n_{\gamma\neq v}\}}\,.$$

7.33 Check that (7.85) reduces to (3.113) in the special case of a common diameter $\sigma_\alpha = \sigma$, regardless of the mole fractions.

7.34 Compare the BMCSL EoS (7.85) with the generic proposals (3.126b), (3.127b), and (3.146) when in the three latter the CS EoS (3.113) is used for $Z_{\mathrm{oc}}(\eta)$. As numerical tests, consider, for instance, the mixtures of Figs. 3.26, 7.9, and 7.10.

7.35 Check the entries of Table 7.3 not derived explicitly in Sect. 7.3.2.

7.36 Compare Table 7.3 with Tables 3.12 and 7.1.

7.37 Check that (7.89) is the solution to (7.88).

7.38 Prove from (7.89) that, in the limit of weak stickiness, $\Lambda^{(2)} = \tau^{-1}(1 + \eta/2)/12(1-\eta)^2 + \mathcal{O}(\tau^{-2})$. Then, by applying that limit to (7.93) and (7.94), recover the HS results (7.38).

7.39 Derive (7.92).

7.40 Derive (7.94) and (7.95).

7.41 Check the correctness of (7.98)–(7.100).

7.42 Derive (7.101)–(7.103).

7.43 Check from (7.101) that

$$\chi_{T,\mathrm{PY}}^{-1}\Big|_{\eta=\eta_{\mathrm{th}},\tau=\tau_{\mathrm{th}}} = \frac{\partial \chi_{T,\mathrm{PY}}^{-1}}{\partial \eta}\Big|_{\eta=\eta_{\mathrm{th}},\tau=\tau_{\mathrm{th}}} = 0 \,.$$

7.44 Check from (7.101) that, if $\tau < \tau_{\mathrm{th}}$, then $\chi_{T,\mathrm{PY}}^{-1} = 0$ at $\eta = \eta_0(\tau)$ where

$$\eta_0(\tau) = \frac{3 - \tau^{-1} - \frac{9}{2}\sqrt{4 - \frac{8}{3}\tau^{-1}(1 - \frac{1}{12}\tau^{-1})}}{12 - 7\tau^{-1}} \,.$$

Does that mean that the scaled pressure $\eta Z_{\mathrm{PY}}^{(c)}(\eta, \tau)$ presents a local minimum at $\eta = \eta_0(\tau)$?

7.45 Take carefully the limit $\tau^{-1} \to 0$ to check that (7.93), (7.94), (7.95), and (7.102) reduce to (7.38a), (7.38b), (7.42), and (7.45), respectively.

7.46 Check (7.105)–(7.107).

7.47 Derive (7.108) and (7.109).

7.48 Check that, in the limit $\xi \to 1$, (7.108) and (7.109) are consistent with (7.87), (7.88), (7.93), and (7.94).

7.49 Prove that (7.95), (7.98b), (7.102), and (7.105) are all consistent with the first three exact virial coefficients, i.e.,

$$Z = 1 + (4 - \tau^{-1})\eta + \left(10 - 5\tau^{-1} + \tau^{-2} - \frac{\tau^{-3}}{18}\right)\eta^2 + \mathcal{O}(\eta^3) \,.$$

7.50 Plot the scaled pressure $\eta Z_{\mathrm{PY}}^{(e)}(\eta, \tau)$ for $\tau = 0.20, 0.15$, and 0.10 from (7.98b) with the choice $Z_{\mathrm{HS}}(\eta) = Z_{\mathrm{PY}}^{(v)}(\eta)$ given by (7.42). Do those isotherms exhibit a physically acceptable behavior?

7.51 Reproduce the numerical values in Table 7.4 and in Fig. 7.15 corresponding to the PY approximation.

7.52 Derive (7.113).

7.53 Obtain (7.116).

7.54 Check (7.118) and (7.119).

7.55 Check (7.122).

7.56 Prove that (7.122) reduces to (7.13a) either in the limit $\sigma' \to 1$ or in the limit $T^* \to \infty$.

7.57 Check (7.124)–(7.126).

7.58 Check (7.127)–(7.130).

7.59 Derive (7.138).

References

1. R. Balescu, *Equilibrium and Nonequilibrium Statistical Mechanics* (Wiley, New York, 1974)
2. M.S. Wertheim, Phys. Rev. Lett. **10**, 321 (1963)
3. E. Thiele, J. Chem. Phys. **39**, 474 (1963)
4. M.S. Wertheim, J. Math. Phys. **5**, 643 (1964)
5. J.L. Lebowitz, Phys. Rev. **133**, A895 (1964)
6. J.S. Høye, L. Blum, J. Stat. Phys. **16**, 399 (1977)
7. C. Freasier, D.J. Isbister, Mol. Phys. **42**, 927 (1981)
8. E. Leutheusser, Physica A **127**, 667 (1984)
9. R.J. Baxter, J. Chem. Phys. **49**, 2770 (1968)
10. J.W. Perram, E.R. Smith, Chem. Phys. Lett. **35**, 138 (1975)
11. B. Barboy, Chem. Phys. **11**, 357 (1975)
12. B. Barboy, R. Tenne, Chem. Phys. **38**, 369 (1979)
13. S.B. Yuste, A. Santos, Phys. Rev. A **43**, 5418 (1991)
14. S.B. Yuste, M. López de Haro, A. Santos, Phys. Rev. E **53**, 4820 (1996)
15. M. López de Haro, S.B. Yuste, A. Santos, in *Theory and Simulation of Hard-Sphere Fluids and Related Systems*, ed. by A. Mulero. Lecture Notes in Physics, vol. 753 (Springer, Berlin, 2008), pp. 183–245
16. J.R. Solana, *Perturbation Theories for the Thermodynamic Properties of Fluids and Solids* (CRC Press, Boca Raton, 2013)
17. R.D. Rohrmann, A. Santos, Phys. Rev. E **76**, 051202 (2007)
18. L. Carlitz, Duke Math. J. **24**, 151 (1957)
19. J. Abate, W. Whitt, Queueing Syst. **10**, 5 (1992)
20. A. Santos, Radial distribution function for hard spheres. Wolfram Demonstrations Project (2013), http://demonstrations.wolfram.com/RadialDistributionFunctionForHardSpheres/
21. H. Reiss, H.L. Frisch, J.L. Lebowitz, J. Chem. Phys. **31**, 369 (1959)
22. E. Helfand, H.L. Frisch, J.L. Lebowitz, J. Chem. Phys. **34**, 1037 (1961)
23. J.L. Lebowitz, E. Helfand, E. Praestgaard, J. Chem. Phys. **43**, 774 (1965)
24. A. Santos, Phys. Rev. Lett. **109**, 120601 (2012)
25. K.W. Kratky, J. Stat. Phys. **27**, 533 (1982)
26. S. Labík, J. Kolafa, A. Malijevský, Phys. Rev. E **71**, 021105 (2005)

27. N. Clisby, B.M. McCoy, J. Stat. Phys. **122**, 15 (2006)
28. R.J. Wheatley, Phys. Rev. Lett. **110**, 200601 (2013)
29. C. Zhang, B.M. Pettitt, Mol. Phys. **112**, 1427 (2014)
30. A.J. Schultz, D.A. Kofke, Phys. Rev. E **90**, 023301 (2014)
31. J. Kolafa, S. Labík, A. Malijevský, Phys. Chem. Chem. Phys. **6**, 2335 (2004). See also http://www.vscht.cz/fch/software/hsmd/
32. S.B. Yuste, A. Santos, J. Stat. Phys. **72**, 703 (1993)
33. S.B. Yuste, A. Santos, Phys. Rev. E **48**, 4599 (1993)
34. S.B. Yuste, A. Santos, M. López de Haro, J. Chem. Phys. **108**, 3683 (1998)
35. A. Santos, S.B. Yuste, M. López de Haro, J. Chem. Phys. **109**, 6814 (1998)
36. A. Malijevský, S.B. Yuste, A. Santos, J. Chem. Phys. **125**, 074507 (2006)
37. A. Santos, R.D. Rohrmann, Phys. Rev. E **87**, 052138 (2013)
38. M. Mandell, H. Reiss, J. Stat. Phys. **13**, 113 (1975)
39. Y. Rosenfeld, J. Chem. Phys. **89**, 4272 (1988)
40. M. Heying, D. Corti, J. Phys. Chem. B **108**, 19756 (2004)
41. T. Boublík, J. Chem. Phys. **53**, 471 (1970)
42. E.W. Grundke, D. Henderson, Mol. Phys. **24**, 269 (1972)
43. L.L. Lee, D. Levesque, Mol. Phys. **26**, 1351 (1973)
44. G.A. Mansoori, N.F. Carnahan, K.E. Starling, J.T.W. Leland, J. Chem. Phys. **54**, 1523 (1971)
45. M. Barošová, A. Malijevský, S. Labík, W.R. Smith, Mol. Phys. **87**, 423 (1996)
46. M.A. Miller, D. Frenkel, Phys. Rev. Lett. **90**, 135702 (2003)
47. M.A. Miller, D. Frenkel, J. Chem. Phys. **121**, 535 (2004)
48. J. Largo, M.A. Miller, F. Sciortino, J. Chem. Phys. **128**, 134513 (2008)
49. M.A. Miller, D. Frenkel, J. Phys. Condens. Matter **16**, S4901 (2004)
50. A.J. Post, E.D. Glandt, J. Chem. Phys. **84**, 4585 (1986)
51. N.A. Seaton, E.D. Glandt, J. Chem. Phys. **84**, 4595 (1987)
52. N.A. Seaton, E.D. Glandt, J. Chem. Phys. **86**, 4668 (1987)
53. N.A. Seaton, E.D. Glandt, J. Chem. Phys. **87**, 1785 (1987)
54. R.D. Rohrmann, A. Santos, Phys. Rev. E **89**, 042121 (2014)
55. R.O. Watts, D. Henderson, R.J. Baxter, Adv. Chem. Phys. **21**, 421 (1971)
56. M.G. Noro, D. Frenkel, J. Chem. Phys. **113**, 2941 (2000)
57. H.E. Stanley, *Introduction to Phase Transitions and Critical Phenomena* (Oxford University Press, Oxford, 1971)
58. L.E. Reichl, *A Modern Course in Statistical Physics*, 2nd edn. (Wiley, New York, 1998)
59. S. Fishman, M.E. Fisher, Physica A **108**, 1 (1981)
60. A. Malijevský, A. Malijevský, S.B. Yuste, A. Santos, M. López de Haro, Phys. Rev. E **66**, 061203 (2002)
61. A. Malijevský, S.B. Yuste, A. Santos, M. López de Haro, Phys. Rev. E **75**, 061201 (2007)
62. S.B. Yuste, A. Santos, M. López de Haro, J. Chem. Phys. **128**, 134507 (2008). Erratum: **140**, 179901 (2014)
63. R.D. Rohrmann, A. Santos, Phys. Rev. E **83**, 011201 (2011)
64. R.D. Rohrmann, A. Santos, Phys. Rev. E **84**, 041203 (2011)
65. A. Malijevský, A. Santos, J. Chem. Phys. **124**, 074508 (2006)
66. A. Malijevský, S.B. Yuste, A. Santos, Phys. Rev. E **76**, 021504 (2007)
67. R. Fantoni, A. Giacometti, A. Malijevský, A. Santos, J. Chem. Phys. **131**, 124106 (2009)
68. S.B. Yuste, A. Santos, J. Chem. Phys. **101**, 2355 (1994)
69. L. Acedo, A. Santos, J. Chem. Phys. **115**, 2805 (2001)
70. J. Largo, J.R. Solana, S.B. Yuste, A. Santos, J. Chem. Phys. **122**, 084510 (2005)
71. S.B. Yuste, A. Santos, M. López de Haro, Mol. Phys. **109**, 987 (2011)
72. A. Santos, S.B. Yuste, M. López de Haro, Condens. Matter Phys. **15**, 23602 (2012)
73. A. Santos, S.B. Yuste, M. López de Haro, M. Bárcenas, P. Orea, J. Chem. Phys. **139**, 074503 (2013)
74. R. Fantoni, A. Santos, Phys. Rev. E **84**, 041201 (2011)
75. R. Fantoni, A. Santos, Phys. Rev. E **87**, 042102 (2013)

76. R. Fantoni, A. Santos, J. Chem. Phys. **140**, 244513 (2014)
77. M.A.G. Maestre, R. Fantoni, A. Giacometti, A. Santos, J. Chem. Phys. **138**, 094904 (2013)
78. M. Robles, M. López de Haro, J. Chem. Phys. **107**, 4648 (1997)
79. A. Díez, J. Largo, J.R. Solana, J. Chem. Phys. **125**, 074509 (2006)
80. S.B. Yuste, A. Santos, M. López de Haro, Mol. Phys. **98**, 439 (2000)
81. E. Waisman, Mol. Phys. **25**, 45 (1973)
82. D. Henderson, L. Blum, Mol. Phys. **32**, 1627 (1976)
83. J. Largo, J.R. Solana, L. Acedo, A. Santos, Mol. Phys. **101**, 2981 (2003)
84. A. Lang, G. Kahl, C.N. Likos, H. Löwen, M. Watzlawek, J. Phys. Condens. Matter **11**, 10143 (1999)
85. S. Zhou, J.R. Solana, J. Chem. Phys. **131**, 204503 (2009)

References

This is an integrated list of the references cited in the chapters of the book. The numbers here are for enumeration purposes only and do not correlate with the reference numbers in each chapter.

1. J. Abate, W. Whitt, The Fourier-series method for inverting transforms of probability distributions. Queueing Syst. **10**, 5–88 (1992)
2. M. Abramowitz, I.A. Stegun (eds.), *Handbook of Mathematical Functions* (Dover, New York, 1972)
3. L. Acedo, A. Santos, A square-well model for the structural and thermodynamic properties of simple colloidal systems. J. Chem. Phys. **115**, 2805–2817 (2001)
4. L. Acedo, A. Santos, The penetrable-sphere fluid in the high-temperature, high-density limit. Phys. Lett. A **323**, 427–433 (2004). Erratum: **376**, 2274–2275 (2012)
5. M.P. Allen, D.J. Tildesley, *Computer Simulation of Liquids* (Clarendon Press, Oxford, 1987)
6. N.W. Ashcroft, D.C. Langreth, Structure of binary liquid mixtures. I. Phys. Rev. **156**, 685–692 (1967)
7. R. Balescu, *Equilibrium and Nonequilibrium Statistical Mechanics* (Wiley, New York, 1974)
8. P. Ballone, G. Pastore, G. Galli, D. Gazzillo, Additive and non-additive hard sphere mixtures. Monte Carlo simulation and integral equation results. Mol. Phys. **59**, 275–290 (1986)
9. B. Barboy, Solution of the compressibility equation of the adhesive hard-sphere model for mixtures. Chem. Phys. **11**, 357–371 (1975)
10. B. Barboy, R. Tenne, Distribution functions and equations of state of sticky hard sphere fluids in the Percus–Yevick approximation. Chem. Phys. **38**, 369–387 (1979)
11. J.A. Barker, D. Henderson, Square-well fluid at low densities. Can. J. Phys. **44**, 3959–3978 (1967)
12. J.A. Barker, D. Henderson, What is "liquid"? Understanding the states of matter. Rev. Mod. Phys. **48**, 587–671 (1976)
13. M. Barošová, A. Malijevský, S. Labík, W.R. Smith, Computer simulation of the chemical potentials of binary hard-sphere mixtures. Mol. Phys. **87**, 423–439 (1996)
14. C. Barrio, J.R. Solana, A new analytical equation of state for additive hard sphere fluid mixtures. Mol. Phys. **97**, 797–803 (1999)
15. C. Barrio, J.R. Solana, Theory and computer simulation for the equation of state of additive hard-disk fluid mixtures. Phys. Rev. E **63**, 011201 (2001)

© Springer International Publishing Switzerland 2016
A. Santos, *A Concise Course on the Theory of Classical Liquids*,
Lecture Notes in Physics 923, DOI 10.1007/978-3-319-29668-5

16. C. Barrio, J.R. Solana, Binary mixtures of additive hard spheres. Simulations and theories, in *Theory and Simulation of Hard-Sphere Fluids and Related Systems*, ed. by A. Mulero. Lecture Notes in Physics, vol. 753 (Springer, Berlin, 2008), pp. 133–182

17. M. Baus, J.L. Colot, Thermodynamics and structure of a fluid of hard rods, disks, spheres, or hyperspheres from rescaled virial expansions. Phys. Rev. A **36**, 3912–3925 (1987)

18. M. Baus, C.F. Tejero, *Equilibrium Statistical Physics. Phases of Matter and Phase Transitions* (Springer, Berlin, 2008)

19. R.J. Baxter, Direct correlation functions and their derivatives with respect to particle density. J. Chem. Phys. **41**, 553–558 (1964)

20. R.J. Baxter, Percus–Yevick equation for hard spheres with surface adhesion. J. Chem. Phys. **49**, 2770–2774 (1968)

21. E. Beltrán-Heredia, A. Santos, Fourth virial coefficient of additive hard-sphere mixtures in the Percus–Yevick and hypernetted-chain approximations. J. Chem. Phys. **140**, 134507 (2014)

22. A. Ben-Naim, *Molecular Theory of Solutions* (Oxford University Press, Oxford, 2006)

23. A. Ben-Naim, *A Farewell to Entropy: Statistical Thermodynamics Based on Information* (World Scientific, Singapore, 2008)

24. A. Ben-Naim, A. Santos, Local and global properties of mixtures in one-dimensional systems. II. Exact results for the Kirkwood–Buff integrals. J. Chem. Phys. **131**, 164–512 (2009)

25. C.M. Bender, S.A. Orszag, *Advanced Mathematical Methods for Scientists and Engineers* (McGraw-Hill, Auckland, 1987)

26. L. Berthier, G. Biroli, Theoretical perspective on the glass transition and amorphous materials. Rev. Mod. Phys. **3**, 587–645 (2011)

27. A.B. Bhatia, D.E. Thornton, Structural aspects of the electrical resistivity of binary alloys. Phys. Rev. B **8**, 3004–3012 (1970)

28. R. Blaak, Exact analytic expression for a subset of fourth virial coefficients of polydisperse hard sphere mixtures. Mol. Phys. **95**, 695–699 (1998)

29. S.M. Blinder, Second Virial Coefficients Using the Lennard-Jones Potential, Wolfram Demonstrations Project (2010), http://demonstrations.wolfram.com/SecondVirialCoefficientsUsingTheLennardJonesPotential/

30. B. Borštnik, C.G. Jesudason, G. Stell, Anomalous clustering and equation-of-state behavior as the adhesive-disk limit is approached. J. Chem. Phys. **106**, 9762–9768 (1997)

31. T. Boublík, Hard-sphere equation of state. J. Chem. Phys. **53**, 471–472 (1970)

32. S. Braun, J.P. Ronzheimer, M. Schreiber, S.S. Hodgman, T. Rom, I. Bloch, U. Schneider, Negative absolute temperature for motional degrees of freedom. Science **339**, 52–55 (2013)

33. S. Buzzaccaro, R. Rusconi, R. Piazza, "Sticky" hard spheres: Equation of state, phase diagram, and metastable gels. Phys. Rev. Lett. **99**, 098301 (2007)

34. H.B. Callen, *Thermodynamics and an Introduction to Thermostatistics* (Wiley, New York, 1985)

35. L. Carlitz, A note on the Bessel polynomials. Duke Math. J. **24**, 151–162 (1957)

36. H.O. Carmesin, H. Frisch, J. Percus, Binary nonadditive hard-sphere mixtures at high dimension. J. Stat. Phys. **63**, 791–795 (1991)

37. N.F. Carnahan, K.E. Starling, Equation of state for nonattracting rigid spheres. J. Chem. Phys. **51**, 635–636 (1969)

38. L.D. Carr, Negative temperatures? Science **339**, 42–43 (2013)

39. S.H. Chen, J. Rouch, F. Sciortino, P. Tartaglia, Static and dynamic properties of water-in-oil microemulsions near the critical and percolation points. J. Phys.: Condens. Matter **6**, 109855 (1994)

40. N. Clisby, B. McCoy, Negative virial coefficients and the dominance of loose packed diagrams for D-dimensional hard spheres. J. Stat. Phys. **114**, 1361–1392 (2004)

41. N. Clisby, B.M. McCoy, Analytic calculation of B_4 for hard spheres in even dimensions. J. Stat. Phys. **114**, 1343–1360 (2004)

42. N. Clisby, B.M. McCoy, New results for virial coefficients of hard spheres in D dimensions. Pramana **64**, 775–783 (2005)

43. N. Clisby, B.M. McCoy, Ninth and tenth order virial coefficients for hard spheres in D dimensions. J. Stat. Phys. **122**, 15–57 (2006)
44. E.G.D. Cohen, Einstein and Boltzmann: Determinism and Probability or The Virial Expansion Revisited (2013), http://arxiv.org/abs/1302.2084
45. A. Díez, J. Largo, J.R. Solana, Thermodynamic properties of van der Waals fluids from Monte Carlo simulations and perturbative Monte Carlo theory. J. Chem. Phys. **125**, 074509 (2006)
46. J. Dunkel, S. Hilbert, Consistent thermostatistics forbids negative absolute temperatures. Nat. Phys. **10**, 67–72 (2014)
47. E. Enciso, N.G. Almarza, D.S. Calzas, M.A. González, Low density equation of state of asymmetric hard sphere mixtures. Mol. Phys. **92**, 173–176 (1997)
48. E. Enciso, N.G. Almarza, M.A. González, F.J. Bermejo, Virial coefficients of hard-sphere mixtures. Phys. Rev. E **57**, 4486–4490 (1998)
49. E. Enciso, N.G. Almarza, M.A. González, F.J. Bermejo, The virial coefficients of hard hypersphere binary mixtures. Mol. Phys. **100**, 1941–1944 (2002)
50. A. Erdélyi, *Asymptotic Expansions* (Dover, New York, 1956)
51. J.J. Erpenbeck, M.J. Luban, Equation of state of the classical hard-disk fluid. Phys. Rev. A **32**, 2920–2922 (1985)
52. R. Evans, Density functionals on the theory of nonuniform fluids, in *Fundamentals of Inhomogeneous Fluids*, ed. by D. Henderson (Dekker, New York, 1992)
53. R. Evans, Density functional theory for inhomogeneous fluids I: Simple fluids in equilibrium, in *3rd Warsaw School of Statistical Physics*, ed. by B. Cichocki, M. Napiórkowski, J. Piasecki (Warsaw University Press, Warsaw, 2010), pp. 43–85. http://agenda.albanova.se/getFile.py/access?contribId=260&resId=251&materialId=250&confId=2509
54. R. Fantoni, A. Santos, Nonadditive hard-sphere fluid mixtures: A simple analytical theory. Phys. Rev. E **84**, 041201 (2011)
55. R. Fantoni, A. Santos, Multicomponent fluid of nonadditive hard spheres near a wall. Phys. Rev. E **87**, 042102 (2013)
56. R. Fantoni, A. Santos, Depletion force in the infinite-dilution limit in a solvent of nonadditive hard spheres. J. Chem. Phys. **140**, 244513 (2014)
57. R. Fantoni, A. Giacometti, A. Malijevský, A. Santos, Penetrable-square-well fluids: Analytical study and Monte Carlo simulations. J. Chem. Phys. **131**, 124106 (2009)
58. R. Fantoni, A. Malijevský, A. Santos, A. Giacometti, The penetrable square-well model: extensive versus non-extensive phases. Mol. Phys. **109**, 2723–2736 (2011)
59. L.A. Fernández, V. Martín-Mayor, B. Seoane, P. Verrocchio, Equilibrium fluid–solid coexistence of hard spheres. Phys. Rev. Lett. **108**, 165701 (2012)
60. M.J. Fernaud, E. Lomba, L.L. Lee, A self-consistent integral equation study of the structure and thermodynamics of the penetrable sphere fluid. J. Chem. Phys. **112**, 810–816 (2000)
61. L. Ferrari, Boltzmann vs Gibbs: a finite-size match (2015), http://arxiv.org/abs/1501.04566
62. M.E. Fisher, D. Ruelle, The stability of many-particle systems. J. Math. Phys. **7**, 260–270 (1966)
63. S. Fishman, M.E. Fisher, Critical point scaling in the Percus–Yevick equation. Physica A **108**, 1–13 (1981)
64. C. Freasier, D.J. Isbister, A remark on the Percus–Yevick approximation in high dimensions. Hard core systems. Mol. Phys. **42**, 927–936 (1981)
65. D. Frenkel, B. Smit, *Understanding Molecular Simulation: From Algorithms to Applications*, 2nd edn. (Academic, San Diego, 2002)
66. D. Frenkel, P.B. Warren, Gibbs, Boltzmann, and negative temperatures. Am. J. Phys. **83**, 163–170 (2015)
67. H.L. Frisch, N. Rivier, D. Wyler, Classical hard-sphere fluid in infinitely many dimensions. Phys. Rev. Lett. **54**, 2061–2063 (1985)
68. H. Goldstein, J. Safko, C.P. Poole, *Classical Mechanics* (Pearson Education, Upper Saddle River, 2013)
69. E.W. Grundke, D. Henderson, Distribution functions of multi-component fluid mixtures of hard spheres. Mol. Phys. **24**, 269–281 (1972)

70. J.A. Gualtieri, J.M. Kincaid, G. Morrison, Phase-equilibria in polydisperse fluids. J. Chem. Phys. **77**, 521–536 (1982)
71. P. Hänggi, S. Hilbert, J. Dunkel, Meaning of temperature in different thermostatistical ensembles. Philos. Trans. R. Soc. A **374**, 20150039 (2016)
72. J.P. Hansen, I.R. McDonald, *Theory of Simple Liquids*, 3rd edn. (Academic, London, 2006)
73. H. Hansen-Goos, M. Mortazavifar, M. Oettel, R. Roth, Fundamental measure theory for the inhomogeneous Hard-sphere system based on Santos' consistent free energy. Phys. Rev. E **91**, 052121 (2015)
74. E. Helfand, H.L. Frisch, J.L. Lebowitz, Theory of the two- and one-dimensional rigid sphere fluids. J. Chem. Phys. **34**, 1037–1042 (1961)
75. P.C. Hemmer, G. Stell, Fluids with several phase transitions. Phys. Rev. Lett. **24**, 1284–1287 (1970)
76. D. Henderson, A simple equation of state for hard discs. Mol. Phys. **30**, 971–972 (1975)
77. D. Henderson, L. Blum, Generalized mean spherical approximation for hard spheres. Mol. Phys. **32**, 1627–1635 (1976)
78. D. Henderson, P.J. Leonard, One- and two-fluid van der Waals theories of liquid mixtures, I. Hard sphere mixtures. Proc. Natl. Acad. Sci. U. S. A. **67**, 1818–1823 (1970)
79. D. Henderson, P.J. Leonard, Conformal solution theory: Hard-sphere mixtures. Proc. Natl. Acad. Sci. U. S. A. **68**, 2354–2356 (1971)
80. K.F. Herzfeld, M. Goeppert-Mayer, On the states of aggregation. J. Chem. Phys. **2**, 38–44 (1934)
81. D. Heyes, *The Liquid State: Applications of Molecular Simulations* (Wiley, Chichester, 1998)
82. M. Heying, D. Corti, Scaled particle theory revisited: New conditions and improved predictions of the properties of the hard sphere fluid. J. Phys. Chem. B **108**, 19756–19768 (2004)
83. M. Heying, D.S. Corti, The one-dimensional fully non-additive binary hard rod mixture: Exact thermophysical properties. Fluid Phase Equil. **220**, 85–103 (2004)
84. S. Hilbert, P. Hänggi, J. Dunkel, Thermodynamic laws in isolated systems. Phys. Rev. E **90**, 062116 (2014)
85. T.L. Hill, *Statistical Mechanics* (McGraw-Hill, New York, 1956)
86. J.S. Høye, L. Blum, Solution of the Yukawa closure of the Ornstein–Zernike equation. J. Stat. Phys. **16**, 399–413 (1977)
87. J.T. Jenkins, F. Mancini, Balance laws and constitutive relations for plane flows of a dense, binary mixture of smooth, nearly elastic, circular disks. J. Appl. Mech. **54**, 27–34 (1987)
88. H. Kamerlingh Onnes, Expression of the equation of state of gases and liquids by means of series. Commun. Phys. Lab. Univ. Leiden **71**, 3–25 (1901). Reprinted in Expression of the equation of state of gases and liquids by means of series, in *Through Measurement to Knowledge*. Boston Studies in the Philosophy and History of Science, vol. 124 (Springer Netherlands, 1991), pp. 146–163.
89. S.C. Kim, S.H. Suh, Inhomogeneous structure of penetrable spheres with bounded interactions. J. Chem. Phys. **117**, 9880–9886 (2002)
90. J.G. Kirkwood, F.P. Buff, The statistical mechanical theory of solutions. I. J. Chem. Phys. **19**, 774–777 (1951)
91. W. Klein, H. Gould, R.A. Ramos, I. Clejan, A.I. Mel'cuk, Repulsive potentials, clumps and the metastable glass phase. Physica A **205**, 738–746 (1994)
92. J. Kolafa, S. Labík, A. Malijevský, The bridge function of hard spheres by direct inversion of computer simulation data. Mol. Phys. **100**, 2629–2640 (2002). See also http://www.vscht.cz/fch/software/hsmd/
93. J. Kolafa, S. Labík, A. Malijevský, Accurate equation of state of the hard sphere fluid in stable and metastable regions. Phys. Chem. Chem. Phys. **6**, 2335–2340 (2004). See also http://www.vscht.cz/fch/software/hsmd/
94. D.T. Korteweg, On van der Waals's isothermal equation. Nature **45**, 152–154 (1891)
95. K.W. Kratky, A new graph expansion of virial coefficients. J. Stat. Phys. **27**, 533–551 (1982)
96. R. Kubo, *Statistical Mechanics. An Advanced Course with Problems and Solutions* (Elsevier, Amsterdam, 1965)

97. J. Kurchan, The quest for a transition and order in glasses, in *4th Warsaw School of Statistical Physics*, ed. by B. Cichocki, M. Napiórkowski, J. Piasecki (Warsaw University Press, Warsaw 2012), pp. 131–167. https://www.icts.res.in/media/uploads/Program/Files/kurchan.pdf

98. S. Labík, J. Kolafa, Analytical expressions for the fourth virial coefficient of a hard-sphere mixture. Phys. Rev. E **80**, 051122 (2009). Erratum: **84**, 069901 (2011)

99. S. Labík, A. Malijevský, Monte Carlo simulation of the background correlation function of non-spherical hard body fluids. I. General method and illustrative results. Mol. Phys. **53**, 381–388 (1984)

100. S. Labík, J. Kolafa, A. Malijevský, Virial coefficients of hard spheres and hard disks up to the ninth. Phys. Rev. E **71**, 021105 (2005)

101. A. Lang, G. Kahl, C.N. Likos, H. Löwen, M. Watzlawek, Structure and thermodynamics of square-well and square-shoulder fluids. J. Phys.: Condens. Matter **11**, 10143–10161 (1999)

102. J. Largo, J.R. Solana, L. Acedo, A. Santos, Heat capacity of square-well fluids of variable width. Mol. Phys. **101**, 2981–2986 (2003)

103. J. Largo, J.R. Solana, S.B. Yuste, A. Santos, Pair correlation function of short-ranged square-well fluids. J. Chem. Phys. **122**, 084510 (2005)

104. J. Largo, M.A. Miller, F. Sciortino, The vanishing limit of the square-well fluid: The adhesive hard-sphere model as a reference system. J. Chem. Phys. **128**, 134513 (2008)

105. J.L. Lebowitz, Exact solution of generalized Percus–Yevick equation for a mixture of hard spheres. Phys. Rev. **133**, A895–A899 (1964)

106. J.L. Lebowitz, D. Zomick, Mixtures of hard spheres with nonadditive diameters: Some exact results and solution of the Percus–Yevick equation. J. Chem. Phys. **54**, 3335–3346 (1971)

107. J.L. Lebowitz, E. Helfand, E. Praestgaard, Scaled particle theory of fluid mixtures. J. Chem. Phys. **43**, 774–779 (1965)

108. L.L. Lee, D. Levesque, Perturbation theory for mixtures of simple liquids. Mol. Phys. **26**, 1351–1370 (1973)

109. E. Leutheusser, Exact solution of the Percus–Yevick equation for a hard-core fluid in odd dimensions. Physica A **127**, 667–676 (1984)

110. C.N. Likos, Effective interactions in soft condensed matter physics. Phys. Rep. **348**, 267–439 (2001)

111. C.N. Likos, H. Löwen, M. Watzlawek, B. Abbas, O. Jucknischke, J. Allgaier, D. Richter, Star polymers viewed as ultrasoft colloidal particles. Phys. Rev. Lett. **80**, 4450–4453 (1998)

112. C.N. Likos, M. Watzlawek, H. Löwen, Freezing and clustering transitions for penetrable spheres. Phys. Rev. E **58**, 3135–3144 (1998)

113. C.N. Likos, A. Lang, M. Watzlawek, H. Löwen, Criterion for determining clustering versus reentrant melting behavior for bounded interaction potentials. Phys. Rev. E **63**, 031206 (2001)

114. M. López de Haro, A. Santos, S.B. Yuste, A student-oriented derivation of a reliable equation of state for a hard-disc fluid. Eur. J. Phys. **19**, 281–286 (1998)

115. M. López de Haro, S.B. Yuste, A. Santos, Alternative approaches to the equilibrium properties of hard-sphere liquids, in *Theory and Simulation of Hard-Sphere Fluids and Related Systems*, ed. by A. Mulero. Lecture Notes in Physics, vol. 753 (Springer, Berlin, 2008), pp. 183–245

116. M. López de Haro, A. Malijevský, S. Labík, Critical consolute point in hard-sphere binary mixtures: Effect of the value of the eighth and higher virial coefficients on its location. Collect. Czech. Chem. Commun. **75**, 359–369 (2010)

117. Lord Rayleigh, On the virial of a system of hard colliding bodies. Nature **45**, 80–82 (1891)

118. H. Löwen, Density functional theory for inhomogeneous fluids II: Statics, dynamics, and applications, in *3rd Warsaw School of Statistical Physics*, ed. by B. Cichocki, M. Napiórkowski, J. Piasecki (Warsaw University Press, Warsaw, 2010), pp. 87–121. http://www2.thphy.uni-duesseldorf.de/~hlowen/doc/ra/ra0025.pdf

119. M. Luban, A. Baram, Third and fourth virial coefficients of hard hyperspheres of arbitrary dimensionality. J. Chem. Phys. **76**, 3233–3241 (1982)

120. M. Luban, J.P.J. Michels, Equation of state of hard D-dimensional hyperspheres. Phys. Rev. A **41**, 6796–6804 (1990)

121. S. Luding, Global equation of state of two-dimensional hard sphere systems. Phys. Rev. E **63**, 042201 (2001)
122. S. Luding, A. Santos, Molecular dynamics and theory for the contact values of the radial distribution functions of hard-disk fluid mixtures. J. Chem. Phys. **121**, 8458–8465 (2004)
123. S. Luding, O. Strauß, The equation of state of polydisperse granular gases, in *Granular Gases*, ed. by T. Pöschel, S. Luding. Lecture Notes in Physics, vol. 564 (Springer, Berlin, 2001), pp. 389–409
124. J.F. Lutsko, Recent developments in classical density functional theory. Adv. Chem. Phys. **144**, 1–92 (2010)
125. J.F. Lutsko, Direct correlation function from the consistent fundamental-measure free energies for hard-sphere mixtures. Phys. Rev. E **87**, 014103 (2013)
126. I. Lyberg, The fourth virial coefficient of a fluid of hard spheres in odd dimensions. J. Stat. Phys. **119**, 747–764 (2005)
127. M.A.G. Maestre, R. Fantoni, A. Giacometti, A. Santos, Janus fluid with fixed patch orientations: Theory and simulations. J. Chem. Phys. **138**, 094904 (2013)
128. A. Malijevský, A. Santos, Structure of penetrable-rod fluids: Exact properties and comparison between Monte Carlo simulations and two analytic theories. J. Chem. Phys. **124**, 074508 (2006)
129. A. Malijevský, A. Malijevský, S.B. Yuste, A. Santos, M. López de Haro, Structure of ternary additive hard-sphere mixtures. Phys. Rev. E **66**, 061203 (2002)
130. A. Malijevský, S.B. Yuste, A. Santos, How "sticky" are short-range square-well fluids? J. Chem. Phys. **125**, 074507 (2006)
131. A. Malijevský, S.B. Yuste, A. Santos, Low-temperature and high-temperature approximations for penetrable-sphere fluids: Comparison with Monte Carlo simulations and integral equation theories. Phys. Rev. E **76**, 021504 (2007)
132. A. Malijevský, S.B. Yuste, A. Santos, M. López de Haro, Multicomponent fluid of hard spheres near a wall. Phys. Rev. E **75**, 061201 (2007)
133. M. Mandell, H. Reiss, Scaled particle theory: Solution to the complete set of scaled particle theory conditions: Applications to surface structure and dilute mixtures. J. Stat. Phys. **13**, 113–128 (1975)
134. G.A. Mansoori, N.F. Carnahan, K.E. Starling, T.W. Leland, Equilibrium thermodynamic properties of the mixture of hard spheres. J. Chem. Phys. **54**, 1523–1525 (1971)
135. C. Marquest, T.A. Witten, Simple cubic structure in copolymer mesophases. J. Phys. France **50**, 1267–1277 (1989)
136. G.A. Martynov, G.N. Sarkisov, Exact equations and the theory of liquids. V. Mol. Phys. **49**, 1495–1504 (1983)
137. J.E. Mayer, M. Goeppert Mayer, *Statistical Mechanics* (Wiley, New York, 1940)
138. M.A. Miller, D. Frenkel, Competition of percolation and phase separation in a fluid of adhesive hard spheres. Phys. Rev. Lett. **90**, 135702 (2003)
139. M.A. Miller, D. Frenkel, Phase diagram of the adhesive hard sphere fluid. J. Chem. Phys. **121**, 535–545 (2004)
140. M.A. Miller, D. Frenkel, Simulating colloids with Baxter's adhesive hard sphere model. J. Phys.: Condens. Matter **16**, S4901–S4912 (2004)
141. B.M. Mladek, G. Kahl, M. Neuman, Thermodynamically self-consistent liquid state theories for systems with bounded potentials. J. Chem. Phys. **124**, 064503 (2006)
142. T. Morita, Theory of classical fluids: Hyper-netted chain approximation, I. Prog. Theor. Phys. **20**, 920–938 (1958)
143. T. Morita, Theory of classical fluids: Hyper-netted chain approximation. III. Prog. Theor. Phys. **23**, 829–845 (1960)
144. A. Mulero (ed.), *Theory and Simulation of Hard-Sphere Fluids and Related Systems*. Lecture Notes in Physics, vol. 753 (Springer, Berlin, 2008)
145. A. Mulero, F. Cuadros, C. Galán, Test of equations of state for the Weeks–Chandler–Andersen reference system of two-dimensional Lennard-Jones fluids. J. Chem. Phys. **107**, 6887–6893 (1997)

146. A. Mulero, C.A. Galán, M.I. Parra, F. Cuadros, Equations of state for hard spheres and hard disks, in *Theory and Simulation of Hard-Sphere Fluids and Related Systems*, ed. by A. Mulero. Lecture Notes in Physics, vol. 753 (Springer, Berlin, 2008), pp. 37–109

147. J.H. Nairn, J.E. Kilpatrick, van der Waals, Boltzmann, and the fourth virial coefficient of hard spheres. Am. J. Phys. **40**, 503–515 (1972)

148. B.R.A. Nijboer, L. van Hove, Radial distribution function of a gas of hard spheres and the superposition approximation. Phys. Rev. **85**, 777–783 (1952)

149. M.G. Noro, D. Frenkel, Extended corresponding-states behavior for particles with variable range attractions. J. Chem. Phys. **113**, 2941–2944 (2000)

150. V. Ogarko, S. Luding, Equation of state and jamming density for equivalent bi- and polydisperse, smooth, hard sphere systems. J. Chem. Phys. **136**, 124508 (2012)

151. F.W.J. Olver, D.W. Lozier, R.F. Boisvert, C.W. Clark (eds.), *NIST Handbook of Mathematical Functions* (Cambridge University Press, New York, 2010)

152. L. Onsager, Theories of concentrated electrolytes. Chem. Rev. **13**, 73–89 (1933)

153. P. Pajuelo, A. Santos, Classical Scattering with a Penetrable Square-Well Potential, Wolfram Demonstrations Project (2011), http://demonstrations.wolfram.com/ ClassicalScatteringWithAPenetrableSquareWellPotential

154. G. Parisi, F. Zamponi, Mean-field theory of hard sphere glasses and jamming. Rev. Mod. Phys. **82**, 789–845 (2010)

155. G. Pellicane, C. Caccamo, P.V. Giaquinta, F. Saija, Virial coefficients and demixing of athermal nonadditive mixtures. J. Phys. Chem. B **111**, 4503–4509 (2007)

156. J.K. Percus, G.J. Yevick, Analysis of classical statistical mechanics by means of collective coordinates. Phys. Rev. **110**, 1–13 (1958)

157. J.W. Perram, E.R. Smith, A model for the examination of phase behaviour in multicomponent systems. Chem. Phys. Lett. **35**, 138–140 (1975)

158. R. Piazza, *Soft Matter. The Stuff that Dreams are Made of* (Springer, Dordrecht, 2011)

159. R. Piazza, V. Peyre, V. Degiorgio, "Sticky hard spheres" model of proteins near crystallization: A test based on the osmotic compressibility of lysozyme solutions. Phys. Rev. E **58**, R2733–R2736 (1998)

160. D. Pontoni, S. Finet, T. Narayanan, A.R. Rennie, Interactions and kinetic arrest in an adhesive hard-sphere colloidal system. J. Chem. Phys. **119**, 6157–6165 (2003)

161. A.J. Post, E.D. Glandt, Cluster concentrations and virial coefficients for adhesive particles. J. Chem. Phys. **84**, 4585–4594 (1986)

162. P.N. Pusey, The effect of polydispersity on the crystallization of hard spherical colloids. J. Phys. France **48**, 709–712 (1987)

163. P.N. Pusey, E. Zaccarelli, C. Valeriani, E. Sanz, W.C.K. Poon, M.E. Cates, Hard spheres: crystallization and glass formation. Philos. Trans. R. Soc. A **367**, 4993–5011 (2009)

164. F.H. Ree, W.G. Hoover, Reformulation of the virial series for classical fluids. J. Chem. Phys. **41**, 1635–1645 (1964)

165. F.H. Ree, N. Keeler, S.L. McCarthy, Radial distribution function of hard spheres. J. Chem. Phys. **44**, 3407–3425 (1966)

166. L.E. Reichl, *A Modern Course in Statistical Physics*, 1st edn. (University of Texas Press, Austin, 1980)

167. L.E. Reichl, *A Modern Course in Statistical Physics*, 2nd edn. (Wiley, New York, 1998)

168. F. Reif, *Fundamentals of Statistical and Thermal Physics* (McGraw-Hill, Boston, 1965)

169. H. Reiss, H.L. Frisch, J.L. Lebowitz, Statistical mechanics of rigid spheres. J. Chem. Phys. **31**, 369–380 (1959)

170. H. Reiss, H.L. Frisch, E. Helfand, J.L. Lebowitz, Aspects of the statistical thermodynamics of real fluids. J. Chem. Phys. **32**, 119–124 (1960)

171. M. Robles, M. López de Haro, On the contact values of the derivatives of the hard-sphere radial distribution function. J. Chem. Phys. **107**, 4648–4657 (1997)

172. F.J. Rogers, D.A. Young, New, thermodynamically consistent, integral equation for simple fluids. Phys. Rev. A **30**, 999–1007 (1984)

173. R.D. Rohrmann, A. Santos, Structure of hard-hypersphere fluids in odd dimensions. Phys. Rev. E **76**, 051202 (2007)
174. R.D. Rohrmann, A. Santos, Exact solution of the Percus–Yevick integral equation for fluid mixtures of hard hyperspheres. Phys. Rev. E **84**, 041203 (2011)
175. R.D. Rohrmann, A. Santos, Multicomponent fluids of hard hyperspheres in odd dimensions. Phys. Rev. E **83**, 011201 (2011)
176. R.D. Rohrmann, A. Santos, Equation of state of sticky-hard-sphere fluids in the chemical-potential route. Phys. Rev. E **89**, 042121 (2014)
177. R.D. Rohrmann, M. Robles, M. López de Haro, A. Santos, Virial series for fluids of hard hyperspheres in odd dimensions. J. Chem. Phys. **129**, 014510 (2008)
178. V. Romero-Rochín, Nonexistence of equilibrium states at absolute negative temperatures. Phys. Rev. E **88**, 022144 (2013)
179. D. Rosenbaum, P.C. Zamora, C.F. Zukoski, Phase behavior of small attractive colloidal particles. Phys. Rev. Lett. **76**, 150–153 (1996)
180. Y. Rosenfeld, Scaled field particle theory for the structure and thermodynamics of isotropic hard particle fluids. J. Chem. Phys. **89**, 4272–4287 (1988)
181. Y. Rosenfeld, Free-energy model for the inhomogeneous hard-sphere fluid mixture and density-functional theory of freezing. Phys. Rev. Lett. **63**, 980–983 (1989)
182. Y. Rosenfeld, M. Schmidt, H. Löwen, P. Tarazona, Fundamental-measure free-energy density functional for hard spheres: Dimensional crossover and freezing. Phys. Rev. E **55**, 4245–4263 (1997)
183. Y. Rosenfeld, M. Schmidt, M. Watzlawek, H. Löwen, Fluid of penetrable spheres: Testing the universality of the bridge functional. Phys. Rev. E **62**, 5006–5010 (2000)
184. R. Roth, Fundamental measure theory for hard-sphere mixtures: a review. J. Phys.: Condens. Matter **22**, 063102 (2010)
185. R. Roth, R. Evans, A. Lang, G. Kahl, Fundamental measure theory for hard-sphere mixtures revisited: the White Bear version. J. Phys.: Condens. Matter **14**, 12063–12078 (2002)
186. J.S. Rowlinson, F. Swinton, *Liquids and Liquid Mixtures* (Butterworth, London, 1982)
187. D. Ruelle, *Statistical Mechanics: Rigorous Results* (World Scientific, Singapore, 1999)
188. G.B. Rybicki, Exact statistical mechanics of a one-dimensional self-gravitating system. Astrophys. Space Sci. **14**, 56–72 (1971)
189. F. Saija, The fourth virial coefficient of a nonadditive hard-disc mixture. Phys. Chem. Chem. Phys. **13**, 11885–11891 (2011)
190. F. Saija, G. Fiumara, P.V. Giaquinta, Fourth virial coefficient of hard-body mixtures in two and three dimensions. Mol. Phys. **87**, 991–998 (1996). Erratum: **92**, 1089 (1997)
191. F. Saija, G. Fiumara, P.V. Giaquinta, Research note: Fifth virial coefficient of a hard-sphere mixture. Mol. Phys. **89**, 1181–1186 (1996)
192. F. Saija, G. Fiumara, P.V. Giaquinta, Research note: Fifth virial coefficient of a two-component mixture of hard discs. Mol. Phys. **90**, 679–681 (1997)
193. F. Saija, G. Fiumara, P.V. Giaquinta, Virial expansion of a non-additive hard-sphere mixture. J. Chem. Phys. **108**, 9098–9101 (1998)
194. F. Saija, A. Santos, S.B. Yuste, M. López de Haro, Fourth virial coefficients of asymmetric nonadditive hard-disk mixtures. J. Chem. Phys. **136**, 184505 (2012)
195. Z.W. Salsburg, R.W. Zwanzig, J.G. Kirkwood, Molecular distribution functions in a one-dimensional fluid. J. Chem. Phys. **21**, 1098–1107 (1953)
196. A. Santos, On the equivalence between the energy and virial routes to the equation of state of hard-sphere fluids. J. Chem. Phys. **123**, 104102 (2005)
197. A. Santos, Are the energy and virial routes to thermodynamics equivalent for hard spheres? Mol. Phys. **104**, 3411–3418 (2006)
198. A. Santos, Exact bulk correlation functions in one-dimensional nonadditive hard-core mixtures. Phys. Rev. E **76**, 062201 (2007)
199. A. Santos, Thermodynamic consistency between the energy and virial routes in the mean spherical approximation for soft potentials. J. Chem. Phys. **126**, 116101 (2007)

200. A. Santos, Second Virial Coefficients for the Lennard-Jones (2n-n) Potential, Wolfram Demonstrations Project (2012), http://demonstrations.wolfram.com/SecondVirialCoefficientsForTheLennardJones2nNPotential/

201. A. Santos, Radial Distribution Function for Sticky Hard Rods, Wolfram Demonstrations Project (2012), http://demonstrations.wolfram.com/RadialDistributionFunctionForStickyHardRods/

202. A. Santos, Chemical-potential route: A hidden Percus–Yevick equation of state for hard spheres. Phys. Rev. Lett. **109**, 120601 (2012)

203. A. Santos, Class of consistent fundamental-measure free energies for hard-sphere mixtures. Phys. Rev. E **86**, 040102(R) (2012)

204. A. Santos, Note: An exact scaling relation for truncatable free energies of polydisperse hard-sphere mixtures. J. Chem. Phys. **136**, 136102 (2012)

205. A. Santos, Radial Distribution Function for Hard Spheres, Wolfram Demonstrations Project (2013), http://demonstrations.wolfram.com/RadialDistributionFunctionForHardSpheres/

206. A. Santos, Virial Coefficients for a Hard-Sphere Mixture, Wolfram Demonstrations Project (2014), http://demonstrations.wolfram.com/VirialCoefficientsForAHardSphereMixture/

207. A. Santos, Playing with marbles: Structural and thermodynamic properties of hard-sphere systems, in *5th Warsaw School of Statistical Physics*, ed. by B. Cichocki, M. Napiórkowski, J. Piasecki (Warsaw University Press, Warsaw, 2014). http://arxiv.org/abs/1310.5578

208. A. Santos, Radial Distribution Function for One-Dimensional Square-Well and Square-Shoulder Fluids, Wolfram Demonstrations Project (2015), http://demonstrations.wolfram.com/RadialDistributionFunctionForOneDimensionalSquareWellAndSqua/

209. A. Santos, Radial Distribution Functions for Nonadditive Hard-Rod Mixtures, Wolfram Demonstrations Project (2015), http://demonstrations.wolfram.com/RadialDistributionFunctionsForNonadditiveHardRodMixtures/

210. A. Santos, M. López de Haro, A branch-point approximant for the equation of state of hard spheres. J. Chem. Phys. **130**, 214104 (2009)

211. A. Santos, A. Malijevský, Radial distribution function of penetrable sphere fluids to the second order in density. Phys. Rev. E **75**, 021201 (2007)

212. A. Santos, G. Manzano, Simple relationship between the virial-route hypernetted-chain and the compressibility-route Percus–Yevick values of the fourth virial coefficient. J. Chem. Phys. **132**, 144508 (2010)

213. A. Santos, R.D. Rohrmann, Chemical-potential route for multicomponent fluids. Phys. Rev. E **87**, 052138 (2013)

214. A. Santos, M. López de Haro, S.B. Yuste, An accurate and simple equation of state for hard disks. J. Chem. Phys. **103**, 4622–4625 (1995)

215. A. Santos, S.B. Yuste, M. López de Haro, Radial distribution function for a multicomponent system of sticky hard spheres. J. Chem. Phys. **109**, 6814–6819 (1998)

216. A. Santos, S.B. Yuste, M. López de Haro, Equation of state of a multicomponent d-dimensional hard-sphere fluid. Mol. Phys. **96**, 1–5 (1999)

217. A. Santos, S.B. Yuste, M. López de Haro, Virial coefficients and equations of state for mixtures of hard discs, hard spheres, and hard hyperspheres. Mol. Phys. **99**, 1959–1972 (2001)

218. A. Santos, M. López de Haro, S.B. Yuste, Equation of state of nonadditive d-dimensional hard-sphere mixtures. J. Chem. Phys. **122**, 024514 (2005)

219. A. Santos, R. Fantoni, A. Giacometti, Penetrable square-well fluids: Exact results in one dimension. Phys. Rev. E **77**, 051206 (2008)

220. A. Santos, R. Fantoni, A. Giacometti, Thermodynamic consistency of energy and virial routes: An exact proof within the linearized Debye–Hückel theory. J. Chem. Phys. **131**, 181105 (2009)

221. A. Santos, S.B. Yuste, M. López de Haro, Rational-function approximation for fluids interacting via piece-wise constant potentials. Condens. Matter Phys. **15**, 23602 (2012)

222. A. Santos, S.B. Yuste, M. López de Haro, M. Bárcenas, P. Orea, Structural properties of fluids interacting via piece-wise constant potentials with a hard core. J. Chem. Phys. **139**, 074503 (2013)

223. A. Santos, S.B. Yuste, M. López de Haro, G. Odriozola, V. Ogarko, Simple effective rule to estimate the jamming packing fraction of polydisperse hard spheres. Phys. Rev. E **89**, 040302(R) (2014)
224. A. Santos, S.B. Yuste, M. López de Haro, V. Ogarko, Equation of state of polydisperse hard-disk mixtures (2016, in preparation)
225. W. Schirmacher, *Theory of Liquids and Other Disordered Media. A Short Introduction.* Lecture Notes in Physics, vol. 887 (Springer, Cham, 2014)
226. M. Schmidt, An ab initio density functional for penetrable spheres. J. Phys.: Condens. Matter **11**, 10163–10169 (1999)
227. M. Schmidt, M. Fuchs, Penetrability in model colloid-polymer mixtures. J. Chem. Phys. **117**, 6308–6312 (2002)
228. A.J. Schultz, D.A. Kofke, Fifth to eleventh virial coefficients of hard spheres. Phys. Rev. E **90**, 023301 (2014)
229. N.A. Seaton, E.D. Glandt, Aggregation and percolation in a system of adhesive spheres. J. Chem. Phys. **86**, 4668–4677 (1987)
230. N.A. Seaton, E.D. Glandt, Monte Carlo simulation of adhesive disks. J. Chem. Phys. **84**, 4595–4601 (1987)
231. N.A. Seaton, E.D. Glandt, Monte Carlo simulation of adhesive spheres. J. Chem. Phys. **87**, 1785–1790 (1987)
232. C.E. Shannon, W. Weaver, *The Mathematical Theory of Communication* (University of Illinois Press, Urbana, 1971)
233. Simulations performed by the author with the Glotzilla library (2015), http://glotzerlab.engin.umich.edu/home/
234. J.R. Solana, *Perturbation Theories for the Thermodynamic Properties of Fluids and Solids* (CRC Press, Boca Raton, 2013)
235. P. Sollich, Predicting phase equilibria in polydisperse systems. J. Phys.: Condens. Matter **14**, R79–R117 (2002)
236. P. Sollich, P.B. Warren, M.E. Cates, Moment free energies for polydisperse systems. Adv. Chem. Phys. **116**, 265–336 (2001)
237. H.E. Stanley, *Introduction to Phase Transitions and Critical Phenomena* (Oxford University Press, Oxford, 1971)
238. G. Stell, Sticky spheres and related systems. J. Stat. Phys. **63**, 1203–1221 (1991)
239. R.B. Stewart, R.T. Jacobsen, Thermodynamic properties of argon from the triple point to 1200 K with pressures to 1000 MPa. J. Phys. Chem. Ref. Data **18**, 639–798 (1989). http://www.nist.gov/data/PDFfiles/jpcrd363.pdf
240. F.H. Stillinger, D.K. Stillinger, Negative thermal expansion in the Gaussian core model. Physica A **244**, 358–369 (1997)
241. Y. Tang, On the first-order mean spherical approximation. J. Chem. Phys. **118**, 4140–4148 (2003)
242. P. Tarazona, J.A. Cuesta, Y. Martínez-Ratón, Density functional theories of hard particle systems, in *Theory and Simulation of Hard-Sphere Fluids and Related Systems*, ed. by A. Mulero. Lecture Notes in Physics, vol. 753 (Springer, Berlin, 2008), pp. 247–341
243. The On-Line Encyclopedia of Integer Sequences (OEIS) (1996), http://oeis.org/A002218, http://oeis.org/A013922
244. E. Thiele, Equation of state for hard spheres. J. Chem. Phys. **39**, 474–479 (1963)
245. M. Thiesen, Untersuchungen über die Zustandsgleichung. Ann. Phys. **260**, 467–492 (1885)
246. L. Tonks, The complete equation of state of one, two, and three-dimensional gases of elastic spheres. Phys. Rev. **50**, 955–963 (1936)
247. S. Torquato, F.H. Stillinger, Local density fluctuations, hyperuniformity, and order metrics. Phys. Rev. E **68**, 041113 (2003)
248. S. Torquato, F.H. Stillinger, New conjectural lower bounds on the optimal density of sphere packings. Exp. Math. **15**, 307–331 (2006)
249. G.E. Uhlenbeck, P.C. Hemmer, M. Kac, On the van der Waals theory of the vapor–liquid equilibrium. II. Discussion of the distribution functions. J. Math. Phys. **4**, 229–247 (1963)

250. I. Urrutia, Analytical behavior of the fourth and fifth virial coefficients of a hard-sphere mixture. Phys. Rev. E **84**, 062101 (2011)
251. L. van Hove, Sur l'intégrale de configuration pour les systèmes de particules à une dimension. Physica **16**, 137–143 (1950)
252. J.M.J. van Leeuwen, J. Groeneveld, J. de Boer, New method for the calculation of the pair correlation function. I. Physica **25**, 792–808 (1959)
253. H. Verduin, J.K.G. Dhont, Phase diagram of a model adhesive hard-sphere dispersion. J. Colloid Interf. Sci. **172**, 425–437 (1995)
254. L. Verlet, Integral equations for classical fluids I. The hard sphere case. Mol. Phys. **41**, 183–190 (1980)
255. L. Verlet, Integral equations for classical fluids II. Hard spheres again. Mol. Phys. **42**, 1291–1302 (1981)
256. L. Verlet, D. Levesque, Integral equations for classical fluids III. The hard discs system. Mol. Phys. **46**, 969–980 (1982)
257. J.M.G. Vilar, J.M. Rubi, Communication: System-size scaling of Boltzmann and alternate Gibbs entropies. J. Chem. Phys. **140**, 201101 (2014)
258. A.Y. Vlasov, A.J. Masters, Binary mixtures of hard spheres: how far can one go with the virial equation of state? Fluid Phase Equil. **212**, 183–198 (2003)
259. E. Waisman, The radial distribution function for a fluid of hard spheres at high densities. Mean spherical integral equation approach. Mol. Phys. **25**, 45–48 (1973)
260. R.O. Watts, D. Henderson, R.J. Baxter, Hard spheres with surface adhesion: The Percus–Yevick approximation and the energy equation. Adv. Chem. Phys. **21**, 421–430 (1971)
261. M.S. Wertheim, Exact solution of the Percus–Yevick integral equation for hard spheres. Phys. Rev. Lett. **10**, 321–323 (1963)
262. M.S. Wertheim, Analytic solution of the Percus–Yevick equation. J. Math. Phys. **5**, 643–651 (1964)
263. R.J. Wheatley, The fifth virial coefficient of hard disc mixtures. Mol. Phys. **93**, 665–679 (1998)
264. R.J. Wheatley, Phase diagrams for hard disc mixtures. Mol. Phys. **93**, 965–969 (1998)
265. R.J. Wheatley, On the virial series for hard-sphere mixtures. J. Chem. Phys. **111**, 5455–5460 (1999)
266. R.J. Wheatley, The sixth virial coefficient of hard disc mixtures. Mol. Phys. **96**, 1805–1811 (1999)
267. R.J. Wheatley, Calculation of high-order virial coefficients with applications to hard and soft spheres. Phys. Rev. Lett. **110**, 200601 (2013)
268. R.J. Wheatley, F. Saija, P.V. Giaquinta, Fifth virial coefficient of hard sphere mixtures. Mol. Phys. **94**, 877–879 (1998)
269. B. Widom, Some topics in the theory of fluids. J. Chem. Phys. **39**, 2808–2812 (1963)
270. Z. Yan, S.V. Buldyrev, N. Giovambattista, H.E. Stanley, Structural order for one-scale and two-scale potentials. Phys. Rev. Lett. **95**, 130604 (2005)
271. S.B. Yuste, A. Santos, Radial distribution function for hard spheres. Phys. Rev. A **43**, 5418–5423 (1991)
272. S.B. Yuste, A. Santos, Radial distribution function for sticky hard-core fluids. J. Stat. Phys. **72**, 703–720 (1993)
273. S.B. Yuste, A. Santos, Sticky hard spheres beyond the Percus–Yevick approximation. Phys. Rev. E **48**, 4599–4604 (1993)
274. S.B. Yuste, A. Santos, A model for the structure of square-well fluids. J. Chem. Phys. **101**, 2355–2364 (1994)
275. S.B. Yuste, M. López de Haro, A. Santos, Structure of hard-sphere metastable fluids. Phys. Rev. E **53**, 4820–4826 (1996)
276. S.B. Yuste, A. Santos, M. López de Haro, Structure of multicomponent hard-sphere mixtures. J. Chem. Phys. **108**, 3683–3693 (1998)
277. S.B. Yuste, A. Santos, M. López de Haro, Direct correlation functions and bridge functions in additive hard-sphere mixtures. Mol. Phys. **98**, 439–446 (2000)

278. S.B. Yuste, A. Santos, M. López de Haro, Structure of the square-shoulder fluid. Mol. Phys. **109**, 987–995 (2011)

279. S.B. Yuste, A. Santos, M. López de Haro, Depletion potential in the infinite dilution limit. J. Chem. Phys. **128**, 134507 (2008). Erratum: **140**, 179901 (2014)

280. M.W. Zemansky, *Heat and Thermodynamics* (McGraw-Hill, New York, 1981)

281. C. Zhang, B.M. Pettitt, Computation of high-order virial coefficients in high-dimensional hard-sphere fluids by Mayer sampling. Mol. Phys. **112**, 1427–1447 (2014)

282. S. Zhou, J.R. Solana, Inquiry into thermodynamic behavior of hard sphere plus repulsive barrier of finite height. J. Chem. Phys. **131**, 204503 (2009)

Index

© Springer International Publishing Switzerland 2016
A. Santos, *A Concise Course on the Theory of Classical Liquids*,
Lecture Notes in Physics 923, DOI 10.1007/978-3-319-29668-5

Printed in the United States
By Bookmasters